Microwave Oven
Service & Repair

DEDICATION

To my friend and instructor, Dr. Harold Tornheim, whose special interests were microwaves and helping his student.

Microwave Oven Service & Repair

By Clayton L. Hallmark

TAB BOOKS
Blue Ridge Summit, Pa. 17214

FIRST EDITION

FIRST PRINTING—NOVEMBER 1977

Printed in the United States
of America

Library of Congress Cataloging in Publication Data

Hallmark, Clayton.
Microwave oven service & repair.

Includes index.
1. Microwave ovens—Maintenance and repair.
I. Title.
TX657.064H34 683'.83 77-20922
ISBN 0-8306-792-6
ISBN 0-8306-6920-0 pbk.

(Cover photo courtesy of Frigidaire Division, General Motors Corporation)

Preface

The microwave oven is probably the biggest breakthrough in food preparation since primitive man first learned to cook his food over an open fire. A far greater advance than the electric oven, which offered flameless cooking, the electronic microwave oven offers what can be considered heatless cooking. Unlike all previous cooking methods which cooked food by producing heat and applying it to the food, the microwave oven cooks food by producing heat within the food itself. The microwave oven itself runs cool; heat is transferred to it from the food as the food heats. In fact, in microwave cooking, it can be said that the food heats the oven, instead of the other way around!

The microwave oven is as revolutionary in economic impact as it is in concept. Retail sales of microwave ovens tripled between 1973 and 1976, and by 1980 it is estimated that sales will triple again—to almost $2 billion. The microwave oven surpassed the gas range in sales in 1975 and the electric range in 1976; and by 1980, microwave sales will outstrip gas and electric sales *combined*—by almost a half-billion dollars. Soon, the largest of all industries, the food industry, will jump on the microwave bandwagon by introducing foods packaged and prepared expressly for microwave cooking, and the impact on sales may be greater than anyone can yet imagine.

The microwave oven may become as much a symbol of the 1970s as the television is of the 1950s.

The microwave is the wave of the future, and the time to join the microwave revolution is now—the sooner the better. Technicians that made the most of the television boom in the 1950s were the ones that jumped on the bandwagon early, and technicians that make the most of the microwave boom will be the ones that jump on the bandwagon now.

Heretofore the manufacturers of microwave ovens have been reticent about providing servicing information on these products, presumably because of the highly publicized hazards of microwave radiation—which turned out to be more apparent than real. Now the Bureau of Radiological Health, the federal agency that establishes regulations pertaining to microwave radiation, says that only qualified persons should be permitted to service microwave ovens, but that manufacturers must make the service information available to anyone on request. Here in this book, for the first time anywhere, is servicing information on all the leading makes of microwave ovens. Here is specific, pinpointed, illustrated, and authoritative data on troubleshooting, maintaining, and repairing the most popular models of ten manufacturers—using the specific procedures and techniques recommended by the manufacturers themselves.

You could get this same information by purchasing dozens of separate service manuals from ten different sources—at a considerable expenditure of time, effort, and money. But the information is all here in one convenient volume—the most needed information on the most popular models, logically organized in a uniform format, manufacturer by manufacturer, for fast, certain results. All of this is made possible through the courtesy of the firms and persons listed on the following pages. I gratefully acknowledge their help and that of my typist, Gloria Cains.

Clayton Hallmark

Acknowledgements

AMANA REFRIGERATION, INC.
Richard Mayes
Michael Humbert

FRIGIDAIRE DIVISION, GENERAL MOTORS
Rex Smith

LITTON MICROWAVE COOKING PRODUCTS
Richard Parrish
Nancy Williams

PANASONIC COMPANY
Joseph Lacovara

SHARP ELECTRONICS CORPORATION
R. F. Liguori

TAPPAN COMPANY
Lee Snavely

CUNNINGHAM AND WALSH, INC.
Edgar Falk

GENERAL ELECTRIC COMPANY
Helen Gibson

MAGIC CHEF, INC.
Joseph Bagwell

ROPER SALES
Donald Ball

SIMPSON ELECTRIC COMPANY
Stewart Cudworth

WHITE-WESTINGHOUSE COMPANY
Donald Magee

Contents

History of the Microwave Oven. 1953: The first mass-produced microwave oven, offered by Raytheon Company for restaurant and hospital use, cost $3000. **1958**: Models with pushbutton controls for self-service in vending operations were introduced, and the price came down to $2000. **1967:** Amana introduced a countertop unit designed for use on the standard 115V wiring circuit found in homes, and the microwave oven became a household appliance. **1974: Microwave** had become a household word, and this computerized Amana model ushered in the beginning of the microwave boom. (Courtesy Amana Refrigeration, Inc.)

Chapter 1
Advantages of Microwave Cooking

In the early days of microwave ovens, people were truly amazed to see a hot dog cook in 30 seconds or a baked potato cook in a mere 4 minutes, but few were willing to pay $500 for what they preceived to be a hot dog cooker and potato baker. They had to be convinced that microwave ovens could do much more before the microwave market really began to take off—and they had to be convinced that microwave ovens were safe. Today the microwave oven is finally attaining its rightful status in the market place, but it is still a relatively new phenomenon, and there are still many persons who are uninformed or misinformed about it. It is the duty of anyone who seeks to sell or service microwave ovens to be well versed in their many virtues and to help educate the public about these virtues.

SAFETY

Safety is the number 1 virtue of the microwave oven. Ironically, it has also been the number 1 concern about the microwave oven. The first microwave ovens for the home were introuced by Tappan and Litton in the 1950s, and *Popular Electronics* predicted skyrocketing sales.

The fact is microwave ovens were, are, and always will be inherently safe. The fact is, also, that there will always be some persons who, because of ignorance or a desire to create a sensation, will exaggerate the dangers of any important new development. There have always been barnburners, who would burn down a barn to kill the rats, and bookburners, who would burn books to kill ideas. And now we have, in seemingly unprecedented numbers, calendar burners, who would burn all the calendars and take us all back to the last century. It is not likely that they will ever again be able to stand in the way of microwave cooking. Nothing is more powerful than an idea whose time has come, and if there is any idea whose time has clearly come, it is microwave cooking.

The microwave oven is probably the safest cooking device ever invented. Countless persons have suffered asphyxiation, burns, or other injuries because of gas leaks or explosions in gas ranges. Others have been killed or injured when an electric range, which draws several times more current than a microwave oven, overloaded a house circuit and started a fire. And almost everyone who has ever used a gas or electric range has been burned by accidental contact with a hot surface of a range. But according to former U. S. Surgeon General Jesse Steinfeld, "There hasn't been any documented case of human injury...[and] it is doubtful that any significant health damage has occurred to anyone from microwave ovens." The radiation hazard from the operation of a microwave oven has been expressed by Dr. James A. Van Allen, the radiation expert after whom the Van Allen belt in the outer atmosphere is named, as "about the same as the likelihood of getting a skin tan from moonlight."

Microwave sources do not present a radiation hazard at all in the same sense as radioactive sources do, because microwave radiation and radiation from radioactive sources are very different in their effects. Radiation from radioactive sources is *ionizing* radiation. Ionizing radiation—which includes X-rays, gamma rays, cosmic rays, and ultraviolet rays—can cause chemical changes to take place with little or no temperature change. Microwave radiation is nonionizing radiation, a relatively innocuous form of radiation that includes radio waves, infrared waves, and light waves. Nonionizing radiation of sufficient intensity will cause a rise in temperature, but it does not cause any cumulative damage to tissues and it cannot be stored by the body as ionizing radiation can.

Under present standards, radiation leakage from microwave ovens cannot exceed 1 milliwatt per square centimeter at 2 inches from the oven prior to factory release, and 5 milliwatts per square centimeter when the oven is in the customer's possession. Before these standards went into effect, industry standards for radar sites allowed workers to be exposed to a microwave intensity of up to 10 mW/cm^2 (milliwatts per square centimeter). Even this higher standard has a considerable margin of safety. It would take a long exposure to 10 mW/cm^2 of microwave radiation to raise the body temperature even 1°F—even if every square centimeter of the body was exposed. There is probably a greater danger to health from a warm bath.

We are exposed constantly to nonionizing radiation without harm. Solar energy reaching the earth on a sunny day may reach 60–100 mW/cm^2. Microwave diathermy treatment of the eye—one of the organs most suscepticable to radiation damage—is carried out at levels of over 100 mW/cm^2 with the eye positioned just 3 inches from the microwave source. Since radiation intensity diminishes by the square of the distance from the source, a microwave oven that leaks the allowable 5 mW/cm^2 at 2 inches from the oven will produce an intensity of just 0.05 mW/cm^2 (a hundredth as much) at a normal operating distance of 20 inches. If 100 mW/cm^2 can be considered therapeutic, microwave critics should feel no more threatened by 0.05 mW/cm^2 than by moonlight, although they may be more affected than most persons by moonlight. For most persons the main danger from a microwave oven is that it might fall off a countertop and onto their foot.

SPEED OF COOKING

A microwave oven can save up to 70% of cooking time in many cases—sometimes even more. You don't have to wait 10 or 15 minutes for the oven to preheat. The microwaves instantly penetrate the food to a depth of about 1.5 inches for meats and 3 inches for other foods, producing the full cooking effect as soon as you turn the oven on. (Beyond the depth of microwave penetration, the foods cook by conduction heating as in conventional ovens.) This means that you can cook a 3-pound beef roast in about 18 minutes, compared to 1.5 hours

■ Microwave Oven ■ Conventional Gas Oven Room Temperature Defrosting

Menu	Amount	Purpose	1 min. / 15 min. / 30 min. / 45 min. / 1 hr. / 1 hr. 15 min. / 1 hr. 30 min.
MEATS			
Roast Beef	per pound	Cooking *	35 min.
Frozen Beef	3 lbs	Defrosting *	14 min. (7 hr.)
POULTRY			
Baked Chicken	3 lbs	Cooking	1 hr. 20 min.
Frozen Chicken	3 lbs	Defrosting	12 min. (7 hr.)
SEAFOOD			
Frozen Lobster	10 oz	Defrosting and Cooking	1 hr. 10 min.
Brook Trout	10 oz	Cooking	30 min.
VEGETABLES			
Baked Potato	7 oz	Cooking	30 min.
Frozen Spinach	10 oz	Defrosting	4 min. 1 hr
DESSERTS			
Pineapple Upside-Down Cake	9½ oz	Cooking	40 min.
Frozen Pastry	10 oz	Defrosting	3-½ min. 1 hr. 15 min.

* Times for NE-5300 will be slightly longer.

Fig. 1-1. Comparison of cooking and defrosting times. (Courtesy Panasonic Company)

in a conventional oven or 1 hour in a pressure cooker. The difference between microwave defrosting and conventional room-temperature defrosting is even greater—almost startling—a mere 14 minutes versus 7 hours. Figure 1-1 shows some other examples of microwave speed.

Cleanup is quick, too, because food spatters can't bake onto oven surfaces. All you do is wipe spills occasionally with a damp cloth.

ECONOMY

The energy crisis has brought new dimensions to this advantage of microwave cooking. Tests by both utility companies and appliance manufacturers show that microwave cooking results in an energy saving of up to 75%. The value of this to consumers in a time of escalating gas and electric bills is obvious. The social advantage can also be great, since the energy that can be saved by microwave cooking is needed to run factories and schools. If even half of the nation's 65 million households converted to microwave cooking, think of the jobs that could be saved during a gas or electric shortage.

The main reasons for the microwave oven's low cost of operation are that it uses relatively little current during operation (about the same as an electric skillet) and it need be in operation for only a short time. The work area stays cooler, which results in additional savings of air conditioning. Since the microwave oven operates for only a short period each time it is used, it has a long service life and a long warranty—up to 5 years or even more.

Microwave ovens prepare leftovers so fast they don't have time to dry out. Foods taste freshly cooked each time. Now leftovers you wouldn't have saved can make a delicious meal for your family. And since the oven thaws foods so rapidly, you can take advantage of food sales to stock your freezer.

CLEANLINESS

Since microwaves have no heating effect on the oven surfaces, food spills and spatters can't mar the oven or your cooking utensils. You can wipe the oven with a damp cloth—no messy, smelly oven cleaners are needed. The door seal of the oven should be kept clean at all times, but this is out of the way of most spills and spatters. Browning elements, when present, are often self-cleaning, and glass parts detach for cleaning.

NUTRITION

Since microwave cooking allows vegetables to be cooked with much less water and more quickly than by conventional means, there is a savings in nutritional value. Most frozen vegetables require no additional water, and most fresh fruits and vegetables require very little water. Since most vitamins are destroyed by heat, long cooking by slower means can reduce the nutritional value of foods.

SIMPLICITY

Microwave ovens are so simple and safe to use that even a child can heat his own lunch. You just place the food in the oven cavity, close the door, set the timer, and press the button. When the timer rings, the food is cooked. You can even reheat foods or finish cooking foods not quite done enough.

CONVENIENCE

Many foods can be cooked and served in the same utensil, resulting in fewer dishes to wash. You can even cook on paper plates. Dishes and cookware are easier to clean because foods do not bake on them. Quick defrosting is another convenience of microwave ovens; foods can be defrosted as needed, giving the homemaker new freedom from the kitchen.

DISADVANTAGE

As you can see, there are many advantages to microwave cooking. There is only one significant disadvantage, however. The cool-running microwave oven does not cause enough surface heating to produce the brown surface color expected of certain foods. With microwave cooking only, a steak or hamburger patty might have a gray, unappetizing appearance. Baked goods would not have a brown crust (these are often covered with frosting, however). The appearance of such items can be enhanced by placing them in a hot oven for a few minutes, or in the case of a steak or meat pattie, on a grill, under a broiler, or in a skillet. Foods that because of their size take longer than 10 or 12 minutes to cook in a microwave oven will brown by microwave cooking alone, particularly beef roasts and roasting chickens and turkeys, in which surface fats aid the browning process. Some microwave ovens incorporate a browning element or some other way to produce sufficient surface heating to brown foods in the microwave oven itself. A microwave oven with this feature and defrosting capability is a nearly universal means of food preparation—one that is almost ideally suited to modern living.

Chapter 2
Microwaves: Their Nature, Generation & Effects

When a stone is thrown into a quiet pond, waves are formed that radiate outward from the point at which the stone broke the surface of the water. There are other kinds of waves that we cannot see, but we know that they exist because we observe their effects. They are known collectively under the term *radiation*. Radiation refers to the process of emitting energy in the form of waves or particles, or to the waves or particles themselves.

WAVE CHARACTERISTICS

Radiant waves, or electromagnetic waves, as they are often called, radiate outward from their source like the waves on the surface of a pond. They travel at a much greater speed, of course, the speed of light—182,000 miles or 300 million meters per second. Electromagnetic waves are characterized by their length (Fig. 2-1A) and by their *frequency*, or the number of waves that occur in a second.

Wavelength (λ) and frequency are inversely proportional. That is, the higher the frequency, the shorter the wavelength; and the lower the frequency, the longer the wavelength. The relationship between wavelength and frequency may be expressed by the formula

$$\text{Frequency} = \frac{\text{speed of light}}{\text{wavelength}}$$

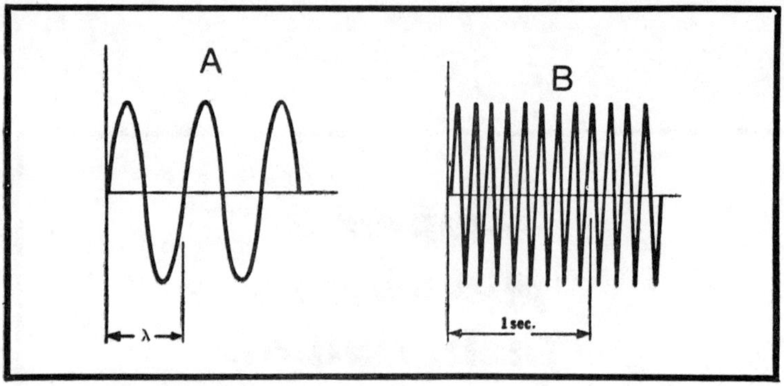

Fig. 2-1. (A) Wavelength, symbolized by the Greek letter λ (lamda). (B) Frequency: the number of cycles in one second.

You can get a feeling for this formula by considering the analogy of a freight train. If two freight trains, one consisting of short coal cars and the other consisting of long piggyback (trailer on flat car) cars, pass by at the same speed, more coal cars will pass in a minute than piggyback cars. The piggyback cars can be said to have a higher frequency and a shorter length. The relationship can be expressed as

$$\text{Frequency of cars} = \frac{\text{speed of train}}{\text{car length}}$$

The relationship for electromagnetic waves is shown in Fig. 2-2.

The frequencies used in electronics, sometimes called radio frequencies, occupy a small portion of the electromagnetic spectrum between infrared and heat radiations. The radio-frequency (RF) spectrum itself may be subdivided into a number of bands as in Fig. 2-3. Microwave ovens operate in the ultrahigh-frequency or UHF band. Since the frequency of the oven waves is so high, the waves are very short—microwaves. By convention, waves shorter than 30 cm are considered microwaves. Since a 30 cm wavelength corresponds to a frequency of 1000 MHz (megahertz, one million waves per second), frequencies higher than 1000 MHz are considered to be in the microwave region.

The almost universally used frequency for microwave ovens today is 2450 MHz, which corresponds to a wavelength of about 12 cm or 5 inches. At least one oven was produced that operated at 915 MHz, which is considered to be in the microwave region since the breakpoints between portions of electromagnetic waves and RF are not considered precise or abrupt.

The term *hertz* is used to honor Heinrich Hertz, the German physicist who first succeeded in producing and detecting electromagnetic (radio) waves. The abbreviation

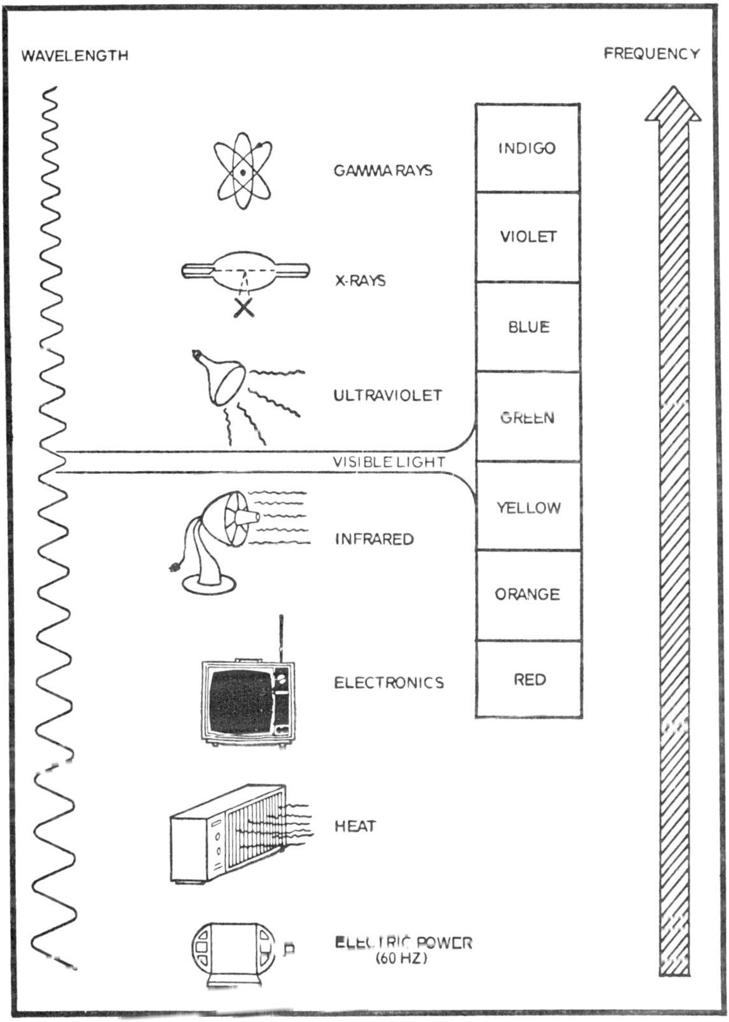

Fig. 2-2. The frequency spectrum.

WAVE-LENGTH	BAND	USES		FREQUENCY
	EXTREMELY HIGH FREQUENCIES (30-300 GHz, 10-1 mm)		RADAR	
	SUPERHIGH FREQUENCIES (3-30 GHz, 10-1 cm)	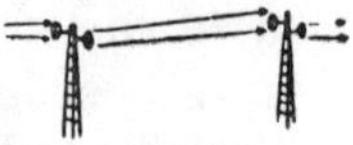	MICROWAVE RELAYS, RADAR	
	ULTRAHIGH FREQUENCIES (300-3000 MHz, 1-0.1 m)		TV CHANNEL 14-83, AIR CRAFT NAVIGATION AND DISTANCE MEASURING EQUIPMENT, MICROWAVE OVENS	
	VERY HIGH FREQUENCIES (30-300 MHz, 10-1m)		TV CHANNELS 2-13; FM BROADCASTING, TAXI, POLICE, AND OTHER TWO-WAY RADIO COMMUNICATIONS	
	HIGH FREQUENCIES (3-30 MHz, 100-10m)		INTERNATIONAL SHORT WAVE BROADCASTING, AMATEUR RADIO, CB RADIO	
	MEDIUM FREQUENCIES (300-3000 kHz, 100-1000m)		AM BROADCASTING, POLICE, SHIP, AND AVIATION COMMUNICATIONS	
	LOW FREQUENCIES (30-300 kHz, 10-1 km)	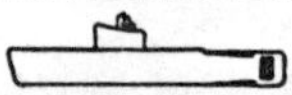	SHIP COMMUNICATIONS AND NAVIGATION	
	VERY LOW FREQUENCIES (3-30 kHz, 100-10 km)		LONG DISTANCE STATION-TO STATION COMMUNICATIONS	

Fig. 2-3. The divisions of the radio-frequency spectrum.

GHz stands for *gigahertz*, or billions (thousands of millions) of hertz. Microwave-oven vibrations occur at the rate of 2.45 billion or 2,450 million times per second, that is, 2.45 GHz.

HOW FOOD IS COOKED BY MICROWAVES

In ordinary cooking, heat is applied to food from an external source. In microwave cooking, the heat is generated inside the food. The microwaves cause the millions of molecules of food to rub against each other. The rubbing of the molecules together generates heat by friction in the same way that heat is generated in two sticks of wood when a Boy Scout rubs them together to start a fire without matches.

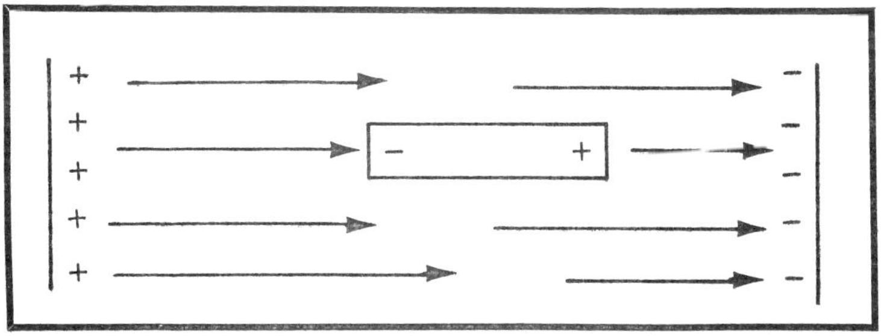

Fig. 2-4. A polarized molecule in electrostatic field.

Polarized Molecules in an Electrostatic Field

Materials which can be heated by microwaves have *polarized* molecules, which means one end of each molecule has a positive charge and the other end has a negative charge. Water, for instance, which is a compound of hydrogen and oxygen, has a positive charge on the hydrogen end of each molecule and a negative charge on the oxygen end.

Figure 2-4 shows how a polarized molecule aligns itself in an electrostatic field. The positive end of the molecule is repelled from the external positive charge and attracted to a negative charge, and vice versa.

Under ordinary circumstances the molecules in a glass of water or in a bit of food are arranged in a helter-skelter fashion as shown in Fig. 2-5. (These drawings, of course, are not intended to be actual pictorial drawings of molecules.)

When the water or food is placed in an electrostatic field of sufficient strength, the molecules are all turned so that their positive charges are pointed in the same direction. (see Fig. 2-6)

Flipping Molecules

When the direction of the electrostatic field is reversed, the molecules also reverse their directions, as shown in Fig.

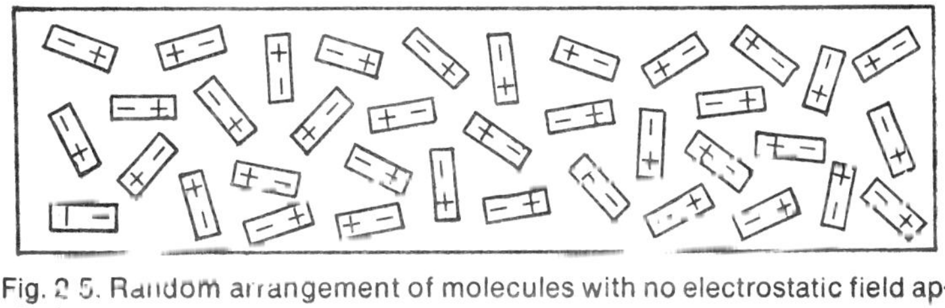

Fig. 2-5. Random arrangement of molecules with no electrostatic field applied.

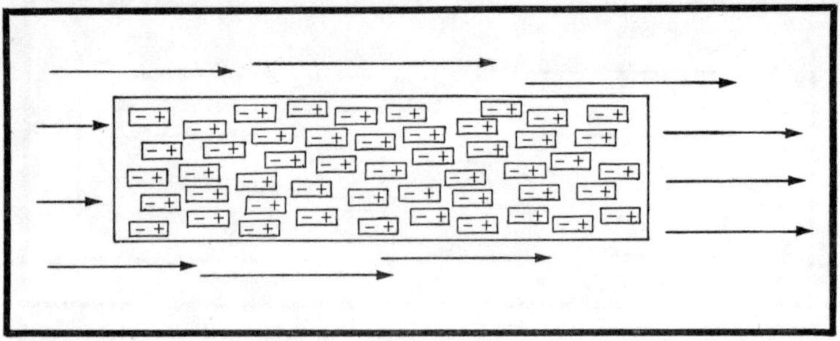

Fig. 2-6. Polarized molecules lined up in same direction by electrostatic field.

2-7. If the field reverses rapidly and has sufficient strength, the molecules will be flipped over rapidly and will rub together enough to generate heat, as described above.

The frequency of the microwaves in microwave ovens is 2.45 GHz, which means that the electrostatic field is reversed and then changed back to its original direction 2.45 billion times each second.

More Heat Near the Outside

Each successive layer of molecules absorbs energy from the electrostatic field penetrating the food. The intensity of the field, therefore, is less at the center of the food than it is at the surface. (See Fig. 2-8.) This causes the food to become hotter near the surface than at the center.

At 2.45 GHz, the power will be cut in half for every three-quarters of an inch of penetration if beef or a similar food is being cooked. Thick cuts of meat therefore become more done on the outside than on the inside.

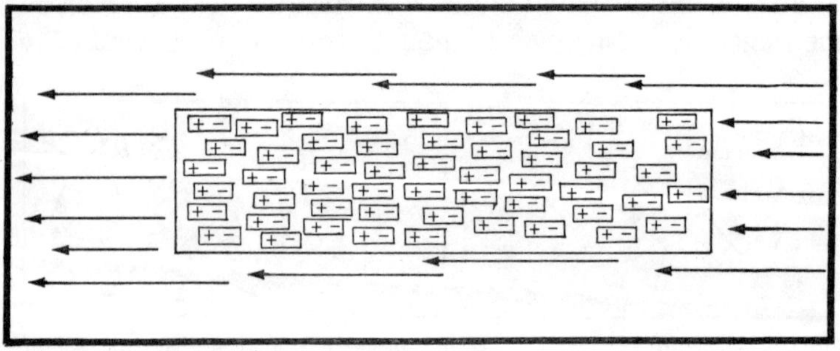

Fig. 2-7. When field is reversed, molecules are also reversed.

HOW AN ALTERNATING ELECTROSTATIC FIELD IS GENERATED

The generating of microwaves can best be explained by first studying the generation of lower frequency waves. The electronic generating process of alternating current (AC) takes place in a tuned or resonant circuit, consisting of a coil of wire and a capacitor. The coil and the capacitor each can store energy, but in different ways; and when they are connected together and connected to a source of energy, AC is generated. To see how this happens, we must first examine the properties of the coil and capacitor separately.

Coils Store Energy in a Magnetic Field

A coil of wire carrying an electric current will have a magnetic field around it exactly like that around a permanent magnet. (See Fig. 2-9). This field will have a north pole and a south pole at opposite ends of the coil. If the direction of current through the coil is reversed, the direction of the magnetic field also will be reversed.

If the voltage source supplying the electric current is cut off, the magnetic field around the coil collapses. This collapsing of the field generates a voltage in the coil which, for a short time, keeps the current flowing in the same direction.

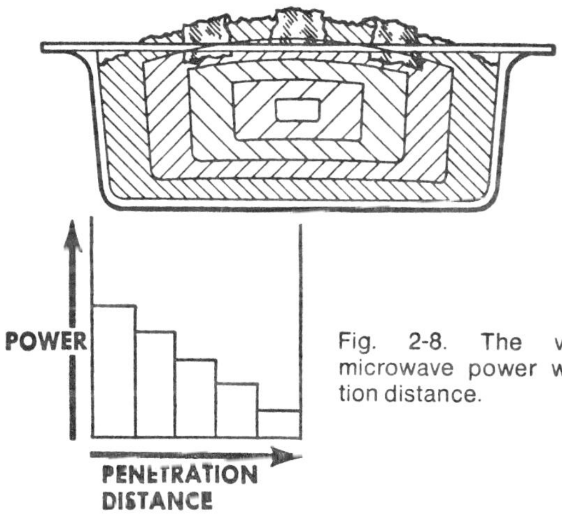

Fig. 2-8. The variation of microwave power with penetration distance.

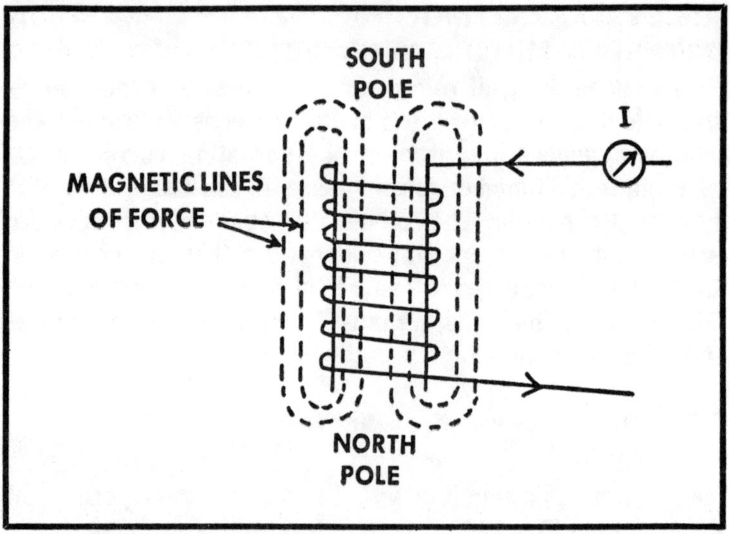

Fig. 2-9. Magnetic field surrounding a current-carrying coil.

In this way the energy stored in the magnetic field is returned to the circuit.

This ability of a coil to store energy is called *inductance*. A straight piece of wire does the same thing, but it cannot store as much energy as a coil.

Capacitors Store Electric Charges

A capacitor is usually made of two metallic plates close together separated by air, paper, oil, mica, or other

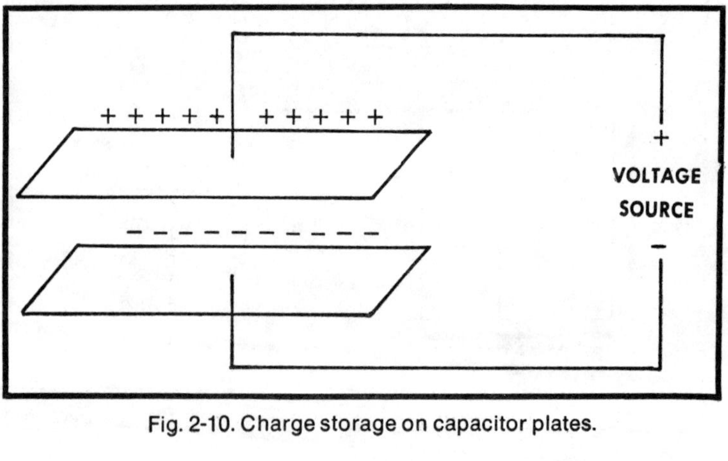

Fig. 2-10. Charge storage on capacitor plates.

nonconducting material. If the two plates are connected to the terminals of a battery or other source of voltage the plates of the capacitor will become charged, with negative electrons on one plate and positive charges on the other plate, as in Fig. 2-10.

When the connections to the battery, or other voltage source, are reversed the electrons will travel through the external circuit until the polarity of the charges on the capacitor is reversed. This current through the external circuit can be used in the generation of AC as described in the following section.

Resonant Circuit Storing Energy

When a coil and a capacitor are connected in parallel (see Fig. 2-11), energy is stored at one instant in the charge on the capacitor, then in current flow through the coil. The sequence is as follows:

1. Taking as a starting point the instant when the capacitor is fully charged, as in Fig. 2-11, the current in the circuit is zero. This is shown in a plot of current versus time as point *1* in Fig. 2-12.
2. Immediately the electrons on the negative plate of the capacitor start to flow through the coil to the other plate, where they neutralize the positive charges. Eventually all the positive charges are neutralized, meaning that the charge on the capacitor is reduced to

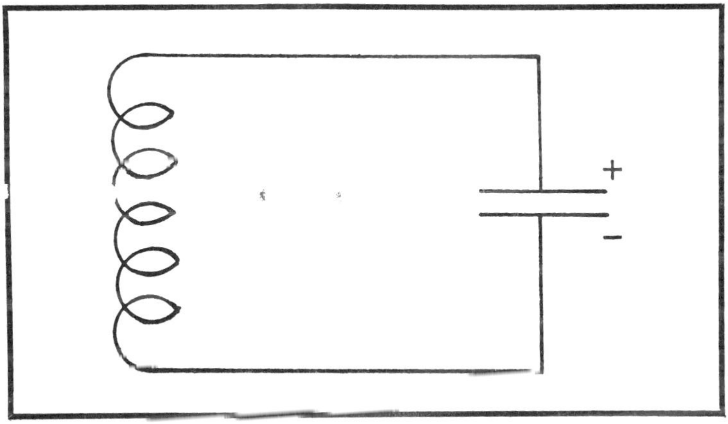

Fig. 2-11. Resonant circuit when capacitor is fully charged.

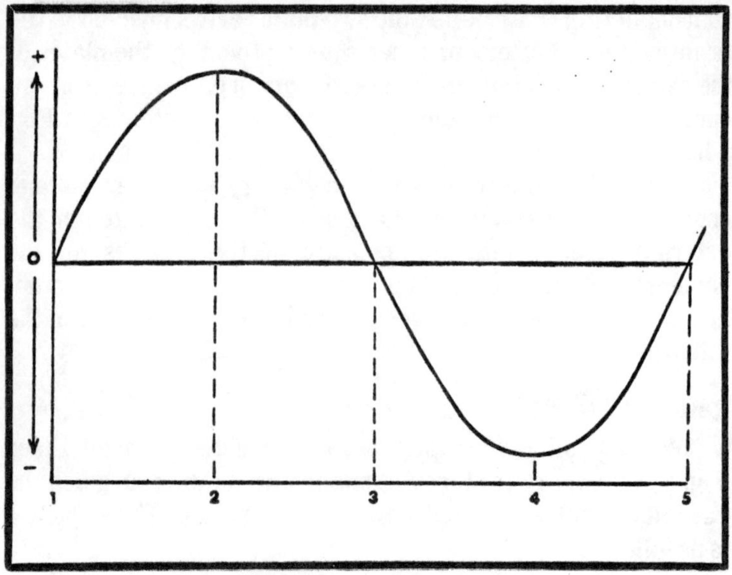

Fig. 2-12. One cycle of a sine wave.

zero. At this point the current in the coil is a maximum and energy is stored in the magnetic field around the coil as shown in Fig. 2-13. This is point 2 in Fig. 2-12.

3. The capacitor can no longer furnish electrons to the circuit so the flow of electrons starts to decrease. The resulting collapse of the magnetic field around the coil keeps the electrons flowing, charging the capacitor with a polarity opposite to its previous charge; that is, the plate which formerly had a negative charge now

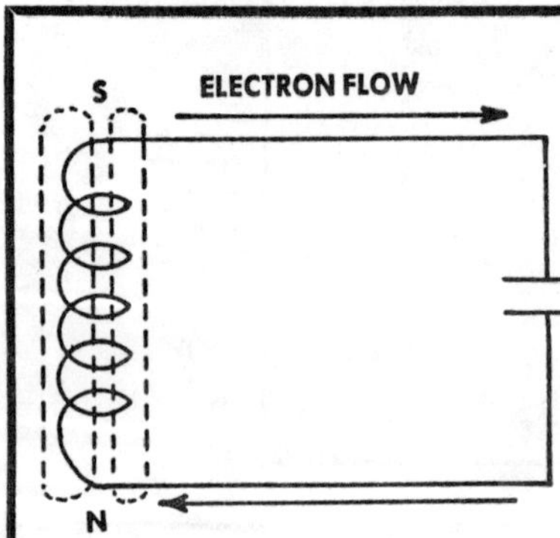

Fig. 2-13. Resonant circuit when electron flow is maximum.

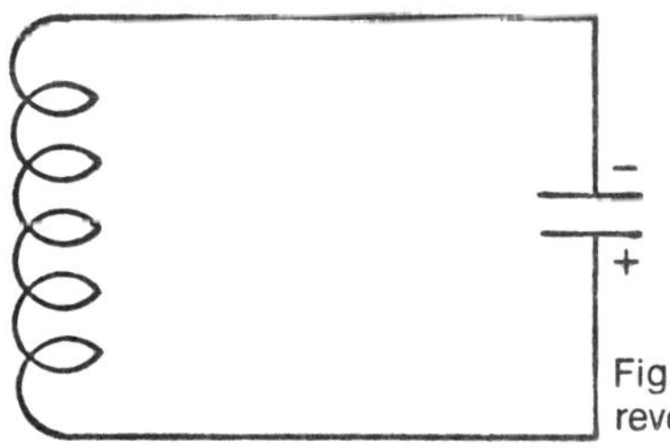

Fig. 2-14. Resonant circuit with reversed charge on capacitor.

becomes positively charged and vice versa. See Fig. 2-14. At point *3* in Fig. 2-12 the current has decreased to zero.

4. Now the capacitor starts to discharge again, sending electrons flowing in a direction opposite to that of their previous flow. The magnetic field around the coil again builds up but it also is in a direction opposite to its previous direction. Figure 2-15 shows the circuit at the time when electron flow has reached its maximum and the charge on the capacitor has been reduced to zero. This is point *4* in Fig. 2-12.
5. With the capacitor discharged the electron flow decreases and the magnetic field around the coil collapses, keeping the electrons flowing long enough to recharge the capacitor with the polarity shown in Fig. 2-11. At point *5* in Fig. 2-12 the electron flow has dropped to zero and the capacitor is recharged as it was at point *1*. The cycle then is repeated.

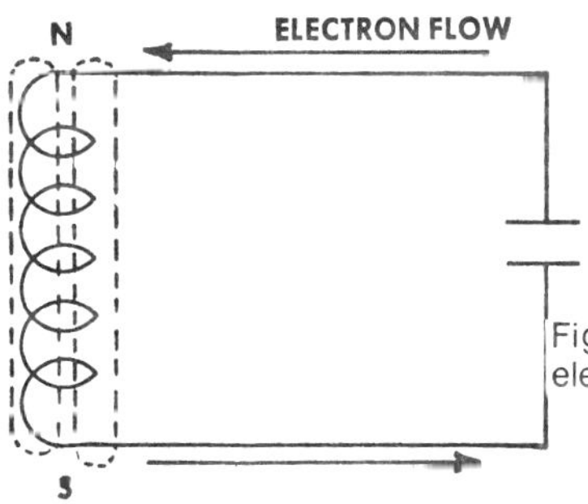

Fig. 2-15. Resonant circuit with electron flow reversed.

In this way AC in the circuit is produced, flowing first in one direction, then in the other. The plot of the current (Fig. 2-12) is called a *sine* wave from the mathematical expression for its shape.

AC in a resonant circuit would decrease rapidly because of resistance in the circuit if energy were not supplied from some other source. In actual use, the circuit must be connected to a vacuum tube or a solid-state circuit which must then be powered by a voltage source.

Coil and Capacitor Size Determine Frequency

A larger capacitor will take longer to discharge than a small one. Also a large coil with many turns will store more energy in its magnetic field than a small one, and the collapse of the field will keep current flowing longer. Therefore, the larger the coil and the capacitor the lower will be the frequency of oscillation, which is the number of cycles per second. *Cycles per second* are now dubbed *hertz.*

The converse also is true—the smaller the coil and capacitor the higher will be the frequency of oscillation. The tuning circuits of some television sets for VHF (very high frequency) use resonant circuits consisting only of a short piece of wire instead of a coil and, for a capacitor, the wire acts as one plate and the chassis of the television set acts as the other.

HOW MICROWAVES ARE GENERATED

The frequency of microwaves is much too high to be produced by a resonant circuit like that described above. Instead, the inductance of the coil and the capacitance of the capacitor must be lumped together in a single unit. A *cavity* can be used for this purpose. *Magnetrons* such as those used in microwave ovens use several cavities.

Oscillation in a Cavity

The anode of the magnetron is a hollow piece of iron with cavities in its outside wall. As shown in Fig. 2-16, the open side of the cavity acts as a capacitor while the wall of the cavity carries current and therefore has inductance. The amount of inductance and capacitance, however, is very low, which is the reason for the ultra-high frequency of oscillation. To maintain oscillation energy must be supplied to the cavities. The next several paragraphs describe how this is done.

A simplified diagram of a magnetron is shown in Fig. 2-17A. The magnetron is essentially a diode that has its plate at ground potential and its cathode at a high negative potential during the time it is oscillating. The diode is placed in a powerful magnetic field produced by a permanent magnet.

When a negative pulse is applied to the cathode and there is no magnetic field present, electrons move in straight lines from the cathode to the plate, as shown in part 1 of Fig. 2-17B. When a weak magnetic field (part 2) is applied, the electron paths become curved; and as the magnetic field becomes stronger (parts 2, 3, 4, and 5), the electron paths become progressively more curved. Finally, the paths become so curved that the electrons are moving in closed circular orbits that miss the plate entirely, and no plate current flows. The plate in Fig. 2-17C is a copper cylinder, the internal surface of which is separated into a number of segments by holes in the cylinder that serve as tuned circuits. As the electrons move in circles past the plate segments, they induce currents electrostatically in the walls of the cavities. The energy of the magnetron's output pulse is contained in the field associated with these currents. The smaller the circles the electrons make, the higher the frequency of the oscillations. The frequency depends on the size of the cylinder, the strength of the magnetic field, and the difference in potential between the cathode and plate. Energy is coupled out of the magnetron by

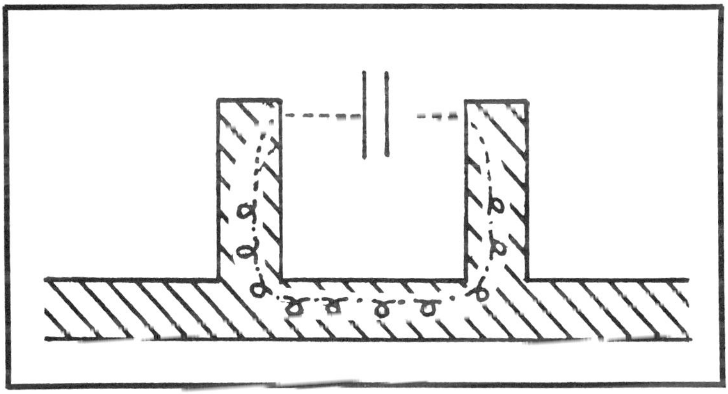

Fig. 2-16. Inductance and capacitance in a cavity.

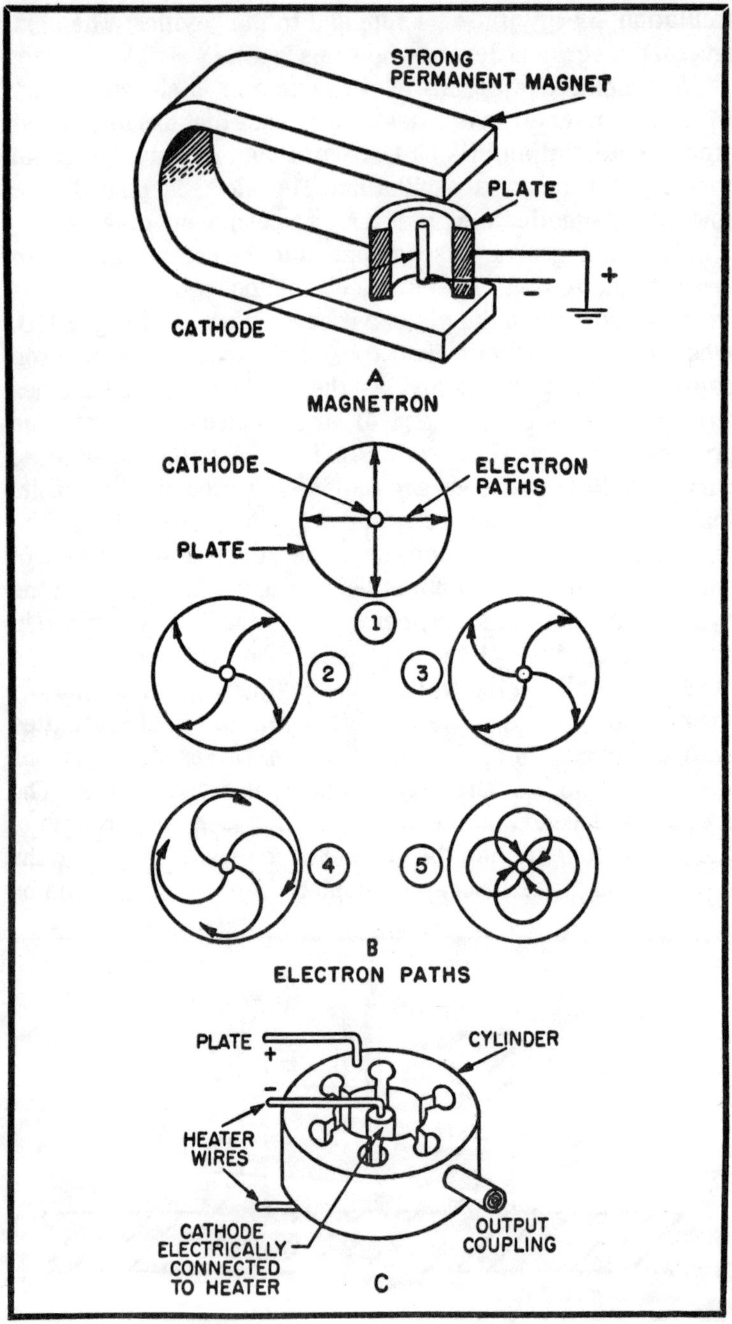

Fig. 2-17. Simplified diagram illustrating magnetron operation.

means of a loop or probe. In microwave ovens, this element is called the antenna.

A metal strip, the antenna, is welded to the anode in such a manner that the microwave energy generated in the magnetron is conducted to the upper portion, where it is radiated down the waveguide to the oven.

Once inside the oven cavity, the microwaves are either absorbed by the food at once or are reflected by the steel sides until they enter the food and are absorbed. There is an air-driven fan-like device called a stirrer at the top of the oven cavity. See Fig. 2-18. As the stirrer rotates it causes the microwaves to reflect randomly (stirs them up so that they do not all follow the same paths). If the waves are not randomly reflected, certain areas of the food would be overcooked and other areas would have barely started to cook.

MICROWAVES ARE SIMILAR TO LIGHT WAVES

Microwaves, like light waves, are waves of electromagnetic energy and have many of the characteristics of light waves. They travel in a straight line; they can be transmitted or absorbed. However, the materials which reflect, transmit, or absorb are different for light and microwaves. For example, the perforated holes in the door of the microwave oven will reflect microwave power while they transmit light. Opaque paper transmits microwaves but absorbs or reflects light waves. Water absorbs microwave energy but transmits light.

The microwave oven cavity has metallic walls, ceiling, and floor which reflect microwaves. Reflecting the microwave energy, rather than absorbing it, causes the walls of the oven cavity to remain comparatively cool. When the walls do get warm it is because of heat transferred from the food, not because they were heated by the microwave energy.

CHOKE CAVITY PREVENTS MICROWAVE LEAKAGE

It is very important that microwaves are not allowed to escape from the oven. The edge of the door must therefore be carefully sealed. The primary door seal is a device known as a *wave choke*. This choke is in the form of a cavity built around the periphery of the door. This cavity has a depth equal to a

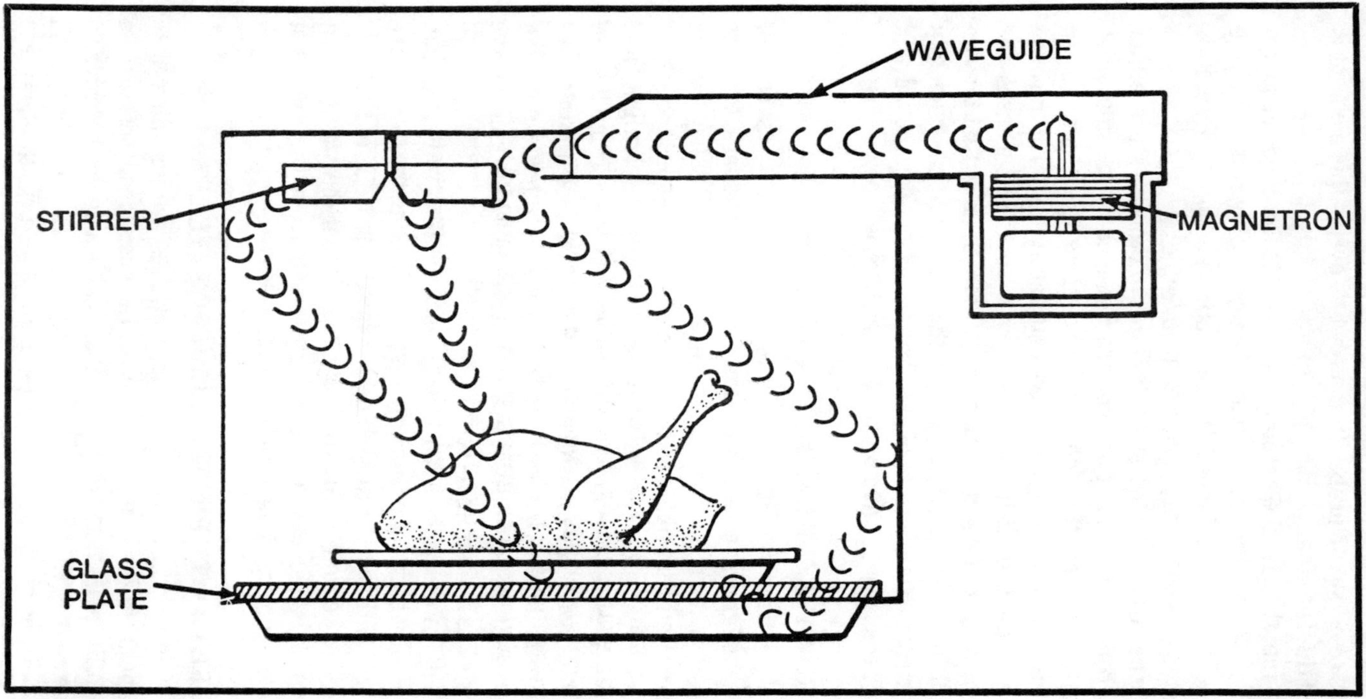

Fig. 2-18. How microwaves are delivered to food load.

quarter-wavelength of the microwaves. The full wavelength is slightly less than 5 inches. A characteristic of a quarter-wavelength cavity is that it has a very high impedance to the microwaves.

Any microwaves that leave the oven cavity will have to reflect back and forth in the space between the tapered flange of the oven cavity and the door. As they reach the gap at the front part of the choke they will enter the choke cavity. Once a wave enters the choke cavity, it travels to the rear of the cavity, is reflected off the back surface of the cavity, then travels to the front of the cavity. This wave has now traveled a half-wavelength and is positioned so that it will cancel another oncoming wave. Because of this canceling effect, no 2.45 GHz energy gets past this cavity.

The choke cavity is filled with polypropylene, which has no effect on microwaves. Polypropylene keeps the interior of the choke clean and helps it maintain its exact shape. The choke cavity cannot be seen, because the parts are all welded together into the die-cast door assembly.

The only waves which the choke cavity will trap are those with a frequency of 2.45 GHz. Most magnetrons also generate low-intensity waves at harmonics of this frequency—that is, at two times three, four, five, etc. times 2.45 GHz. To stop these higher frequencies a secondary door seal is provided. The secondary seal is the door gasket, which is made of a special grade of vinyl. Ordinary vinyl has no effect on microwaves, so the vinyl in the door gasket is loaded with carbon black. It therefore has the ability to absorb any microwave energy that might bypass the choke. This seal is shown in Fig. 2-19.

In some ovens a special filter is used in the waveguide to prevent harmonic frequencies from even reaching the oven. Since the harmonic waves are kept out of the oven, no secondary seal on the door is needed to keep them from escaping around the door.

Microwave energy cannot leak through the small holes in the oven door because of the small size of the holes as compared to the wavelength of the microwaves. These waves could escape, however, through a slit such as would exist around the edge of the door if it were not tightly closed or if the choke and vinyl seal were damaged and did not block the

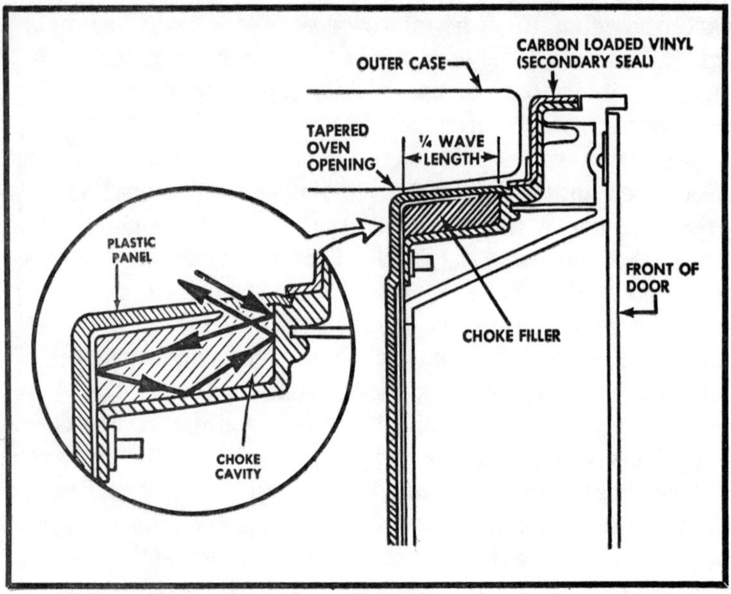

Fig. 2-19. How the oven door prevents microwave leakage.

passage of the waves through the door gap. But if the microwave oven is in good condition, only extremely small amounts of energy could ever be radiated in this way. Also the radiation will diminish in intensity according to the square of the distance from the source.

The normal position of a person using the oven places the eyes and other sensitive areas of the body at distances where traces of energy leaking past the door seal are fully dissipated.

Chapter 3 Basics of Microwave Servicing

The basic design of the microwave oven makes it an inherently safe device to use or service. There are certain precautions that should be followed to maintain this safety. These and various other matters that are of consequence to the serviceman regardless of the make and model of oven are the focus of this chapter.

SAFETY IN SERVICING

The following may be regarded as the Ten Commandments of Microwave Servicing:

1. Always operate the unit from an adequately grounded outlet. Do not operate on a 2-wire extension cord.
2. The oven should never be operated if the door does not fit properly against the seal, if the hinges or hinge bearings are damaged or broken, if the gasket or choke is damaged (cut, pieces missing, etc.), or if any other visible damage can be noted. Check the choke and gasket areas to ensure that they are clean and free of all foreign matter.
3. Always have the unit disconnected when the panels are removed, except when making the live tests called for in the service data. Do not reach into the equipment area while the unit is energized. Make all connections

for the test and check them for tightness before plugging the cord into the outlet.

4. Always ground the capacitors on the filter box with an insulated-handle screwdriver before working in the high-voltage area of the equipment compartment. Some types of failures will leave a charge in these capacitors, and the discharge could cause a reflex action, resulting in a self-inflicted injury.
5. Always remember that in the area of the transformer there is high voltage. When the unit is operating, keep this area clear and free of anything which could possibly cause an arc or short.
6. Do not for any reason defeat the interlock switches. There is no valid reason for this action at any time.
7. Before returning a range to a customer, be sure to check for proper interlock action. The two interlock switches should open in about the first quarter-inch of door motion when measured at the right side of the door. The sensing switch should open beyond about a quarter-inch. Check the manufacturer's specifications for the exact distance.
8. Before returning a range to a customer, be sure that the door spacing is reasonably uniform along the top, bottom, and sides, and that it measures an eighth-inch or less.
9. When applicable, always install a new pop rivet and seal when closing up the range, to protect against unauthorized entry and possible modification.
10. To ensure that the unit does not emit excessive radiation and to meet the Department of Health, Education, and Welfare guidelines, check the unit for microwave leakage, using an approved radiation meter as outlined later in the chapter.

CUSTOMER EDUCATION

Throughout this book you will find many troubleshooting tables that show problems, their causes, and possible remedies. Often the indicated remedy is customer education in the use of the microwave oven because many problems are caused by misuse of the oven by the customer rather than by an oven defect. Table 3-1 shows some of the problems that experience shows appear with some regularity. In suggesting

Table 3-1. Problems Caused by Misuse of Microwave Oven.

CONDITION	CAUSE	REMEDY
A little longer time required than specified by the cooking book. 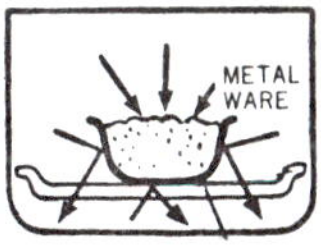	1. Metal ware used: In this case, microwaves from the bottom side of the container are not absorbed by the food, because of metal reflection, so the cooking time becomes longer. Cooking without the glass cook tray yelds the same results. (*)	Do not use metalware. Use the glass cook tray when cooking.
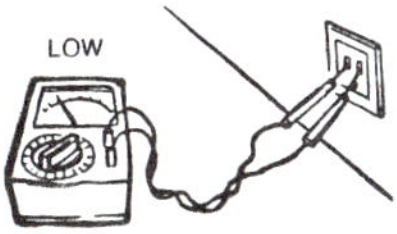	2. Lowering of the powerline voltage: In this case, the output generated by the magnetron becomes low and the cooking time becomes longer. (*)	Use the microwave oven at adequate electrical power.
 UREA RESIN	3. Using a cooking container which has high-absorption characteristics: For instance, using urea resin or phenol resin as cooking ware, cooking time will become longer. (*)	Do not use these types of cooking ware in the microwave oven.
	4. Lowering of the food temperature: Some foods in winter, have a relatively low temperature as compared with the same foods in summer, thereby causing the cooking time to become longer. Food preserved in the refrigerator will have the same condition. (**)	This is not, in reality, a problem. Just cook a little longer.
Spark occurs inside of the oven while cooking. 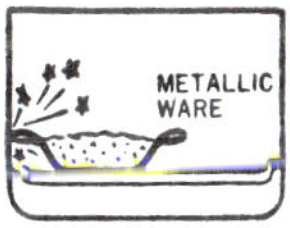	1. A metallic ware contacting the oven wall is used. (***)	Do not use metallic ware for cooking.
	2. Ceramic ware trimmed in gold or silver powder is used. (***)	Do not use this type of ware.
	3. Cooking food incompletely wrapped in aluminum foil. (*)	Wrap the aluminum foil completely.

Table 3-1 cont.

CONDITION	CAUSE	REMEDY
Food or containter in the oven burned or scorched.   	1. A small volume of food is cooked too long.(**) 2. Microwave oven is used to dry clothes, towels, or papers. In this case, these materials gradually become carbonized and finally burn. (**) 3. Bamboo or wooden containers are used for cooking. Wooden or bamboo containers with metal parts may become damaged or scorched in a short time. (*)	Take care not to cook small volume of food too long. Do not use the microwave oven for drying. Do not use bamboo or wooden containers for cooking.
Large meat could not be cooked inside evenly. 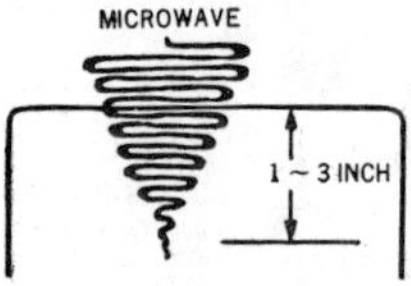 	If meat is relatively large, microwaves will not reach the center of the meat, causing the inside of the meat not to be cooked. Microwave penetration depends on the absorption ratio of the object. Usually the penetration of the microwaves into the food is approximately 1 inch from all sides.(**)	Utilize carryover heating. That is, after finishing cooking the meat, leave the meat for about 10 or 15 minutes until heated inside evenly.
A film forms on the surface of the soup or cocoa. 	A film forms, at times on the surface of soup or cocoa due to the dehydration of the surface. (***)	Stir the food occasionally while cooking.
Food unevenly heated.  	Because the food in the microwave oven is heated by microwaves it is impossible to maintain a constant intensity of microwaves due to microwave characteristics. In order to make up for this deficiency, a stirrer blade is rottated during cooking, thereby agitating the microwaves for more even distribution. If completely even cooking is desired, open the door and give the dish a half-turn, turn the food over during cooking or 1 or 2 times while cooking. This helps maintain more even cooking. (***)	A. Turn 180° once or twice while cooking food such as a large roast pork. B. Change the position once or twice when cooking several pieces of food simultaneously. C. Turn the food over once or twice while cooking food such as roast pork or chicken. D. Wrap the thinner part with aluminum foil when cooking or defrosting large fish or roast chicken. E. Use plastic wrap or cover with a lid. F. Stir once or twice while cooking soup, cocoa, or milk.

Table 3-1 cont.

CONDITION	CAUSE	REMEDY
The microwave oven does not operate when the timer is set for less than 2 minutes.	If the cook timer is set for less than 2 minutes, the timer contact will not turn on, due to the gap in the cam assembly in the timer.(**)	To set the timer for less than 2 minutes, turn the cook timer a little past the 2 minutes position and then return to the desired time.
Glass cook tray becomes cracked. 	If the glass cook tray is exposed to a sudden change in temperature, it might become cracked. If the glass cook tray is subjected to a sudden impact, it also may become broken.(***)	Do not expose the glass cook tray to a sudden change in temperature or to sudden impacts. Be careful not to put the glass cook tray in water when it is very hot.
The ceiling of the microwave oven is burned. 	If a piece of food or spilled liquids adheres to the inside ceiling, the ceiling may become burned.(***)	Wipe the inside ceiling with a damp cloth if a piece of food adheres to the ceiling.
Food does not become browned.	Because the food is cooked inside and outside at the same time, it is not browned. This is one of the features of the microwave oven. To brown the food follow the method on the right.(**)	A. Brown meats slightly in a gas or electric oven, a frying pan, or a skillet before cooking in the microwave oven. B. Another method is to brush meats with a mixture of brown bouquet sauce and water before cooking.
Noise occurs.	This noise is normally caused by the fan motor cooling the magnetron. It is sometimes caused by a defect in the fan motor.(***)	This is not a problem unless the noise is obtrusive, in which case the motor may be defective.
An odor remains in the oven. 	Odors and steam in the oven are exhausted by the air from the fan motor. However, sometimes a remaining odor is detected in the oven. This is the odor of spilled liquid, oil, or a piece of food attached to the inside of the oven.( )	Wipe the oven wall and inside ceiling with a damp cloth to eliminate the odor. If the odor is unusual, wipe with kitchen detergent.

Legend:
(***)—denotes many calls
(**)—denotes occasional calls
(*)—denotes a few calls

the remedies listed in the table, and in any effort at customer education, avoid coming off like the detergent pitchman posing as an appliance repairman in a TV commerical. Use tact in making suggestions.

The following paragraphs provide some helpful tips for users of microwave ovens.

Easy Control Operation

Operating a microwave oven is simplicity itself. There is nothing required but to load the oven, close the door, set the cooking time, and push the START button.

The timer will turn counterclockwise from the time you set it until it reaches the off position. The oven will shut itself off, both oven light and oven signal light will go out, and the solenoid will release the door-locking mechanism, permitting the door to be opened and the food to be removed or tested.

The timer is typically a 12-minute control, numbered counterclockwise from 0 to 11 (see Fig. 3-1). Each mark between the minute numbers indicates 30 seconds. To set the time required, turn the indicator dial to the time you want. The dial can be turned in either direction.

To interrupt a cooking operation, or if you wish to open the door to stir, turn, or check food, push in STOP button. This will release the door lock and permit the door to be opened. To resume cooking, close the door and push the START button. The timer will resume operation from that point without resetting.

The cooking signal light and the oven light will be on when the oven is in operation. The lights will go out when the cooking time has been completed or the door is opened.

Cooking Utensils

There are several different kinds and types of cookware that can be used in the microwave oven.

Glass. There are many heat-tempered, oven-proof, clear-glass utensils in assorted sizes and shapes available for use in the microwave oven. Each recipe states the utensil may cause a variation in cooking time.

Paper. Paper products may be used in recipes where directed. Paper plates should not be used with foods of high moisture content unless the paper has a plastic coating. Paper towels may be used to cover foods that spatter or absorb grease.

Examine frozen-food packages of paper, which may have some metal parts. These should not be used in the microwave oven.

China, Ceramics, Glass Ceramics, Opaque Glass, Pottery. If any utensils of these materials contain metal in the construction, composition, or decoration, they cannot be used

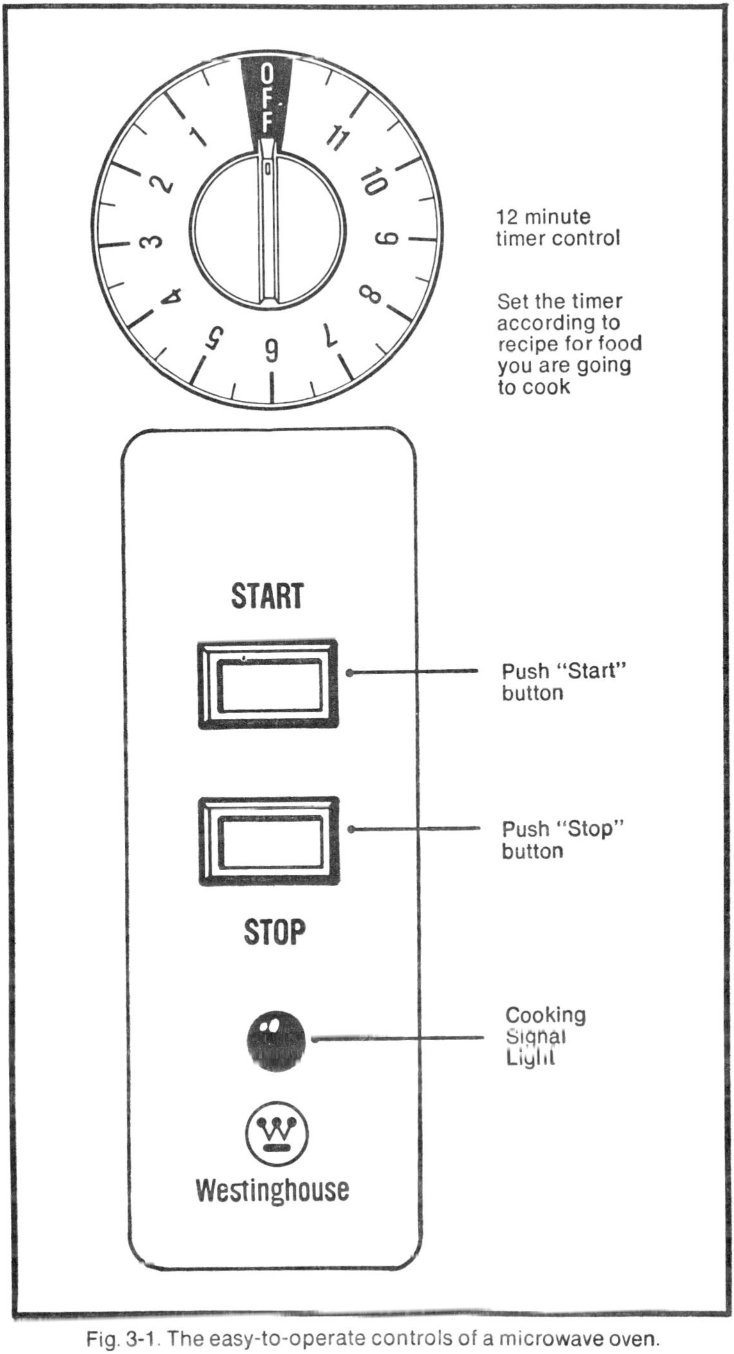

Fig. 3-1. The easy-to-operate controls of a microwave oven.

in the microwave oven. When the body of a utensil becomes quite hot during a short period of time in the oven, it should be eliminated from those utensils suitable for microwave cookery.

Some manufacturers use a code printed on utensils to designate those which can be used with microwaves. Information supplied with the utensil should supply this distinction.

Plastics. Plastics which hold their shape in the presence of heat may be used. The microwaves do not melt plastic, but the heated food conducts heat that may cause shape distortion.

Foods can be thawed in plastic to the just-thawed stage. Plastic cooking pouches of frozen vegetables and main dishes may be used if a small hole is pierced to permit steam to escape during cooking.

Metal Cookware. Metal cookware cannot be used in the microwave oven. Even small amounts of metal present in the oven reflect the microwaves so that they will not penetrate the food to be cooked. The only variance from this is the use of foil on turkey or chicken legs.

Be careful not to use dinnerware or glass cookware with a painted edge, metal trim, or a metal attachment. If arcing should occur, turn off the oven and check cookware carefully for metal.

A meat thermometer has metal parts and cannot be used in the microwave oven. Check the doneness of meat by removing the meat from the oven before inserting the thermometer. Remove thermometer before continuing the cooking of meat.

Special Help

When determining cooking times for a particular food, use the minimum amount of time given. Check the doneness at that time and at intervals until food is done. The microwave oven cooks so quickly that it is easy to overcook.

All cooking times are approximated. The length of cooking time will vary with differences in foods from various parts of the country.

Items requiring more than a few minutes of cooking time should be given a quarter-turn every few minutes, or at least a one-quarter, one-half, and three-quarters turn during the cooking period.

Do not salt meat prior to cooking it in the microwave oven. Salt draws the moisture from the meat and tends to toughen the outer layer of meat. Salt may be added after the meat is cooked. Other seasonings, such as pepper or paprika, may be used during cooking.

Small steaks or pieces of meat which cook faster than they brown have a more acceptable color when a basting agent is brushed on the meat prior to cooking.

For best results with frozen meat, fish, and poultry, the thawing and cooking should be two separate processes. Thaw first, allow to stand 10–15 minutes; then cook the recommended length of time for fresh meat.

Leftovers can be reheated in a very short time. They taste freshly prepared because there is no drying or reheated taste.

Cleaning the oven is a very easy task if a cup of water is allowed to boil for a minute or two in the oven. Wipe the moisture away from the oven interior with a clean, dry cloth or paper towel.

Taking Care of a Microwave Oven

There is little to do to keep the microwave oven in tip-top condition. The inside of the unit is made of coated steel. If it should become splattered, it should be wiped with a paper towel or cleaned with a mild detergent in warm water and a soft sponge or cloth.

To clean the glass shelf, lift up the shelf and remove. Wash in warm water and detergent. Replace with rough side down.

Never pour water into the bottom of the unit.

Do not use an abrasive to clean the inside. It might damage the coated steel.

Do not operate unit when empty. Whenever the oven is operated it should have food in it. Otherwise, the magnetron might be damaged.

Hints and General Information

The recipes should be followed exactly to gain experience in microwave cooking. Follow the recipe, use the specified utensil, measure carefully, use the same ingredients listed, and combine ingredients as directed. Later, a favorite recipe may be tried by substituting the directions for a similar recipe. Remember to check food doneness at the minimum time, followed by interval checking until the desired stage of doneness is reached.

Cooking times are very important. The microwave oven cooks rapidly, and food can overcook easily if the wrong times are used. Always refer to the cooking-time guide or the recipe instructions.

Carryovers occur after food is removed from the oven. Allow for this additional cooking from internal heat within the food.

The cooking-time requirement increases when the amount of food to be cooked is increased. The time increase is not always proportional to the food increase. Prepare two separate dishes and bake separately when double quantities are desired.

Some foods of large volume cook long enough to acquire a desirable brown color. Other recipes call for egg white, sauces, or quick searing to obtain the expected food coloring. Conventional cooking may be called for in a few recipes to aid the microwaves in preparing a more conventional food product.

Foods with a skin or covering which prevents interior steam from escaping should be pierced once or as directed in recipe. Whole potatoes, eggs with unbroken yolks, and apples are examples of such foods. The cook-in-bag pouches of frozen foods should be pricked once to make a steam-escape opening.

When foods are cooked rapidly by microwaves, steam which normally forms during cooking is created faster than it is dissipated. The condensation on walls of the oven is common with many food preparations. Wiping the moisture away following cooking is an easy way to clean the oven interior.

The glass shelf in the bottom of the microwave oven may be removed for cleaning. Handle it carefully and replace with the rough side down. This permits the microwaves to be reflected up into the bottom of foods being cooked. It also makes cleaning easier. The glass shelf is a special glass designed for use in the microwave oven. Ordinary glass will not function properly and should not be substituted.

Important Things Not To Do!

Do not operate the oven when it is empty, because the microwave energy has no food to absorb it and it can cause damage to the magnetron tube.

Do not use metal cookware of any kind. It is possible to damage the magnetron.

Do not put *TV dinners* on aluminum trays in the oven to heat. Transfer such food to a serving plate or use proper utensil.

Do not use a meat thermometer to check food when oven is operating. It may cause arcing and malfunction.

Do not use wax paper or paper with grease on in the oven. When cooking bacon or other meats, always use fresh paper towels or plates with each load.

It is not recommended to make popcorn in the oven because of the load factor. Once the corn is popped, there isn't a load.

Do not wrap foods in aluminum foil to heat. One exception is the use of lightweight foil to cover spots of a whole turkey, which may have a tendency to overbrown. Only small pieces should be used.

Do not operate oven when door is damaged.

Where to Use a Microwave

You can put it on a stand and move it from area to area. Almost every place you go, the microwave oven will serve your family—on the patio, in a boat, or in a summer cottage. And it is ideal for a small apartment.

It may be more convenient to place the microwave oven on a counter that will put it right in the regular food flow in your kitchen. The handsome outer case will harmonize in most kitchen decors. It can add a touch of glamour, too! An ideal place for the microwave oven is at the end of a countertop on a cabinet or stand with wheels. This allows the unit to move from place to place in the kitchen or other areas. Be sure the cabinet is sturdy enough to support the weight of the unit.

MEASURING MICROWAVE POWER

We measure microwave power by converting the microwave energy into heat, then measuring the heat in a simple calorimeter. All we need are two measuring cups or beakers, a thermometer, and some water from the cold-water faucet.

Metric Measure

Pour 2 liters (2.11 quarts) of water into beakers and stir the water with a thermometer. Measure the initial temperature. Place the beakers at the center of the glass cook

$P = 4.18\, m \frac{\Delta T}{\Delta t}$ WATTS | m = gms | ΔT = °C | Δt = sec

plate in the oven and heat for 2 minutes. When the heating is finished, again stir the water with the thermometer and measure its temperature. Determine the temperature difference; then multiply it by 39. The product is the actual power output of the oven. The temperature rise of the water will be greatly affected by the voltage at which the oven is operated. The voltage at the time of the test should be accurately known to properly understand the test results. If the *nominal* output of the oven is 600W (watts), an *actual* output of 540W to 690W may be expected.

U.S. Customary Measure

Pour two cups of cold tap water into a Pyrex measuring cup. Note the temperature. Heat the water in the microwave oven for 1 minute (as noted by a clock or watch with a second hand) and measure the temperature again. The difference in temperature times 18 is the power in watts.

Quickie Test

Place a cup of water at room temperature in the oven. Close the oven door and set the timer to 3 minutes. The water temperature should rise to 212°—boiling temperature—in 2.5 to 3 minutes. This is a simple test to make, and one that the customer can readily understand and appreciate.

In evaluating any of these tests, note that the actual output power will depend greatly on the supply voltage of the house circuit. Furthermore, if the result is at all close to the nominal output, do not replace the magnetron. Magnetrons are expensive, and it is a simple matter to adjust cooking times to correct for low and high output.

MICROWAVE LEAKAGE CHECK

After any service to the latch assembly (primary interlock, secondary interlock, or monitor switch) and the door, the following leakage test must be performed to ensure compliance with federal regulations. The measurements are made with a microwave-leakage tester, one of which is shown in Fig. 3-2. Some recommended instruments are:

- Simpson Model 380
- NARDA Model 8100
- NARDA Model 8200
- Holaday Model HI-1500

Proceed as follows:

1. Pour 275 ±15 cc of 20°±5°C (68 ±9°F) water in a beaker and place the beaker in the center of the oven.
2. Press the START button to measure the leakage by using an electromagnetic-radiation monitor.
3. When using the radiation monitor, follow the manufacturer's recommended test procedure to assure correct results.
4. When measuring the leakage, always use the 2-inch spacer with the probe. This is because the value of the radiation is specified to be measured at the distance of

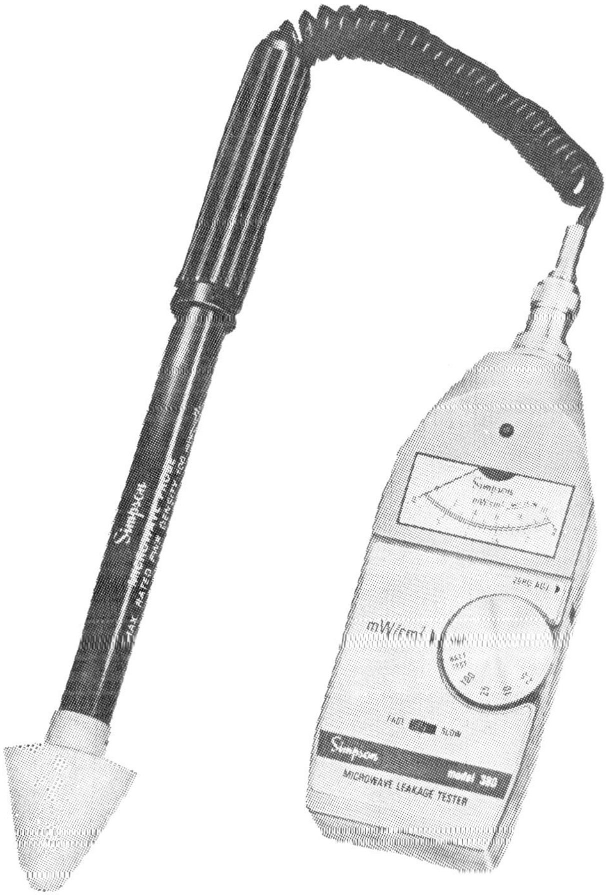

Fig. 3-2. The Simpson Model 380 microwave-leakage tester.

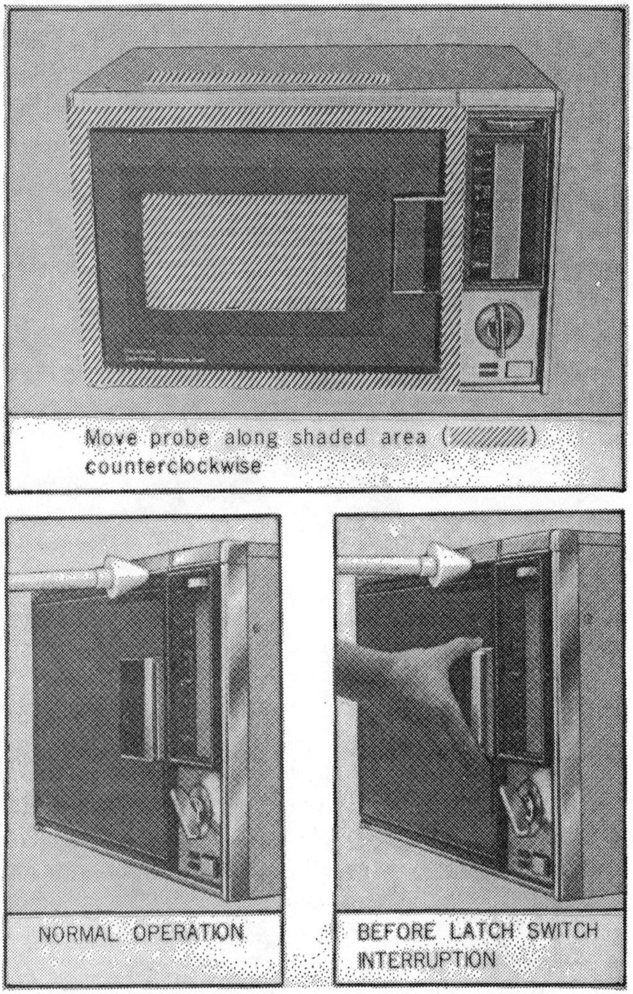

Fig. 3-3. Probing for microwave leakage.

2 inches away from the suspected areas. Hold the probe perpendicular to the cabinet door. Place the spacer cone of the probe on door seam and ventilating port on the cabinet, and move it along the seam. The measurement should be taken in the counterclockwise direction at a rate of 1 inch per second. If the leakage of the door seam is unknown, move the probe slowly. When testing near a corner of the door, keep the probe perpendicular to the areas, making sure the probe end at the base of the cone does not get closer than 2 inches

to any metal (Fig. 3-3). If it does, erroneous reading may result.

5. The measured radiation leakage value must be less than 5 mW/cm^2 after repair or adjustment (Panasonic's voluntary standard is 2 mW/cm^2) with the door in the closed position. Before latch-switch interruption with the door in a slightly opened position, the reading should be less than 5 mW/cm^2 (2 mW/cm^2 for Panasonic).

Simpson 380 Microwave-Leakage Tester

Prior to operation of the instrument, review and perform (where applicable) the following steps and checks. These steps can also be used as a general functional test.

1. Connect the thermocouple probe to the instrument by inserting the probe plug into the connector jack on the front end of the instrument; then tightening the locking collar.
2. Attach the 5 cm (2-inch) spacer cone (see Fig. 3-2) onto the probe tip, using a slight twisting and pushing motion.
3. With batteries installed, turn the range-selector switch from the OFF to BATT TEST position. Observe the meter indication; it should indicate within the BATT OK area on the dial. If it does not, replace batteries.
4. Place the thermocouple probe into a zero-power-density area, or wrap the probe in aluminum foil to avoid stray-power pickup. Set the range-selector switch to the desired range; adjust the ZERO ADJ control for a zero indication.

Note: Rotating the ZERO ADJ thumbwheel clockwise (down) moves the meter pointer in an upscale direction. Rotating the thumbwheel counterclockwise (up) moves the pointer in a downscale direction. The ZERO ADJ potentiometer is a multiturn control; thus several full turns may be required to zero the instrument.

To achieve correct and repeatable measurements, the probe must be postioned at a predetermined distance from the area being checked. The facilitate these measurements, a removable spacer cone is provided, which, when attached to the probe tip, automatically spaces the sensing element to the standardized distance of 5 cm (2 inches).

Because of the presence of mode stirrers in some ovens and excessive ripple voltages from magnetron power supplies, the emitted field densities fluctuate, making accurate readings difficult. To minimize these fluctuations, the instrument provides a slow-operating mode. With the use of the FAST-SLOW switch, the response time of the instrument is changed from a 1-second to a 3-second time constant.

To determine if the probe is defective, zero the indicator with the probe attached and the range switch set to the most sensitive (2.5 mW/cm^2) range. Disconnect the probe assembly from the instrument. If the pointer remains on zero (does not go below zero) the probe is defective.

If the magnitude of the field to be measured is unknown, begin the measurement procedure with the range switch set to the highest (100 mW/cm^2) range.

Check and zero the instrument each time the range is changed, before approaching the microwave field to be measured.

The probe is designed to provide accurate measurements and minimize any perturbations of the field being measured. Probe orientation, whenever possible, should be such that the probe handle is perpendicular to the plane of the radiating source. This natural orientation of the probe also adds to the safety of the operator by maintaining a maximum distance between source of emission and the hand holding the probe, as well as the body of the operator, which is important whenever approaching a device with high unknown emissions.

Slowly introduce the thermocouple probe into the microwave source under survey and rotate the probe around its axis through 180°. Determine the maximum and minimum meter indications. The correct reading then is the average of the two readings. For example, if a minimum reading of 2.3 mW/cm^2 and a maximum of 2.5 mW/cm^2 is obtained, the average power density would be:

$$\frac{2.3 + 2.5}{2} = \frac{4.8}{2} = 2.4\ mW/cm^2$$

If the instrument indication is less than approximately one-fourth of the scale, select the next lower range. Repeat this procedure, as necessary, to obtain the most accurate reading possible. Zero each time the range is changed.

Sources of possible leakage should be thoroughly probed to define the position of maximum leakage; the leakage must be measured when the device is operating at its maximum output.

Caution: If the instrument indication goes to full scale during the initial probing (100 mW/cm^2 plus), discontinue the measurements, as the instrument may be damaged by the excessive microwave power.

The final reading should be taken with the probe tip placed as close to the plane of radiation as the spacer cone permits. Probe about for the maximum indication.

PROPER HANDLING OF MAGNETRON TUBES

A magnetron tube, like a radio or TV tube, must be handled with a reasonable amount of care. Always handle it by the housing area only. Use caution not to touch or strike the glass portion at the top or bottom.

The carton used to ship replacement tubes is reusable. Defective tubes must be returned to factory in this carton.

Magnetron Failure Modes

Magnetron failures that have been identified are generally grouped into the categories listed below:

Air. This is a case where the tube has had the vacuum envelope destroyed and air has entered the tube. This will cause internal arcing and high line current if high voltage is applied. If the tube failed when in use, the copper antenna will not have the characteristic shine but will be oxidized to a dull copper color.

Tipoff Crack. The result of poor handling and striking the glass or of the tipoff being too long and cracking from the thermal or mechanical stress. The crack may be annular around the tipoff or lengthwise as shown in Fig. 3-1.

Suck-in. The result of abnormally high power on the antenna glass which softens the glass. The atmospheric pressure pushes the glass inward toward the vacuum until a hole forms. Improper use can create the fault such as operation with the oven empty or with metal cooking utensils. This fault is usually at the antenna glass-to-metal seal, approximately an eighth-inch in diameter.

Insulation Breakdown. At the end of the leads, a breakdown of insulation may occur from degradation of the

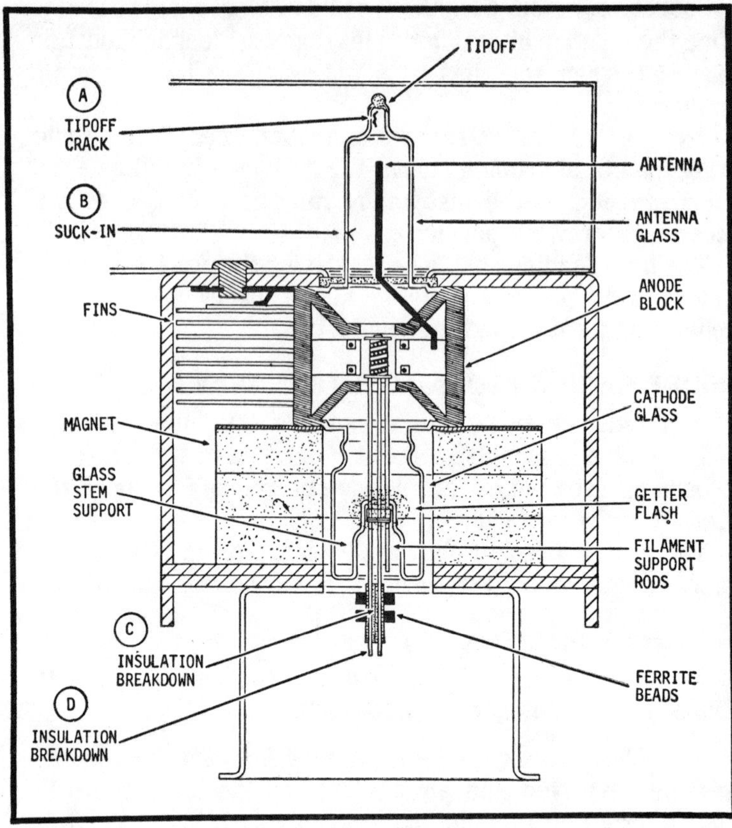

Fig. 3-4. Magnetron failure modes.

material. This is rare but appears to result from high temperatures for an extended time. In the region on the ferrite beads, a breakdown of the insulation may occur if the beads were against the metal filter box. The ferrites are conductive and the tube installer should attempt to keep the beads a half-inch from the metal walls.

Operational Faults

An Open Heater can be determined by ohmmeter when the transformer leads are disconnected. Resistance is normally less than 1Ω (ohm); the filament does not short internally. Any tube removed should be checked since usage tends to make the filament more fragile. Later handling or shipping may open the filiment and thereby mask the true failure when checked at the factory.

Low emission is caused by *wearout* of the emission characteristics of the directly heated cathode. Symptoms are: 1) The tube current will take longer to get to the operating point of 275 mA than a new tube (normally about 2–3 seconds). 2) The tube current does not get high enough to cause the tube to oscillate with normal line voltage (the current will be below 200 mA. 3) The oven produces low power into a load, two-thirds normal or less.

Chapter 4
Servicing
Amana Microwave Ovens

The story of the microwave oven is largely the story of Amana Refrigeration, Inc.[1], and of its parent company, Raytheon. And probably no industrial company has a more intriguing story or more unlikely origin than Amana. The company was founded in 1934 by George Foerstner, who originally built cold-storage lockers. In 1936, the firm was acquired by the Amana Society, a group of former commune members. This seems a strange beginning indeed for a company whose growth and prosperity epitomize the capitalistic ideal. Today the company is still headed by George Foerstner, whose ancestors were original members of the German religious sect that founded these even communal Amana villages in 1888. Now owned by Raytheon Company—a pioneer in microwave research and the principal maker of microwave tubes for radar during World War II—Amana is synonymous with the latest developments in appliances for modern living.

Harking back to the origin of the microwave tube, the trade name *Radarange* is used by Amana for its microwave ovens. Our consideration of Amana products starts with the original home Radarange, Model RR-4, which is something like the Model T of microwave ovens. This model was the industry leader from its inception in 1967 until it was

1. All of the illustrations in this chapter are courtesy of Amana Refrigeration, Inc.

superseded by Model RR-7 in 1976. Both models have two mechanical timers (5 and 30 minutes) and a maximum power output of 675W. The output of the newer RR-7 is variable by means of a Cookmatic Power Shift control.

Also covered are Touchmatic models RR-6 and RR-9. The Touchmatic models have a smooth-glass electronic touch-panel in place of conventional timers and controls. The panel looks somewhat like the keyboard of a pocket calculator, but it uses capacitive switches instead of mechanical ones. As the pads, or *keys*, of the panel are touched, instruction signals are sent to an electronic memory The instructions are later recalled from the memory as the necessary cooking functions are to be performed. Model RR-9, which superseded Model RR-6, also features Cookmatic Power Shift.

Notice: Amana Refrigeration, Inc., discourages the repair of its products by persons other than service personnel expressly authorized by it, and such repair will void the valuable warranty on the products. To obtain information on becoming an authorized repairman, contact the national service manager at Amana Refrigeration, Inc., Amana, Iowa, 52204.

HOW THEY WORK

Figures 4-1 through 4-4 are the schematics of the four models under consideration.

Transformers

The transformers used in these ovens are of both the step-up and step-down variety.

Plate Transformer. This step-up transformer is rated to function on a primary supply voltage of 115V AC ($\pm$10%). At 115V AC the secondary voltage will be approximately 2400V AC.

The transformer is a saturated type, where the secondary core is operating at magnetic saturation; the input line voltage may fluctuate without changing the constant power output of the oven.

An additional tap is provided on the secondary for capacitor values and is covered later.

Filament and Control Transformer. This dual-purpose step-down transformer is rated to function on a primary voltage of 115V AC ($\pm$10%). At 115V AC the secondary output will be 3.1V AC to supply the magnetron filament (and 24V AC

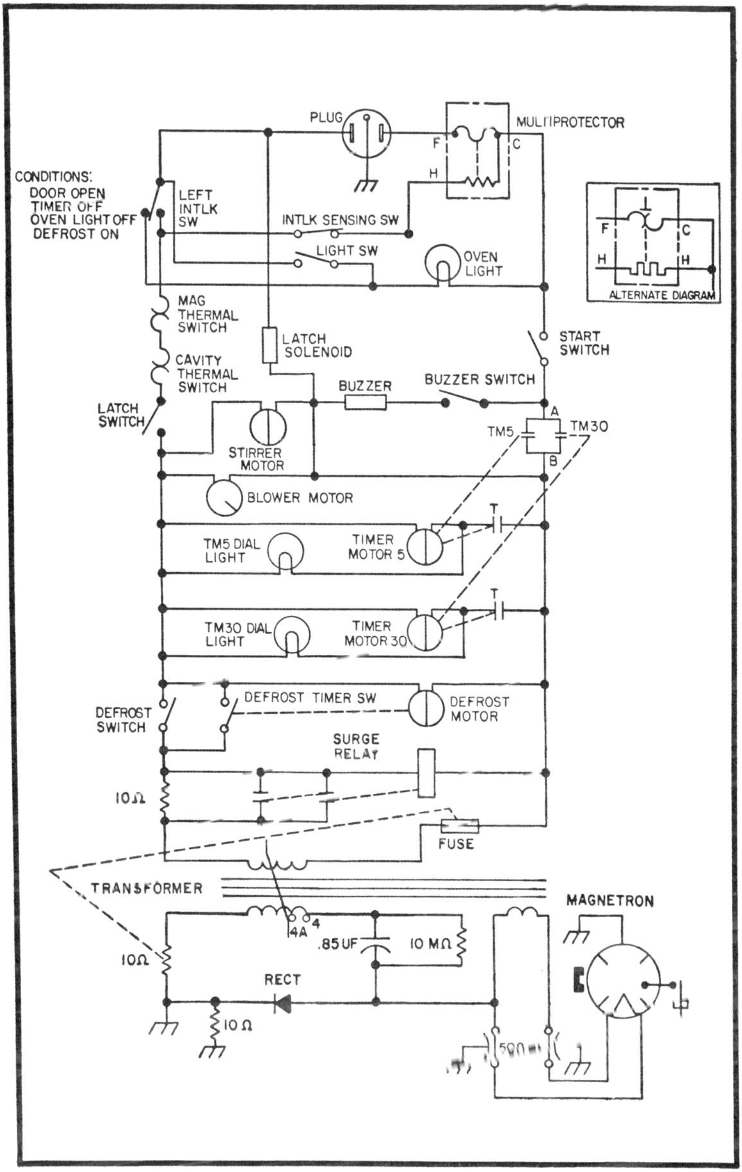

Fig. 4-1. Schematic diagram of Amana Model RR-4D.

to supply the Cookmatic variable-power module in RR-7). The primary leads of this transformer must be polarized (wired per wiring diagram) or a failure of the HV (high-voltage) diode may result.

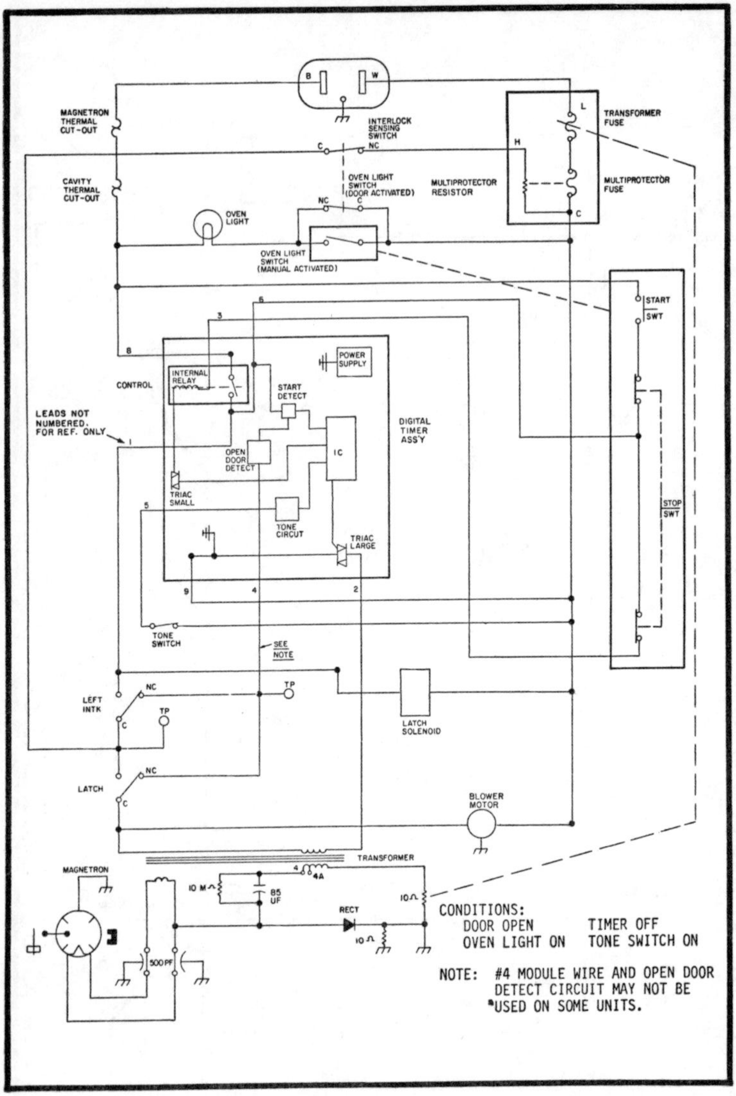

Fig. 4-2. Schematic diagram of Amana Model RR-6.

Transformer-Protect Fuse

The fuse-resistor board contains a 10Ω resistor, which is in the secondary HV circuit. A thermal fuse (in the primary circuit) is placed inside a resistor to protect the plate transformer should a short occur in the secondary (HV) circuit. When a high-voltage short to ground occurs, the

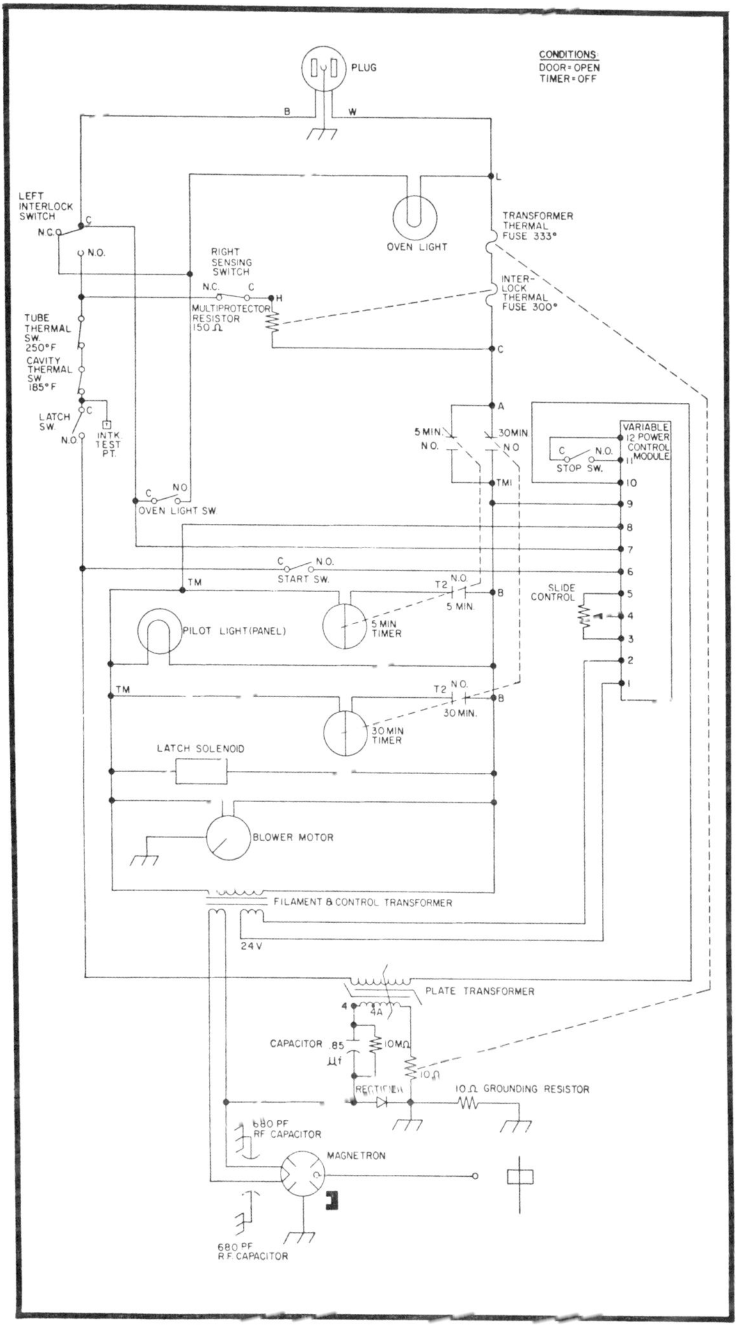

Fig. 4-3. Schematic diagram of Amana Model RR-7.

current heats the resistor, which opens the fuse (333F), in the primary circuit. If the fuse should open, *only* the oven light will operate.

Rectifier

A special rectifier consisting of a single diode is used. The rectifier used conducts high current in one direction, with a low voltage drop, but conducts a low current at a high voltage drop in the other direction.

If the voltage becomes excessively high, this special diode will conduct a high current for a short time and effectively protect the other components from shorting.

Magnetic-Field Circuit

The major components of the magnetic-field circuit are the permanent magnets and the tube housing. The magnetic-field circuit components are an integral part of the magnetron-tube assembly and may not be disassembled. The intensity of the magnetic field imposed on the tube cannot be adjusted.

To prevent possible demagnetization of the magnets, tools and other steel objects should not be allowed to come in contact with the magnets or the tube housing.

Magnetron Tube

The magnetron used in these units utilizes permanent magnets and a half-wave voltage-doubler circuit for tube operation in converting the 60 Hz electricity into microwave energy. The magnetron tube is the heart of the Radarange.

The magnetron needs a high DC voltage (4000V), a low AC voltage (3.1V), and a magnetic field for normal operation. The tube consists of an anode and a filament, which also is the cathode.

For the tube to operate, electrons must flow in the tube from the cathode to the anode. To get the electrons to flow, high voltage has to exist between the cathode and the anode, and the cathode must be heated. The heating of the cathode is done by the filament. The filament or heater power is supplied by the 3.1V AC winding on the transformer.

The above paragraph describes the operation of a simple diode. To convert a diode to a magnetron, a magnetic field is applied parallel to the cathode. The magnetic field causes the electrons to spin around the cathode instead of traveling in a straight line from the cathode to the anode.

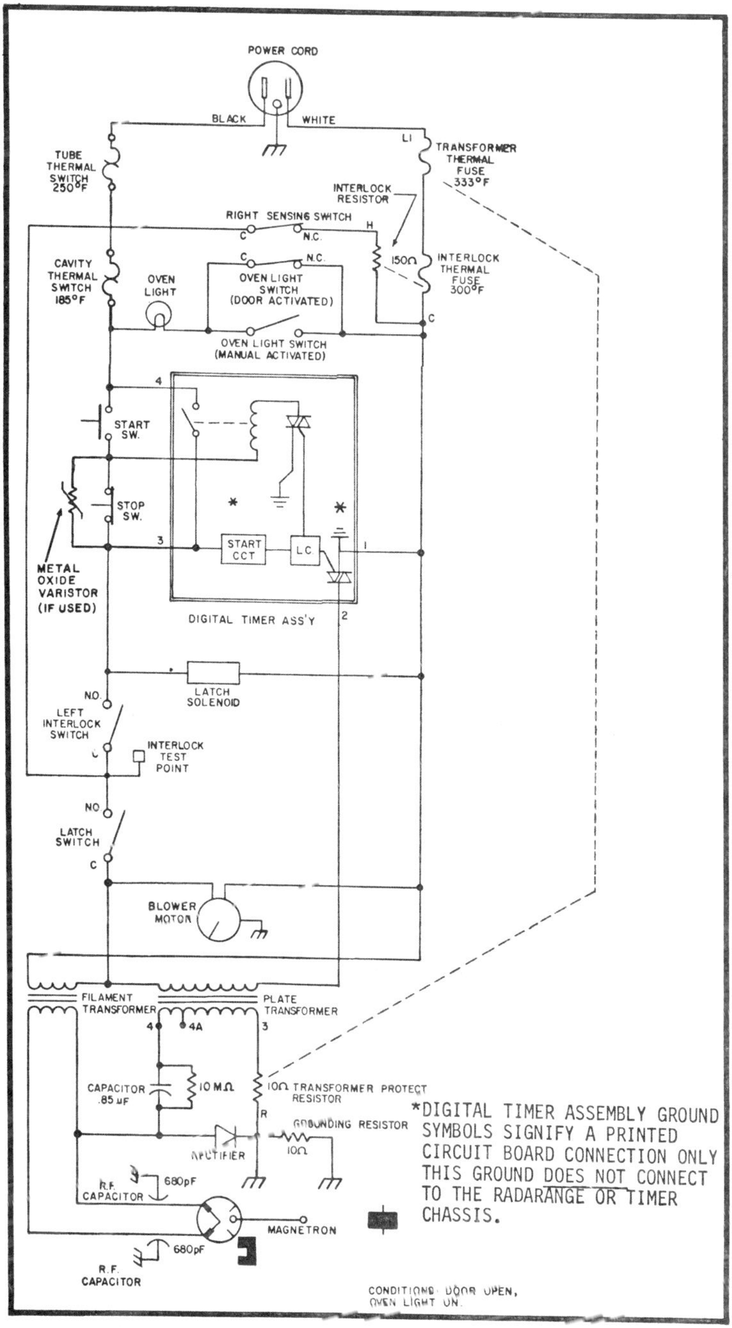

Fig. 4-4. Schematic diagram of Amana Model RR-9.

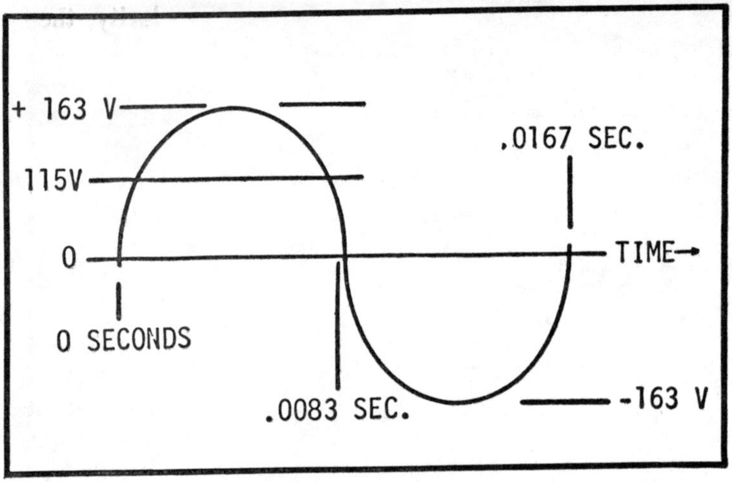

Fig. 4-5. Waveform of power-line voltage.

The spinning action of the electrons and the configuration of the anode causes RF currents to flow on the surface of the anode. An antenna connected to the anode directs the RF power down the waveguide.

Not all energy is transferred by the tube antenna. A small amount of RF current may feed back to the tube filament leads. If this power were allowed to feed back to the house circuit, it could cause TV and radio interference. To prevent this, ferrite rings and bypass capacitors are added to the circuit.

High-Voltage Doubler

Warning: The circuit's 4000V potential cannot be measured without special equipment. *Do not attempt to read secondary voltages.* To attempt to read these voltages serves no useful purpose in servicing the equipment and subjects the technician to an unnecessary hazard.

The high-voltage circuit covered here is called a *voltage-doubler* circuit. A simple explanation of its operation starts with the half-wave voltage doubler.

The 60 Hz AC voltage at the receptacle is shown in Fig. 4-5.

The voltage in AC circuits measured with a standard voltmeter will be an RMS (root-mean-square) value. The effective or RMS voltage is 115V in a home. The voltage is actually, at any instant of time, varying from zero to a maximum of 163V (115 × $\sqrt{2}$) with a positive polarity; then repeating this with negative polarity.

The relationship of the peak value of 163V to the meter reading of 115V is very important in understanding the voltage-doubler circuit.

In these power transformers, the secondary voltage is approximately 2700V measured with an RMS reading meter. This is a peak voltage of 3800V; but remember, the voltage alternates in polarity each half-cycle, which is 8.3 milliseconds long.

If the switch in Fig. 4-6 is closed at the instant the voltage in the sine wave is zero, no current will flow. When the voltage increases with the polarity shown, current *i* will flow in the direction of the arrow, and the capacitor will be charged through the diode. The voltage on the capacitor will rise with the secondary voltage to the maximum of 3800V (peak). As the voltage of the transformer decreases from the maximum positive value, the capacitor voltage is higher and will try to discharge (in a direction opposite to the arrow of Fig. 4-5). The diode will not conduct in this direction, and the capacitor remains charged to 3800V.

When the secondary voltage becomes a negative 3800V (with the polarity of the circles in Fig. 4-5), the voltage on the capacitor, added to the secondary voltage, is the sum voltage across the nonconducting diode. The voltage across the diode (points *A* and *B*) is now 7600V.

The magnetron tube is connected in parallel with the diode (*A* and *B*). When the tube is not conducting (filament not hot), the tube voltage is 7600V. When the tube is conducting, the

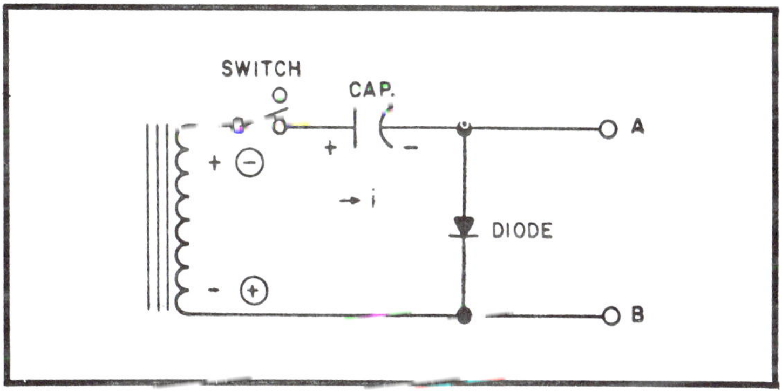

Fig. 4-6. Simple half-wave voltage doubler.

capacitor is discharging, and the voltage will not exceed a peak value of 400V because of the magnetron load.

Interlock-Monitor Circuit

All microwave ovens must by federal law incorporate an interlock-monitor circuit. The monitored circuit—in this case, the left-door interlock—is designated as the *secondary* interlock. The latch-switch interlock, actuated by the door latch, is the *primary* interlock.

Door Interlock & Monitoring-Sensing Switch

Facing the unit, the left-door interlock switch is mounted on a bracket attached to the lower-left front of the oven cavity. The interlock-sensing switch is mounted on a bracket attached to the lower-right front of the oven cavity. Each switch is actuated independently by counterbalanced arms attached to the door.

The normally open (N.O.) terminal of the left-door interlock switch is in series with the normally closed (N.C.) terminal of the monitoring-sensing switch and the time-delay mechanism (a thermal fuse). Refer to Fig. 4-7.

If a malfunction occurs in the door-interlock switch, and the normally open contacts do not open when the door opens, current will flow through the sensing switch, energizing and opening the time-delay mechanism or fuse. Ten seconds is the maximum time to open the time-delay mechanism.

As the monitored left-door interlock is the interlock nearest the power source, the latch-interlock switch and the STOP switch would interrupt power, preventing oven operation during the actuation time of the monitor circuit.

When current flows through the sensing switch, it energizes a hollow resistor in series with the sensing switch. Current through this resistor melts a line-circuit thermal fuse, which is in the center of the resistor.

A faulty door interlock that allows current through the sensing switch *must* be changed before replacing the thermal fuse.

Latch-Switch Interlock

The latch-switch interlock has wires to its common and normally open terminals. It is actuated by the latch arm on the door.

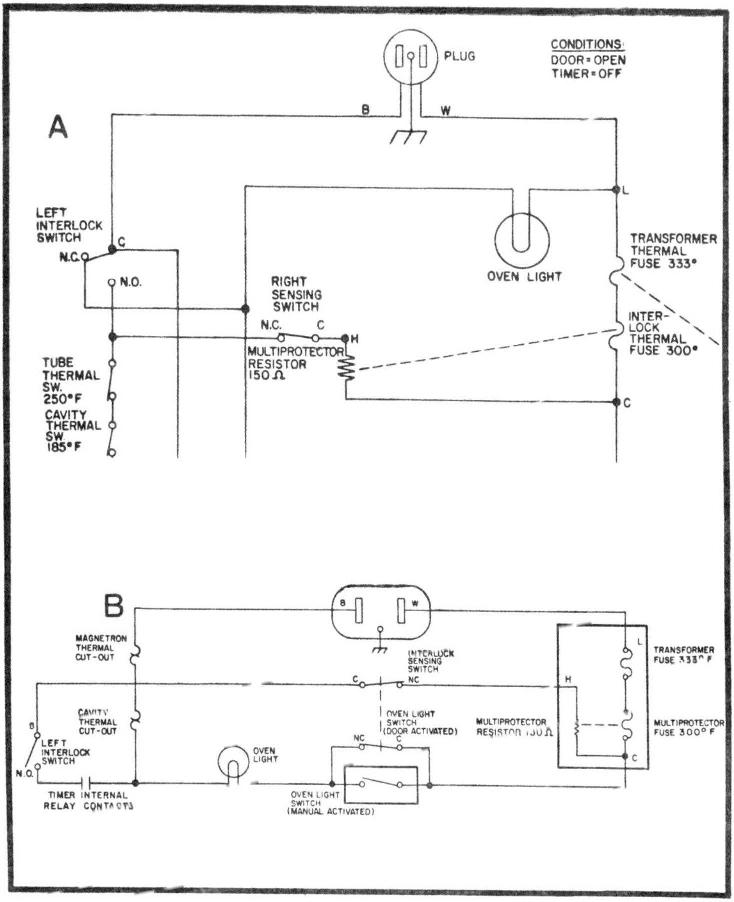

Fig. 4-7. Switch circuits—(A) Models RR-4D and RR-7, (B) Models RR-6 and RR-9.

Tube & Cavity Thermal Protectors

The magnetron-tube protector (opens at approximately 250°F) and oven-cavity protector (opens at approximately 185°F) are self-resetting bimetallic normally closed (N.C.) switches that open when an overtemperature condition is reached on the tube housing or in the oven cavity.

Timers

On models RR-4D and RR-7, the top timer has a scale of 0 to 5 minutes. A second timer is provided for items requiring a longer cooking time. The scale on this timer is from 0 to 30 minutes.

Touchmatic Timer

On models RR-6 and RR-9 the Touchmatic electronic timer is time selectable from 0 to 99 minutes, 99 seconds in 1-second steps. The solid-state light-emitting diodes (LEDs) on the timer panel indicates four digits of selected timer plus cook, hold, defrost, Cookmatic, and timer operation.

Variable-Control Module on Cookmatic Models

The variable-control module (Cookmatic control) is the master-control center of the Cookmatic feature. In conjunction with the Cookmatic slide control, the module will vary the actual *on* time of the magnetron tube from 10% to 100% of each second.

As an example, if the slide control is set at 50% (equal to a middle setting on the slide control) , the plate transformer and magnetron operate for 50% of each second and are off for 50% of each second for the total amount of time selected on the timer. The Cookmatic module controls *only* the voltage to the plate transformer, which supplies high voltage to the magnetron tube. Because this oven has a separate filament transformer, the magnetron filament operates for the total time selected.

Cookmatic Slide Control

The Cookmatic slide control is a specially designed centertapped slide potentiometer. This slide control, in conjunction with the variable-control module, varies the cooking time from 10% to 100% of the time selected on the 5- or 30-minute timer.

Blower

Some heat is generated within the tube, and this heat has to be removed by forced air. Therefore, the unit is provided with a blower. The blower intake is located in the bottom of the unit. The air exhausts from the air-discharge duct, which is located above the timer-control panel.

Microwave Power Transmission Components

From the magnetron, the microwave energy must be directed to the food. One of the basic design considerations which has to be taken into account is that when the energy reaches the food, the food must be heated evenly. Microwaves react much the same way as light, traveling in a straight line An explanation of the transmission components follows.

Waveguide. The magnetron is connected to a waveguide. A waveguide is a rectangular piece of sheet metal, approximately 2 × 4 inches. The waveguide channels the power from the magnetron into the oven. One end of the waveguide is open; this end serves as an antenna to radiate the power into the oven cavity.

Stirrer. The stirrer is a 4-bladed fan-type device, located above the grease shield in the oven cavity. The stirrer blade is driven by moving air through the cavity from the cooling blower. It turns at a very slow speed and is used to distribute the energy to all parts of the oven cavity.

Glass Tray. Food to be cooked in the microwave oven is placed on a glass tray, which rests approximately 1 inch above the bottom center of the oven. The glass tray has no effect on microwaves. The microwaves go through the glass, strike the bottom of the oven, and are reflected back into the food from the bottom side. This allows the microwave energy to cnter the food from all sides.

Containment Components

Once generated, the microwaves must be contained so that only food is cooked. The components are constructed of various materials which reflect, transmit, or absorb power, depending on their intended function.

Oven Walls. The oven enclosure is constructed of stainless steel. Stainless steel reflects microwave energy and does not absorb any power.

Power Door Seal—Choke. The primary microwave seal is the choke (Fig. 4-8). The choke is built on the preriphery of the door and fits the edge of the oven cavity. Microwaves follow the path of least resistance. The microwaves pass by the front edge of the door, strike the back edge of the door, are reflected into the choke cavity, and in turn are reflected back into the oven cavity.

The choke cavity is filled with polypropylene, which has no effect on microwaves. Polypropylene keeps the interior of the choke from becoming filled with dirt and also makes the exterior of the door much easier to clean. The choke seal is so effective in preventing microwave leakage from the oven that these ranges have been specifically and uniquely exempted

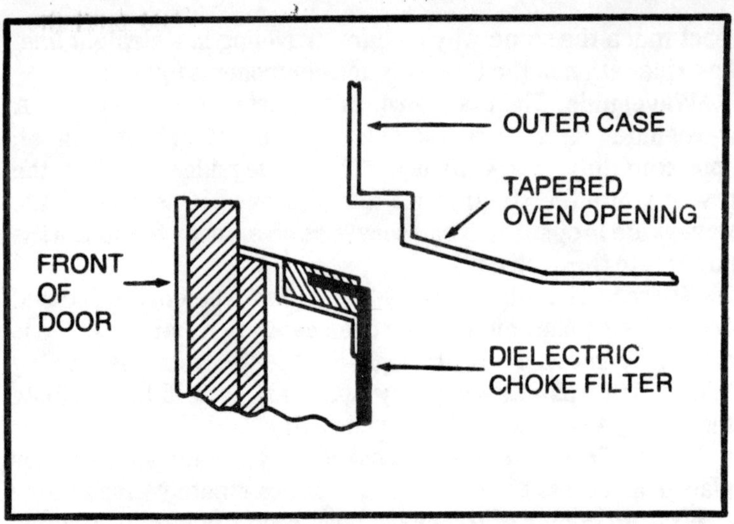

Fig. 4-8. Cavity and inner-door cross section.

from carrying the warning label required on other microwave ovens by federal regulations. Even when the door seal and oven have been subjected to gross abuse the choke seal has prevented excessive leakage.

Secondary Seal. Any power that bypasses the choke and is not reflected back into the oven is absorbed by the door gasket. The door gasket is a special grade of vinyl. Ordinary vinyl passes microwaves, so the vinyl in the door gasket is loaded with carbon black. This makes the vinyl highly absorbent, and therefore it has ability to absorb the microwave power which is passed by the choke.

Oven Light & Switch

The oven light is located at the back of the oven and is a standard 25W appliance bulb. The bulb is removed through an access panel at the top rear of the oven. One screw has to be removed to change this light bulb.

During door opening, the left-door interlock's normally closed terminal, or a right-door sensing light combination switch on some models, energizes the oven light. The oven light may also be operated by the light switch.

Door Latch

The door latch is an electrically operated mechanical device which prevents the door from being opened when the unit is in operation. The door will remain latched until the electrical circuit is interrupted.

Wire Harnesses

To provide easier serviceability, most models incorporate three separate wire-harness connector assemblies. The three assemblies (cavity, power supply, and control module) are each interconnected with multipin connector blocks. This will allow complete removal of each assembly by simply disconnecting the connector block and removing the mounting screws.

Oil-Filled Capacitor

Oil-filled capacitors used in the ovens are supplied with a ±6% tolerance in capacitance. The capacitor supplier color codes the actual value of each capacitor at time of manufacture with a paint dot on the terminal end of the capacitor. An additional transformer tap is provided to compensate for variations in capacitor values. When installing or replacing a capacitor, the technician must check the color code on the capacitor and connect as shown below:

Color Dot	Transformer Tap Used
Red or yellow	4A
White or black	4

Disassembly Procedure

Important: Disconnect power to oven before following any disassembly procedures.

Access to the major electrical components can only be gained by removal of the side panel. (See Figs. 4-9 and 4-10.) To remove the side panel, remove the seal and the three screws from the bottom side. Drill the pop rivet and lift the panel off.

Mandatory: When replacing the end panel after removal, the screws, pop rivets, and seal *must be* replaced.

Caution: During disassembly, some foam gaskets, foam blocks, or aluminum sealing tape may require removal. These parts must be replaced for proper circulation of air over the components and through the oven cavity.

Oven-Light Removal

Disconnect power to oven. The access plate is located in the rear of the unit. Remove one screw and tilt the access

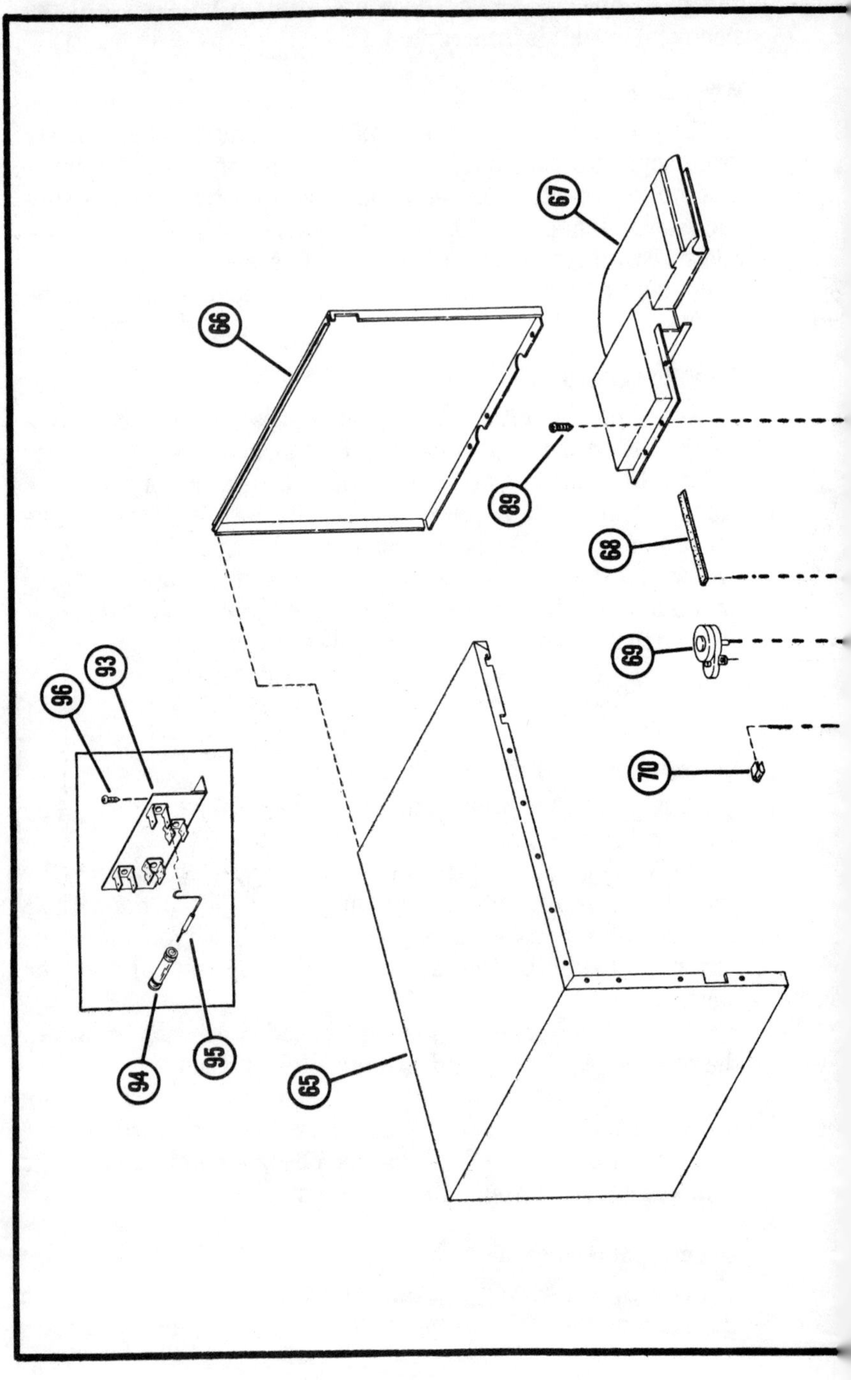
66
67
89
68
69
70
96
93
94
95
65

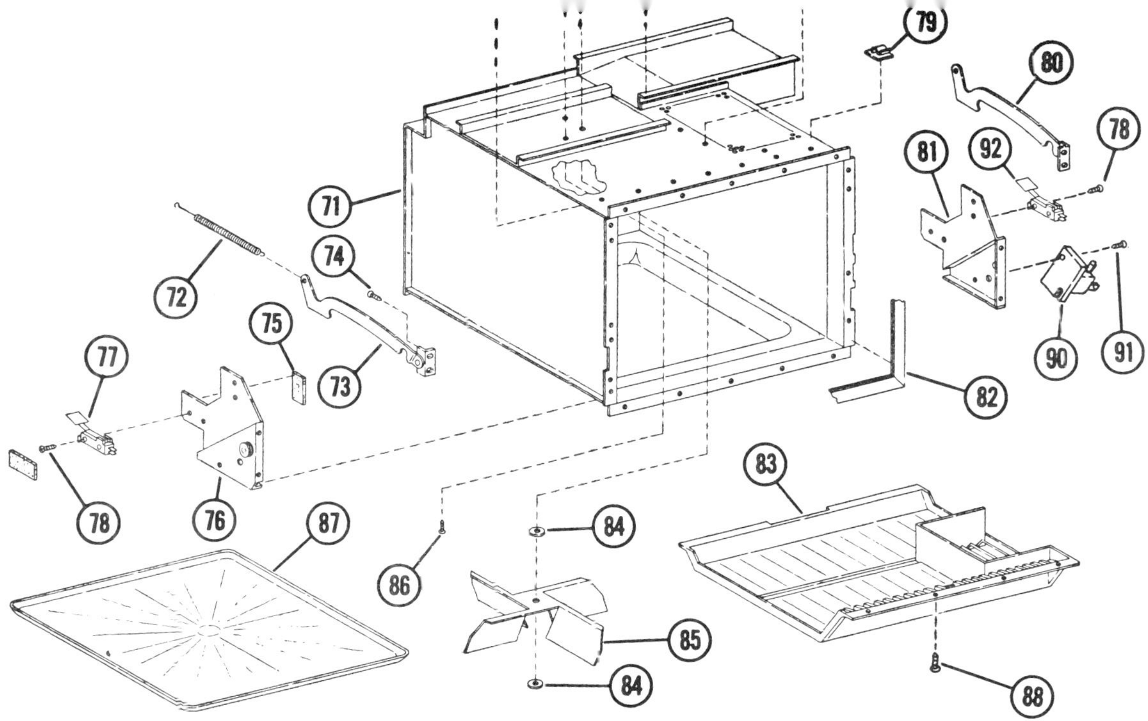

Fig. 4-9. Interior parts—RR-4D (RR-6 similar). (cont. on next page).

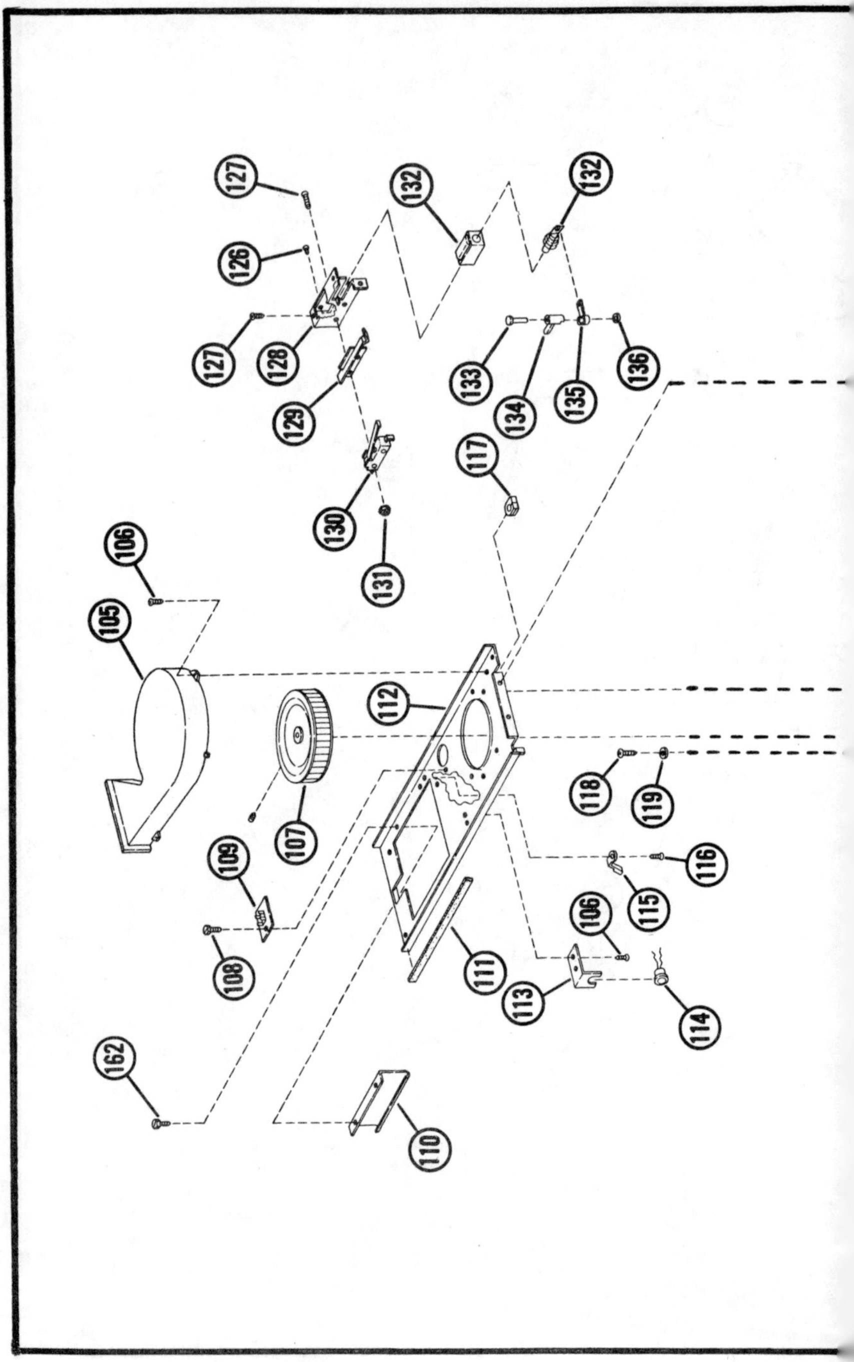
105
106
107
108
109
110
111
112
113
114
115
116
117
118
119
126
127
128
129
130
131
132
133
134
135
136
162

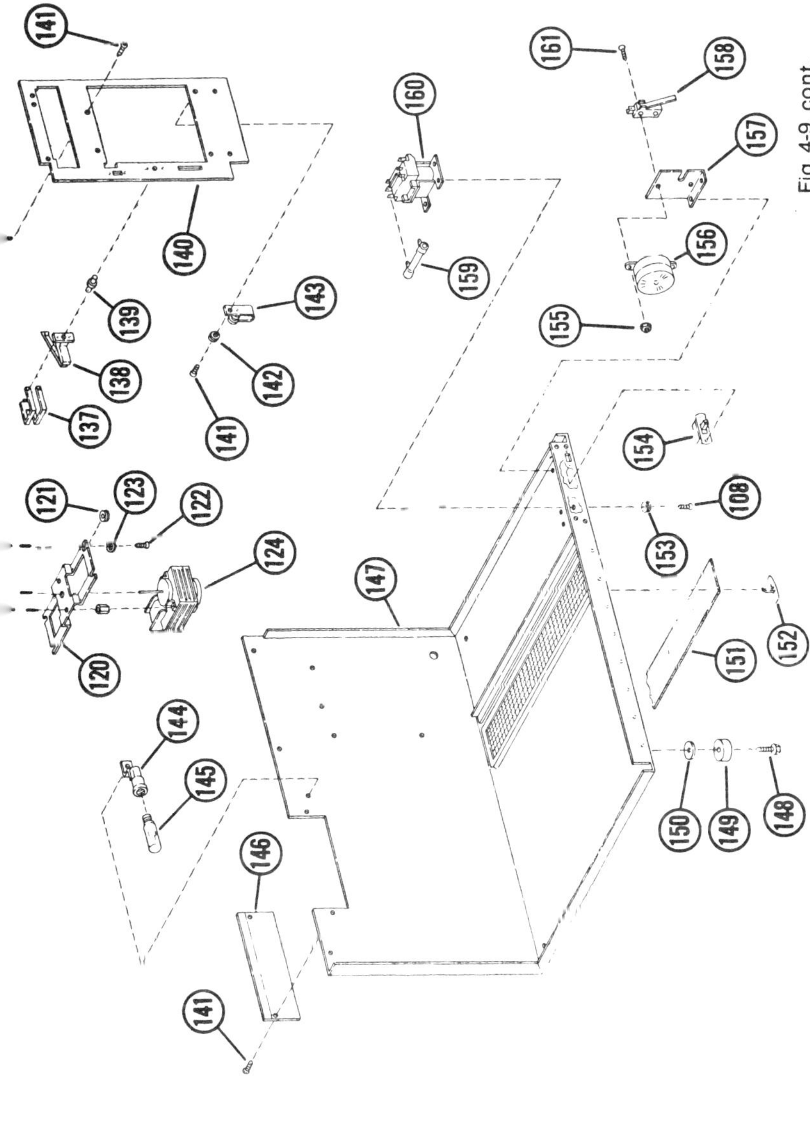
141
140
139
138
137
143
142
141
121
123
122
124
120
144
145
146
141
147
160
159
161
158
157
156
155
154
108
153
151
152
150
149
148

Fig. 4-9. cont.

Fig. 4-9. Parts List.

REF. NO.	PART NUMBER	DESCRIPTION
65	C36884-1	Wrapper, Outer Case
66	C36896-1	Panel, Access
67	D71060-3	Duct, Air Discharge
68	A30303-2	Gasket (Order By The Foot)
69	B56028-3	Motor, Stirrer
70	M1149-1	Clip, Cable
71	D71683-2	Weldment, Oven
72	M7750-1	Spring, Counterbalance Arm
73	B56931-2	Arm, Counterbalance, Left Hand
74	M2280-15	Screw, #6-32 X 5/16", Counterbalance Arm
75	M0509-2	Pad, Insulation
76	B56932-3	Bracket, Counterbalance, Left Hand
77	A31625-3	Switch, Interlock, Left Hand
78	M2275-25	Screw, #6-32 X 7/8", Switch Mounting
79	M1041-2	Clip, Wiring, Adhesive Back
80	B56931-1	Arm, Counterbalance, Right Hand
81	B56932-4	Bracket, Counterbalance, Right Hand
82	C36372-1	Gasket, Oven Cavity
83	D71062-1	Shield, Grease
84	A31091-1	Washer, Clutch, Stirrer Fan
85	C36091-1	Fan Blade, Stirrer
86	M2165-29	Screw, #8-18 X 3/8"
87	D70896-2	Tray, Oven
88	M2300-28	Screw, #8-18 X 5/16", Grease Shield
89	M2300-15	Screw, #6-20 X 1/2", Air Duct
*90	B56887-1	Relay Time Delay
91	M2341-17	Screw, Relay
92	A31625-3	Sensing Switch
**93	B57220-1	Terminal Board
**94	M3054-1	Resistor, Tube, 100 Ohm
**95	M3081-1	Fuse, Thermal, 300° F.
**96	M2300-15	Screw, #6-20 X 5/16", Terminal Board
105	C36673-1	Blower Scroll
106	M2300-15	Screw, Air Duct, Etc. #6-20 X 1/2
107	B56611-1	Blower Wheel
108	M2165-17	Screw, Terminal Board, #6-20 X 1/2"
109	B56070-2	Terminal Board
110	B56916-1	Support, Magnetron
111	A30303-2	Gasket, (Order By The Foot)
112	D71678-1	Bulk Head
113	B56803-1	Bracket, Oven Thermal Cut-Out
114	B56658-3	Thermal Cut-Out, Oven
115	M3209-1	Terminal, Ground
116	M2165-29	Screw, Wiring Clamps #8-18 X 3/8"
117	A45687-2	D, Grommet, Wiring
118	M2026-42	Screw, Motor To Bracket, #10-32 X 5/16"

REF. NO.	PART NUMBER	DESCRIPTION
119	M2726-5	Lock Washer, Motor To Bracket, #10
120	B56473-1	Motor Mount
121	A3815-1	Grommet, Motor Mount, Rubber
121	A3814-1	Eyelet, Motor Mount
122	M2190-18	Screw, Bracket To Bulk Head, #6-20 X 5/8
123	M2705-4	Washer, M2190-18 Screw
124	C36565-1	Blower Motor
126	M2014-14	Screw, Mount Solenoid, #6-32 X 1/8"
127	M2025-21	Screw, Limit Switch, #6-32 X 7/8"
128	C36669-2	Bracket, Latch Switch
129	A31485-1	Spring, Latch
130	A31634-5	Switch, Preform, Latch
131	M2820-5	Nut, Keps #6
132	B56821-1	Solenoid
133	A31826-1	Pin, Bell Crank
134	B56725-2	Bell Crank
135	B56725-3	Bell Crank
136	M2705-30	Washer, Bell Crank .125 ID
137	B56853-2	Spring, Reset Arm
138	C36675-1	Arm, Reset
139	A31831-3	Button, Reset Arm
140	B56602-2	Front, Outer Case
141	M2300-15	Screw, Bracket To Bulkhead, #6-20 X 1/2"
142	M5001-1	Grommet
143	A31525-1	Buzzer
144	B56912-1	Socket, Oven Light
145	A31675-1	Lamp, Oven Light
146	A31981-1	Panel, Bulb Access
147	C36399-9	Weldment, Outer Case Base Pan
148	M2211-32	Screw, Foot, #8-32 X /34"
149	M0110-1	Foot
150	M2705-19	Washer, Foot .219 ID
151	A30949-1	Filter
152	M1112-3	Dart Clip, Filter
153	A3814-1	Eyelet, Motor Mount
154	A31162-1	Switch, Buzzer
155	M2820-5	Nut, Keps, Defrost Switch #6
156	B56998-1	Timer Motor, Defrost
157	B56996-1	Bracket, Defrost Timer
158	A32207-1	Switch, Defrost Timer
159	M3050-7	Resistor
160	C36563-3	Relay, Power
161	M2025-21	Screw, #6-32 X 7/8", Defrost Switch
162	M2165-17	Screw, #6-20 X 1/2", Magnetron Support

plate. Disengage the bulb by carefully pushing inward. Turn it slightly and remove.

Door-Latch Assembly

Refer to Figs. 4-9 and 4-10 for parts locations.

1. Disconnect the power to oven.
2. Remove the end panel.
3. Remove the control panel for easy access.
4. Remove the leads from the switch and solenoid.
5. Remove the exterior cabinet.
6. Remove the air duct and blower housing.
7. Remove the latch-bracket mounting screws.
8. Drill the latch-bracket pop rivet, located in the front flange behind the right-hand door gasket, and remove the assembly from the unit.

Latch-Switch Removal

1. Disconnect the power to oven.
2. Remove the end panel.
3. Remove the control panel (only enough to gain access to switch).
4. The transformer assembly may be removed to gain additional access to switch.
5. Remove the mounting screws.

Caution: Hex-head nuts are used on the back side of the switch. Support them with your fingers to prevent loss while removing mounting screws.

6. Disconnect the leads and remove the switch and latch spring.

Important: After replacement, refer to the interlock-switch procedure for proper switch adjustment.

Latch-Solenoid Removal

1. Disconnect the power to oven.
2. Remove the end panel.
3. Remove the control panel (only enough to gain access to solenoid-mounting screws).
4. Remove the two screws holding solenoid coil to door-latch assembly bracket.
5. Remove the rear latch-switch mounting screw.

Caution: Hex-head nuts are used on the back side of the switch. Support them with your fingers to prevent loss while removing mounting screws.

6. Disconnect the leads and remove the solenoid coil and plunger assembly by moving toward rear of unit and down.

   Note: On reassembly be sure to insert plunger-shaft engagement pin into bell-crank assembly. Check the plunger shaft to be sure it has free return from the locked (in) position. With your fingers, push the plunger in and release it slowly. Check the door to be sure it opens.

Tube Replacement

Caution: See *RF Capacitor Check.* A magnetron tube, like a radio or TV tube, must be handled with a reasonable amount of care. When handling a tube, always handle by the housing only. Use caution not to touch or strike the glass portion at the top or bottom.

The carton used to ship replacement tubes is reusable. All defective tubes *must* be returned to the factory in this carton.

The magnetron assembly is made up of the magnetron, permanent magnets, housing, filter box, and latching device. It can only be replaced as an assembly.

Magnetron Removal

1. Disconnect the power to the oven and remove the access panel.
2. Remove the capacitor cover and disconnect the transformer leads from the RF capacitors after discharging them.

   Caution: Support the capacitor shaft when removing the leads, to prevent damaging the capacitor.
3. Remove the foam air block (or aluminum tape, if present) from the rear of the bulkhead. On models so equipped, remove the bulkhead extension by removing the two screws securing it to the main bulkhead (above plate transformer). The sealing gasket should be cut at the overlap point and the extension removed.
4. Push the tube-assembly latching lever away from yourself toward middle, while supporting the tube assembly at the bottom of the filter can. The tube can

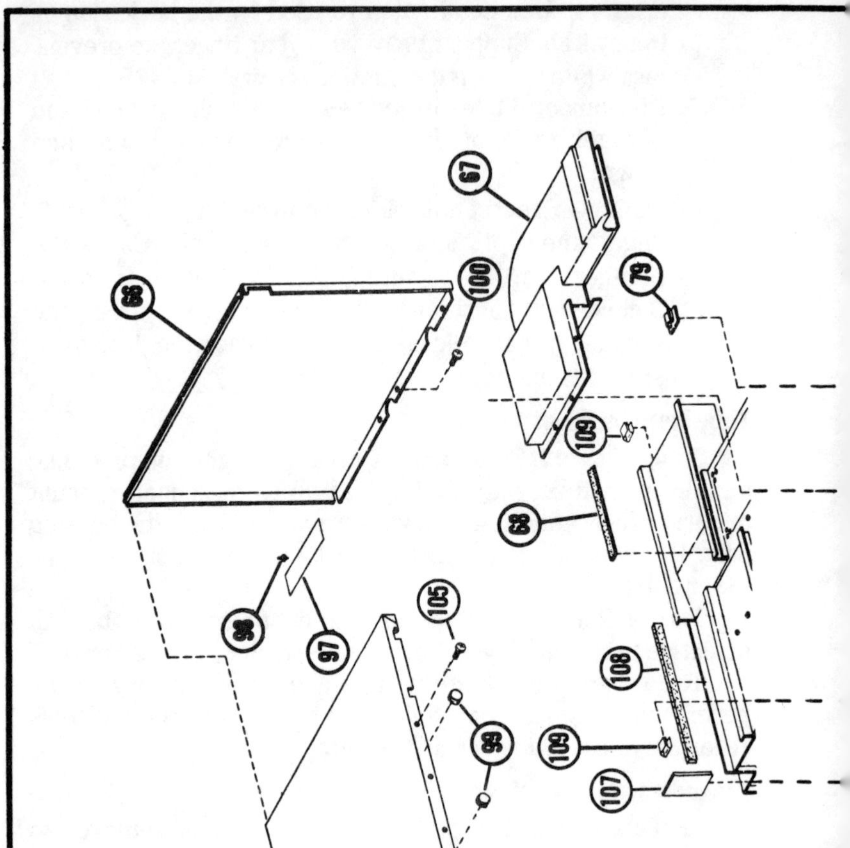
67
66
100
79
109
68
98
97
105
108
99
109
107
65
88
188

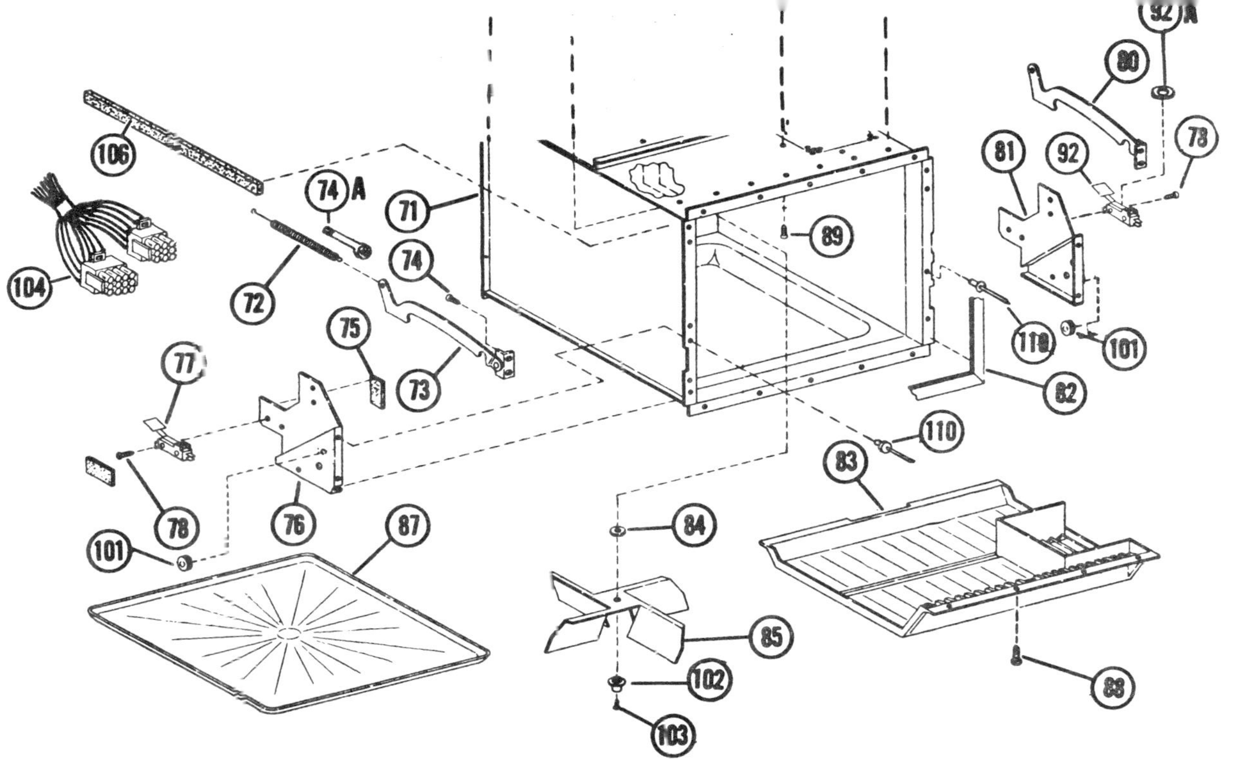

Fig. 4-10. Interior parts—RR-7 (RR-9 similar). (cont. on next page)

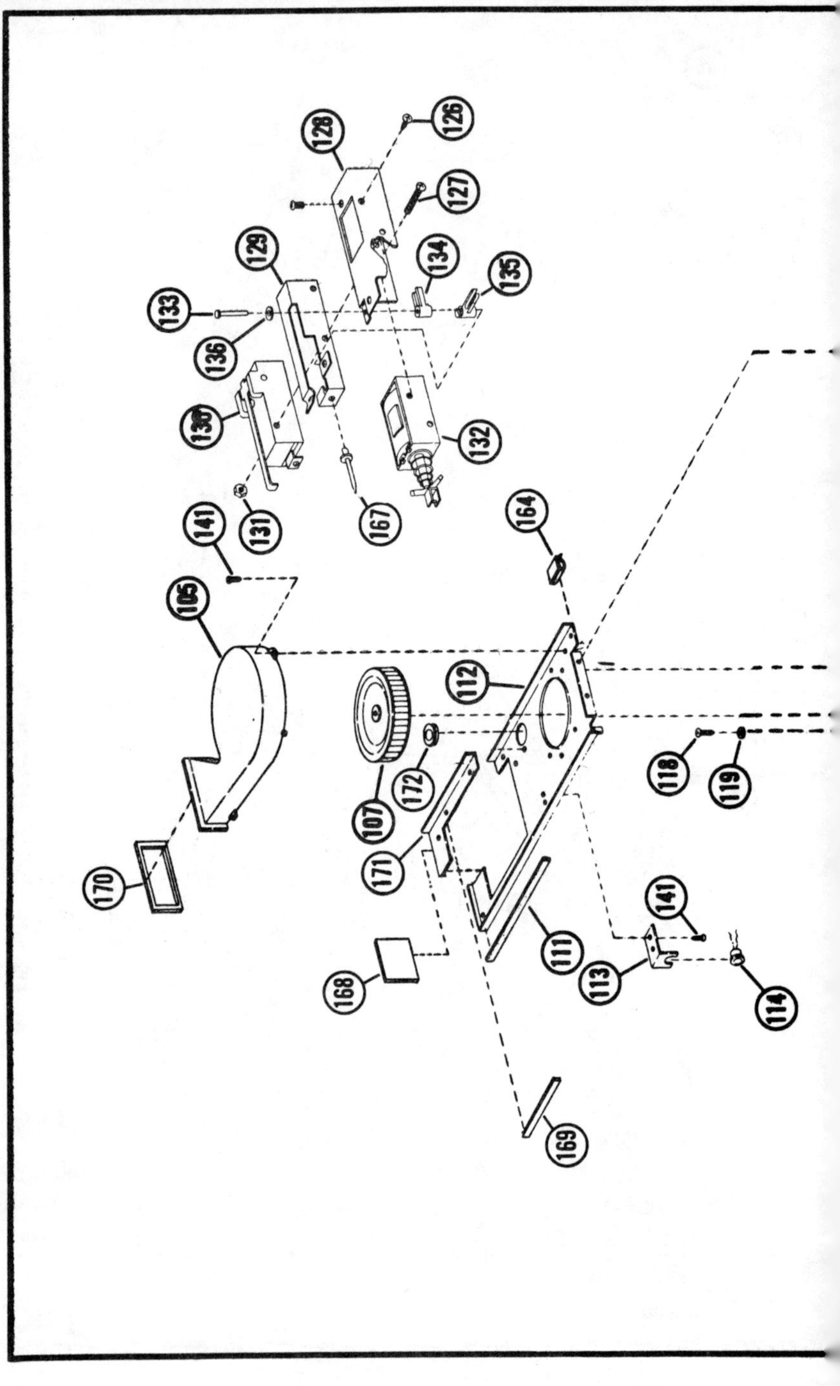
128
126
127
129
134
135
133
136
130
132
141
131
167
164
105
112
107
172
118
119
171
170
111
141
113
168
114
169

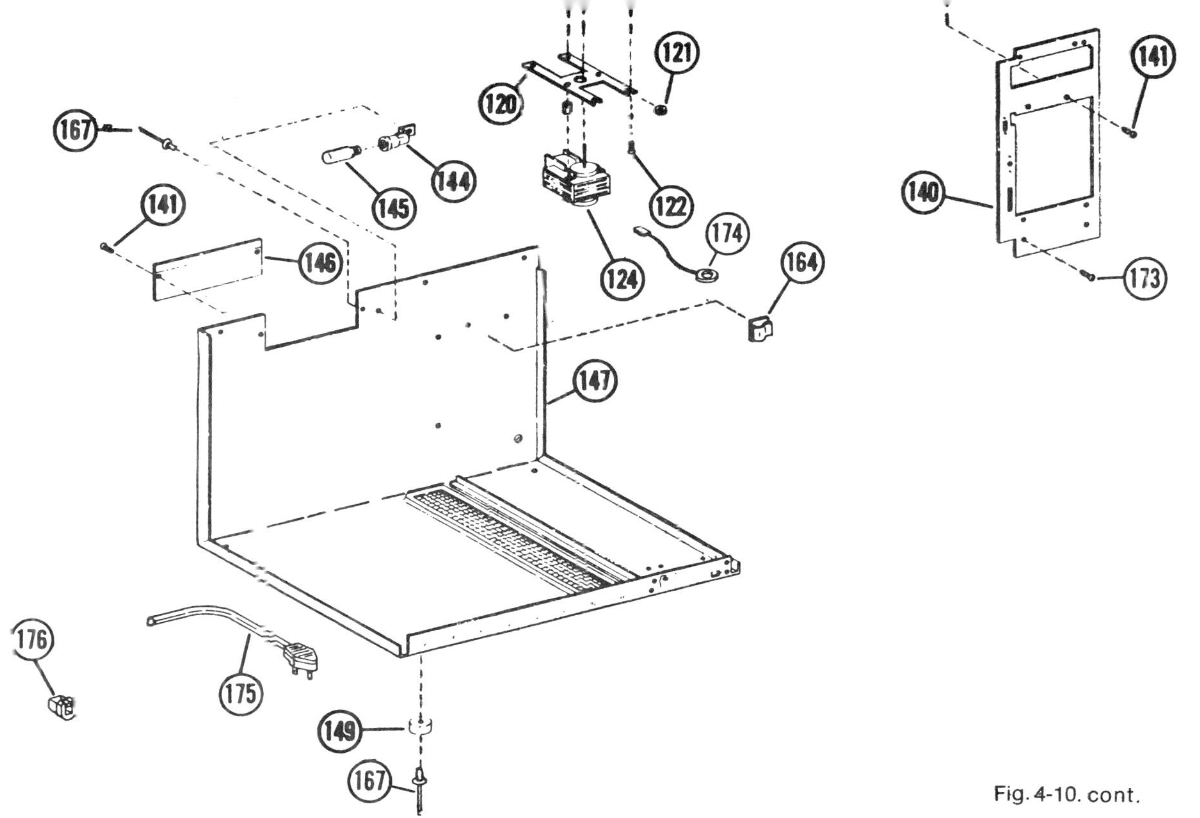
121
120
167
144
145
141
146
122
124
174
164
147
140
141
173
176
175
149
167
Fig. 4-10. cont.

Fig. 4-10. Parts List.

REF. NO.	PART NUMBER	DESCRIPTION
65	D72575-1	Wrapper, Outer Case
66	C87354-1	Panel, Access
67	D72587-3	Duct, Air Discharge
68	A30303-2	Gasket (Order By The Foot)
71	D71683-6	Weldment, Oven
72	M7750-1	Spring, Counterbalance Arm
73	B56931-2	Arm, Counterbalance, Left Hand
74	M2585-18	Screw, Counterbalance Arm (Special Holt Head)
74A	R1930-32	Tool For Holt Head Screw Removal (Not Included With Unit)
75	M0509-2	Pad, Insulation
76	B56932-5	Bracket, Counterbalance, Left Hand With Roller
76	R1571-45	Roller, Mounting Kit
77	A31625-7	Switch, Interlock, Left Hand
78	M2275-25	Screw, #6-32 X 7/8", Switch Mounting
79	M1041-1	Clip, Wiring, Adhesive Back
80	B56931-1	Arm, Counterbalance, Right Hand
81	B56932-6	Bracket, Counterbalance, Right Hand With Roller
81	R1571-45	Roller And Mounting Kit
82	C36372-1	Gasket, Oven Cavity
83	D72797-1	Shield, Grease
84	B57042-3	Bushing, Stirrer
85	C87208-1	Fan Blade, Stirrer
87	D70896-2	Tray, Oven
88	M2300-15	Screw, #8-18 X 5/16", Grease Shield, Cabinet
89	M2300-28	Screw, #6-20 X 1/2", Air Duct
92	C87367-3	Sensing Switch
92A	A31451-1	Washer, Nylon-Switch Plunger
97	B57385-1	Seal, Warranty - Warning
98	A45396-1	Pop Rivet
99	A15115-2	Bumper, Door
100	M2166-16	Screw
101	R1571-45	Roller Kit With Shoulder Screws
102	A32148-1	Bearing, Stirrer
103	A32149-2	Screw, Shoulder, #6-32 X 1-7/16"
104	D72595-1	Wiring Harness, Oven Cavity
105	M2165-15	Screw, Thread Cutting
106	A32639-1	Gasket, Air Sealing
107	B56953-1	Gasket, Air Sealing
108	B56953-2	Gasket, Air Sealing
109	M1188-1	Clip, Cable
110	A45396-1	Pop Rivet
105	C36673-1	Scroll, Blower
107	B56611-1	Wheel, Blower
111	A30303-2	Gasket, (Order By The Foot)
112	D72556-1	Bulk Head
113	B56803-1	Bracket, Oven Thermal Cut-Out
114	B56658-3	Thermal Cut-Out, Oven
118	M2026-42	Screw, Motor To Bracket, #10-32 X 5/16"
119	M2726-5	Lock Washer, Motor To Bracket, #10
120	B57440-1	Motor Mount, Blower
121	A3815-1	Grommet, Motor Mount, Rubber
121	A3814-1	Eyelet, Motor Mount
122	M2165-17	Screw, Bracket To Bulkhead
124	C36565-5	Motor, Blower
126	M2014-14	Screw, Mount Solenoid, #6-32 X 1/8"
127	M2025-21	Screw, Limit Switch, #6-32 X 7/8"
128	C87121-1	Bracket, Latch Switch
129	B57196-1	Spring, Latch
130	A31634-5	Switch, Preform, Latch
131	M2820-5	Nut, Keps #6
132	B56821-3	Solenoid
133	A31826-2	Pin, Bell Crank
134	B56725-2	Bell Crank
135	B56725-3	Bell Crank
136	M2705-30	Washer, Bell Crank, .125 ID
140	D72564-1	Front, Outer Case
141	M2300-15	Screw, #6-20 X 1/2"
144	B57420-1	Socket, Oven Light
145	A31675-1	Lamp, Oven Light
146	A31981-3	Panel, Bulb Access
147	D72567-1	Weldment, Outer Case Base Pan
149	B57325-1	Foot
164	M1041-2	Clip, Cord
167	A45396-1	Rivet, Pop
168	B56953-1	Block, Air
169	A30303-9	Gasket, 5"
170	A30303-1	Gasket, 14-1/2"
171	C87360-1	Extension Bulk Head
172	A32588-1	Grommet
173	M2166-16	Screw
174	C36558-67	Ground Wire, Motor
175	B57035-6	Cord,Power
176	M3700-1	Strain Relief, Power Cord

be removed only when the lever is perpendicular to the edge of the housing. (See Fig. 4-11.)

5. Disconnect the tube thermal-cutout leads.
6. With the latch released, lower the tube assembly to clear the antenna glass to remove it from the oven.
7. If the add-on parts are in good usable condition (top metal gasket, thermal cutout, RF capacitors, nuts, etc.), remove them from the defective tube to install them on the new tube assembly. *Record the tube serial number.* The tube serial number is located on the antenna.

   Important: For correct unit operation, the ferrite air baffle and clips must be installed on the replacement tube.
8. Radarange microwave ovens have a HV circuit used in conjunction with the permanent-magnet (PM) tube assembly. To ensure proper phasing of the complete HV secondary circuit, the transformer and tube leads must be connected as shown in Fig. 4-12. The

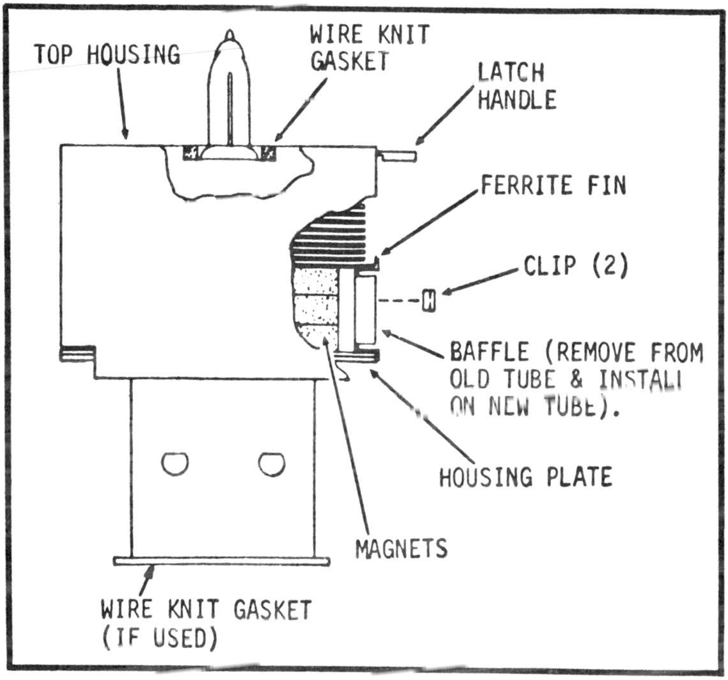

Fig. 4-11. Magnetron assembly.

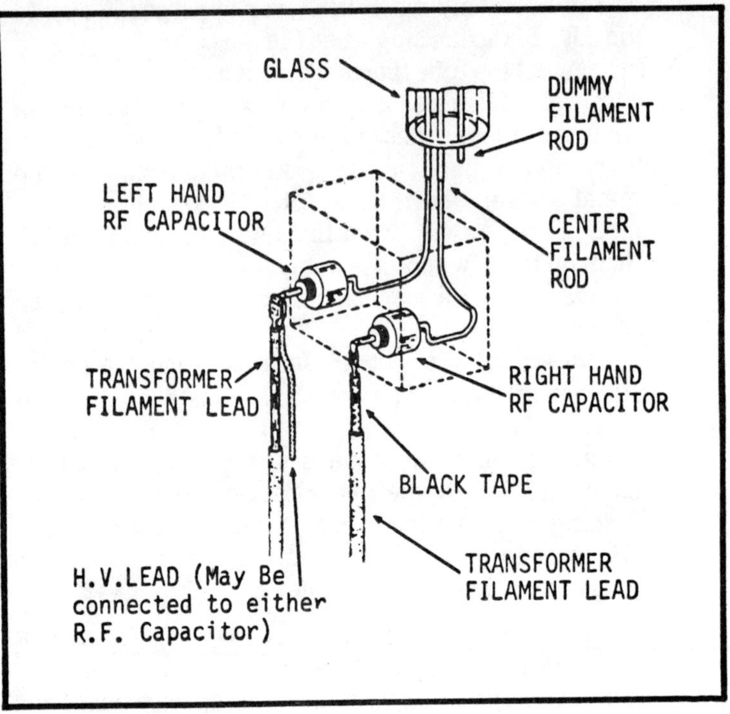

Fig. 4-12. Filament and capacitor connections.

transformer lead marked with the tape or sleeving and the tube lead from the center filament rod must be attached to the right hand RF capacitor. The other outside filament lead is connected to the left RF capacitor. The plate transformer's secondary lead may be connected to either terminal on RR-7 or RR-9 (to left terminal on RR-4D). Failure to connect the components in this manner will cause premature failure of the diode, thereby opening the fuse resistor, causing the oven to be inoperative.

9. To replace the tube, insert the antenna glass portion of the tube assembly into the waveguide and, while holding the tube assembly firmly against the waveguide, pull the latching lever towards yourself to latch tube in place.
10. Ensure that the magnetron assembly is properly locked in place on all four mounting pins. Improper installation will result in malfunctioning of the unit.

11. Connect the tube's filament leads to the RF capacitors, observing step 8, dressing the leads and beads away from sides and bottom of the capacitor box to prevent shorting. A minimum spacing of a quarter-inch is required.
12. Replace the filter, box cover, and clip. On models RR-7 and RR-9, connect the thermal-cutout leads.
13. Install the leads to the outside of the RF capacitors, observing step 8.
14. Install the bulkhead extension if present. Install the foam air block and sealing gaskets, otherwise the stirrer may not rotate.

RF Capacitors

1. Disconnect the power to the oven.
2. Remove the HV cover.

   Caution: The cover's grounding clip must ground the capacitors to ensure against accidental shock.
3. Remove the clip and cover from the bottom of the can.
4. Disconnect the leads on both sides of the capacitor unit while supporting the terminal to prevent breakage.
5. Remove the mounting nut and pull capacitor unit from the can through the bottom.
6. Should the HV grounding clip become too loose to ground the capacitors, bend the clip to less than the quarter-inch space of the capacitor spade.
7. On installation of the capacitor, connect the filament leads (per Fig. 4-12) to the RF capacitors, dressing the leads away from sides and bottom of capacitor box to prevent shorts. A minimum spacing of a quarter-inch is required.
8. Replace the filter, box cover, and clip.
9. Install the transformer leads (per Fig. 4-12) and capacitor cover.

Blower-Wheel Adjustment

Refer to Figs. 4-9 and 4-10 for parts locations.

1. Disconnect the power to the oven
2. Remove the access panel.
3. Drill a quarter-inch hole in the side of the blower housing, approximately midway from top to bottom.

4. Rotate blower to gain access to the blower setscrew. Adjust them with an eighth-inch Allen wrench.
5. Cover the hole in side of the blower housing with tape.

Blower—Motor Removal

1. Disconnect the power to the oven.
2. Remove the access panel.
3. Remove the control-panel assembly.
4. Drill a quarter-inch hole in side of the blower housing, approximately midway from top to bottom.
5. Rotate the blower to gain access to blower setscrew. Loosen them with an eighth-inch Allen wrench.
6. Remove the mounting bracket, motor leads, and ground wire.

Blower-Motor Replacement

Refer to Figs. 4-9 and 4-10.

1. Install the motor mount on the replacement motor and connect leads.
2. Place the rubber grommets and eyelets in their mounts. Slide the shaft into the blower hub.
3. Install the motor mount to the bulkhead with the four screws.
4. Adjust the blower wheel on the motor shaft. Test run it with a jumper cord to ensure that the blower does not strike or rub.
5. Connect the motor leads and *green ground lead.*
6. Cover the hole inside of the blower housing with tape.

Removal of Exterior Cabinet

Refer to Figs. 4-9 and 4-10.

1. Disconnect the power to the oven.
2. Remove the access panel.
3. Remove the cabinet-mounting screws along top front, left side of front, lower-left side, and top-rear flange (when present).
4. When present, remove the mounting screw from far-right upper corner.
5. Lift the cabinet off.

Grease-Shield Removal

Refer to Figs. 4-9 and 4-10.

1. Open the door and remove any oven contents.
2. Remove the four mounting screws from the front of grease shield.
3. Pull it forward to disengage it from the clips in rear. Remove it from the cavity.

Air-Duct Removal

Refer to Figs. 4-9 and 4-10.

1. Disconnect the power to the oven.
2. Remove the end panel.
3. Remove the exterior cabinet per instructions. In some cases the timer escutcheon may have to be removed.
4. Remove the two screws holding the control-panel support frame to the divider bulkhead.
5. Remove the top-mounting screws securing the control-panel support to the cavity flange (under door gasket). In some instances the pop rivet below door-latch hole may need to be removed.
6. Remove the five duct-mounting screws from inside the oven cavity.
7. Bend the support frame forward and lift the air duct out.

Blower Scroll & Wheel

Refer to Figs. 4-9 and 4-10.

1. Disconnect the power to the oven.
2. Remove the end panel.
3. Remove the exterior cabinet per instructions.
4. Remove the sealing tape connecting the scroll waveguide.
5. Remove the air discharge duct.
6. The scroll is held in place by one tab that passes through the bulkhead at the left-front of the scroll and four screws driven from the underside of the bulkhead.
7. Remove the four mounting screws and tilt the scroll up to remove.
8. If removing the blower wheel, rotate the blower to gain access to the blower setscrew; loosen it. Remove the wheel.

9. Replace any sealing tape or gaskets you may have removed.

Door Disassembly

1. Open the door fully.
2. Place a small C-clamp on each counterbalance arm at the cabinet, protecting the metal surface to prevent damage.
3. Remove the three holt-head mounting screws (or two counterbalance screws and one assembly screw) from each side lift off inner door.

Door Glass Removal

1. Follow the steps for door disassembly.
2. Remove the tinnerman nuts holding inner-door molding to inner door and remove molding, glass, and plastic.

Door Handle Removal

1. Follow the steps for door disassembly.
2. Remove the two mounting screws.

Left Interlock Removal

1. Remove the exterior cabinet, or remove the screws along left side of front and lower-left side of unit.
2. Carefully pull the left bottom up to gain access to the switch.

   Caution: Do not pull far enough to deform edge of cabinet.
3. Remove the two mounting screws, disconnect the leads, and remove the switch.

TROUBLESHOOTING & TESTING

Tables 4-1 and 4-2 provide a general troubleshooting guide that can help you isolate the area of trouble in an oven. Use the appropriate schematic to locate the parts referred to in the tables. The following paragraphs describe some of the tests you can make to determine the need for parts replacement.

Output Testing

A grounding resistor (10Ω, 5W) has been added to the HV circuit to protect you while taking milliampere readings. This resistor is attached to the ground side of the fuse-resistor and one of the terminal-board mounting screws. Do not remove the grounding resistor from the circuit except to check meter impedance. If the impedance of the meter being used is known, proceed with the milliampere reading without removing the grounding resistor.

If it is necessary to check the milliammeter impedance, proceed as outlined next.

Checking Meter Impedance

1. Unplug the unit.
2. Remove the unit's end panel.
3. Remove the gray ground lead at terminal-board foot (or black ground lead at bulkhead on some models).
4. Connect the black meter lead to grounding-spade connection.
5. Attach the red meter lead to the gray (or black) ground wire removed from the spade.
6. Set the meter to the next higher scale than 300 mA.
7. Check the meter leads to ensure tightness.
8. Place a water load in the unit (approximately 2 cups).
9. On Cookmatic models, place the Cookmatic control in the far-right position.
10. Start the unit and record the milliampere reading. Unplug the unit.
11. Repeat the test with the grounding resistor removed from the terminal-board mounting screw.
12. Record the meter reading; take the difference between the two readings. Record this figure on the face of the meter for future use. In the future, it will not be necessary to remove the grounding resistor to determine the milliampere output of a unit unless the meter is repaired or recalibrated

Milliampere Reading Example

Meter reading without grounding resistor in circuit	300 mA
Meter reading with grounding resistor in circuit	275 mA
Allowance for meter internal impedance	25 mA

If the actual milliampere reading is lower than 300 mA, change the gray wire coming from the oil-filled capacitor to the No. 4 tap on the plate transformer. If the milliampere reading is high, change the gray lead to the No. 4A tap on the

Table 4-1. RR-4D & RR-6 General Troubleshooting Guide

CONDITIONS	POSSIBLE CAUSES
No voltage at outlet.	Line fuse (external) open. Check for overloaded house circuit.
External line fuse open; condition above okay.	1. Defective terminal or wiring in 120V circuit of unit. 2. Shorted component in 120V circuit of unit. 3. Replace fuse. If fuse blows quickly, do not proceed with step 4; recheck steps 1 and 2, and proceed to step 5. 4. Check amperes and wattage. Compare to specifications. 5. If fuse blows quickly at step 3, disconnect power cord. Disconnect and tape apart transformer primary wires. Connect power cord. 6. If fuse does not blow, disconnect power cord, check transformer primaries for correct ohmage. 7. If fuse blows, check 120V section for loose connections or grounded wires. The timers, light, and motor by themselves draw less than 1A. If the unit blows a fuse while the transformer is disconnected, there are usually burned areas visible.
Power available to power plug; no radarange components operating including oven light.	1. Open power cord; unplug; check the ohmmeter on lowest scale. Reading should be under 1Ω end to end, for each lead. 2. Defective terminal wiring.
No Radarange components operate, except oven light operates on light switch; does not operate when door is opened.	1. Test left-interlock switch. 2. Defective terminal or wiring.
Only oven light operates; operates on light switch or by opening door.	1. Defective terminal or wiring. 2. Open cavity or tube thermal switch. Allow time to reset if cavity or tube is warm. Check for low air flow (motor, fan, etc.). 3. Latch switch not closing. 4. START switch not closing. 5. STOP switch not opening. 6. Multiprotector fuse open. 7. Open transformer-protect fuse 8. Timer contacts not closing. Test by trying other timer. 9. Variable-power module inoperable.
Timers, timer light, latch solenoid, and blower motor operate. Magnetron filament does not glow.	1. Defective terminal or wiring. 2. Open primary or filament winding on filament-control transformer. 3. Open tube tube fulament.
Slide control is to far-right position; timers, timer light, latch solenoid, blower motor, and tube filament operate; unit does not cook. No voltage at plate-transformer's primary leads **Caution**: All other leads are HV.	1. Defective terminal or wiring. 2. Open 24V winding on filament-control transformer. 3. Slide control. 4. Varaible-power module inoperable.
Same as previous condition, but voltage is available at plate-transformer's primary (115–120V) leads. **Caution**: All other leads are at HV.	1. Defective terminal or wiring in HV section. 2. Open transformer-secondary winding. Test ohmage. 3. Open diode, or diode connected backwards. 4. Open oil-filled capacitor. 5. Magnetron tube weak. Check milliampere output.

Table 4-1 cont.

CONDITION	POSSIBLE CAUSES
Open multiprotector fuse in primary circuit.	1. Left-interlock or sensing switch out of sequence. 2. Defective left-interlock sensing switch. 3. If multiprotector fuse is open, also test multiprotector resistor for correct value; inspect for cracks, etc.
Open transformer-protect fuse in primary circuit.	1. Defective terminal or wiring in HV circuit. 2. Defective RF capacitors. 3. Defective rectifier. 4. Defective tube. May show blue-violet glow inside (look through waveguide and perforations) when thermal fuse replaced and oven turned on. This is a 5 – 10 second test, or thermal fuse may open. 5. Defective oil-filled capacitor. 6. Shorted or grunded transformers. Check ohmages.
All unit components operate; unit cooks slowly in all slide-control positons.	1. Check unit to see if it is set at the correct Cookmatic setting for the food being prepared. 2. Run oven-performance test. If okay, customer education in oven use is indicated. 3. Check voltage at outlet. If low, advise customer. 4. Check milliampere output. 5. If milliampere output low, check components in HV circuit; power cord must be disconnected from outlet and RF capacitors discharged. 6. Check capacitor color dot for connection to correct 4 or 4A HV transformer tap. 7. Slide control. 8. Variable-power control module. 9. If above tests are acceptable, replace tube.
All unit components operate; unit cooks per oven-performance test with slide control in far-right position, slide control does not vary cooking speed.	1. Defective terminal or wiring. 2. Slide control. 3. Variable-power module.
Unit cooks food unevenly.	1. Customer is not following cookbook instructions; customer education indicated. 2. Stirrer blade deformed. 3. Stirrer blade not turning. Check screw and blade-clutch bearings for grease or burrs. This may also indicate incorrect air flow; defective blower motor or blower wheel.
Door will not open after STOP switch is pushed. Latch solenoid still operating, blower running, etc., unit cooking or not cooking.	1. STOP switch not closing. 2. START swtch not opening. 3. Variable-power module. 4. Defective terminal or wiring.
Door will not open after STOP switch is pushed. No components operating.	Defective latch-solenoid linkage; adjust.

transformer. If the reading continues to be high or low, the oil-filled capacitor may require changing to correct the problem.

A normal milliampere reading is 300–330 mA. Moving the Cookmatic slide control to the left will cause the reading to decrease and fluctuate.

Diode Check

1. Set the ohmmeter to the R × 10K or higher scale.
2. Check the resistance between *A* and *B* in Fig. 4-13. Reverse the leads and check again. There should be a considerable difference (minimum of one-half of scale) between the two readings, approximately a 10:1 ratio or better. If reversing the leads does not change the reading, replace the diode.

   Note: The diode may be checked in unit, but one lead must be removed. Some high-impedance VTVMs or FET meters may show an acceptable diode as defective. These meters will normally show high ohms (open) in both lead positions.

RF Capacitor Check

Caution: The capacitors are charged to 4000V, and it is extremely important that they be grounded before handling. Capacitors will normally be grounded when the cover is

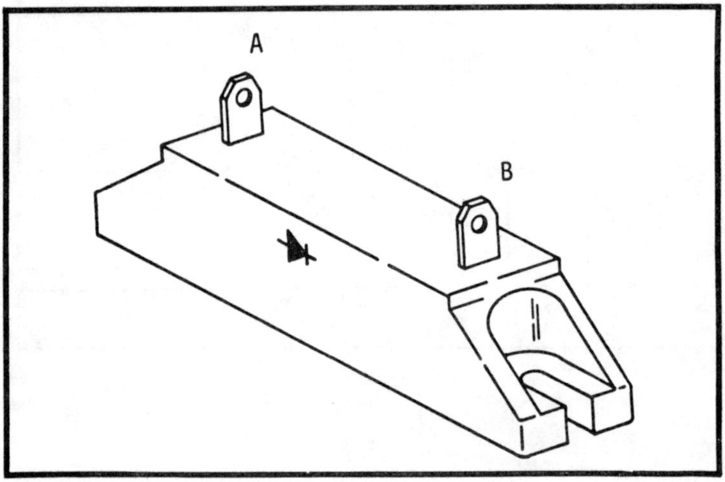

Fig. 4-13. Illustration for diode check.

removed through the cover clip; if not, place a screwdriver blade against the capacitor can; *then* discharge the capacitor. Proceed as follows:

1. Remove the capacitor cover.
2. Remove the capacitor can cover from the bottom of can.
3. Slip the side of the plastic protective cover between the ground clamp and the capacitor while checking the capacitors.
4. Remove both leads from either capacitor.
5. Set the ohmmeter on the R × 10K scale. Check the ohmage from the capacitor to ground. Any reading less than 100K indicates a faulty capacitor.
6. After checking or replacing capacitors, dress the tube leads to ensure that they are not in such a position as to short to the capacitor box. A minimum spacing of a quarter-inch is required between wires or beads and box.

   Important: Wire-knit gasket replacement is required if it is used. Return of the ferrite beads to filament leads is necessary to prevent radio and TV interference.

RR-4 & RR 7 Left-Door Interlock

1. Close the door.
2. Test the oven-cavity light, using the light switch. Shut the light off.
3. Open the door slowly. The oven-cavity light should come on. If it does not, the left-door interlock is defective and must be replaced. For proper switch operation and adjustment, the light must come on before the door opening exceeds 0.3125 inches.

RR-6 Left Interlock

1. Zero the ohmmeter on the lowest scale.
2. Connect the ohmmeter leads to the black lead on the divider-panel terminal-block test point and to blue lead on control-module terminal board.
3. With the door closed, the ohmmeter should indicate zero. This indicates the N.O. contacts of left interlock are closing as designed.

Table 4-2. Touchmatic Models General Troubleshooting Guide

CONDITIONS	POSSIBLE CAUSES
No voltage at outlet.	External line fuse open. Check for overloaded house circuit.
External line fuse open. Condition above okay.	1. Defective terminal or wiring in 120V. 2. Shorted component in 120V. 3. Replace fuse. If fuse blows quickly, **do not** proceed with step 4; check steps 1 and 2 and proceed to step 5. 4. Check amperes or wattage. Compare to specifications. 5. If fuse blows quickly at step 3, disconnect unit power cord, disconnect and tape apart plate- and filament-transformer primary wires; then connect power cord. 6. If fuse does not blow, disconnect power cord and check transformer primaries for correct ohmage. 7. If fuse blows, check 120V section for loose connections or grounded wires. The timer, light, and motor by themselves draw less than 1A. If the unit blows a fuse while the transformer is disconnected, there are usually burned areas visible.
Power available to power plug. No Radarange components operating including oven light.	1. Open power cord; unplug it. Check with ohmmeter on lowest scale. Reading should be under 1μ, end to end, for each lead. 2. Effective terminal or wiring. 3. Open cavity or tube thermal switch. Allow time to reset if cavity or tube is warm. Check for low air flow (motor, fan, etc.). 4. Open multiprotector fuse or open transformer fuse.
Oven light operates. Latch solenoid does not operate; oven does not cook.	1 START switch not closing to activate timer module. 2. STOP switch not closing to activate timer module. 3. Defective terminal or wiring. 4. Defective Touchmatic timer.
Oven light and latch solenoid operate, blower does not operate, oven does not cook.	1. Left-interlock or latch switch not closing. Listen for audible click. An audible click **usually** means the switch is working properly; test. 2. Defective terminal or wiring.
All components operate, but oven does not cook. No voltage at plate-transformer primary (120V) leads. **Caution:** All other transformer leads are at HV.	1. Defective terminal or wiring. 2. Defective Touchmatic timer.
Open multiprotector fuse in primary circuit.	1. Left-interlock or sensing switch out of sequence. 2. Defective left-interlock or sensing switch. 3. If multiprotector fuse is open, also test multiprotector resistor for correct value, cracks, etc.
Open transformer thermal fuse in primary circuit. **Caution**: Some leads are at HV.	1. Defective terminal or wiring in HV circuit. 2. Defective RF capacitor. 3. Defective rectifier. 4. Defective tube. May show blue-violet glow inside (look through waveguide and perforations) when thermal fuse replaced and oven turned on. This is a 5–10 second test, or thermal fuse will open. 5. Defective oil-filled capacitor. 6. Defective plate transformer. Check ohmage.
All oven components operate. Unit cooks slowly in all modes.	1. Check whether unit is in defrost or low Cookmatic mode. 2. Run oven-performance tests. If okay, customer education is required. 3. Check voltage at outlet. If low, customer education is required.

Table 4-2 cont.

CONDITIONS	POSSIBLE CAUSES
	4. Check milliampere output. 5. If milliampere output low, check components in HV circuit; power cord must be disconnected from outlet, and RF capacitors discharged. 6. Check capacitor color dot for connection to correct 4 or 4A HV transformer tap. 7. If above checks are acceptable, replace tube.
Oven does not operate in cookmatic mode or defrost mode or cook mode.	1. Improper usage. 2. Test Touchmatic timer
Touchmatic timer, latch solenoid, and blower motor operate. Magnetron filament does not glow.	1. Defective terminal or wiring. 2. Open primary or filament winding on filament transformer 3. Open tube filament.
Touchmatic timer, latch solenoid, blower motor, and tube filament operate. Unit does not cook. **No** voltage at plate-transformer primary (115 – 120V) leads. **Caution**: All other leads are at HV.	1. Defective terminal or wiring. 2. Touchmatic timer inoperable.
Same as condition above, but voltage is available at plate-transformer primary (115 – 120V) leads. **Caution**: All other leads are at HV.	1. Defective terminal or wiring in HV section. 2. Open transformer secondary winding. Test ohmage. 3. Open diode, or diode connected backwards. 4. Open oil-filled capacitor. 5. Magnetron tube weak. Check milliampere output.
Oven cooks food unevenly.	1. Customer is not following cookbook instructions. 2. Stirrer blade deformed. 3. Stirrer blade not turning. Check screw and blade clutch bearings for grease or burrs. This may also indicate incorrect air flow; defective blower motor or blower wheel.
Door will not open after STOP switch is pushed. Latch solenoid still operating; fan running; oven cooking or not cooking.	1. STOP switch not opening. 2. START switch not opening. 3. Defective terminal or wiring. 4. Defective Touchmatic timer.
Door will not open after STOP switch is pushed.	Defective latch-solenoid linkage; adjust.
No end-of-cycle (EOC) tone.	Defective Touchmatic timer.
Timer will not maintain correct time of day (TOD) or cannot be set.	1. Intermittent 120V supply 2. Customer usage. 3. Defective Touchmatic timer.
Part of 4-digit time display not functioning (LIT). Will not maintain TOD or mode selected cook, hold, Cookmatic, slow cook, or timer indicator not functioning (LIT).	1. Customer usage. 2. Defective Touchmatic timer
Oven emits **loud** buzzing sound or vibration in **all** operation modes.	Loose plate-transformer laminations. Disconnect and tape apart transformer primary leads. Test oven. If loud buzzing is no longer present, replace plate transformer
Oven emits **loud** pulsing sound or vibration in Cookmatic mode (10 – 100% slide setting), but oven is normal with slide control in far-right position.	Defective large triac in TM. Replace TM assembly. **Special Note**: A **slight** pulsing sound or vibration is a normal condition and will not be corrected by replacing the TM assembly.

4. With the unit door opened to 0.3125 inches, the ohmmeter indication must be greater than 300Ω. This indicates the N.O. contacts of the left interlock are opening as designed. These contacts *must* open before the door opening exceeds 0.3125 inches.
5. If steps 3 and 4 show the same reading, replace the switch or adjust switch-actuator arm.

RR-9 Left-Door Interlock

1. Zero the ohmmeter on the lowest scale.
2. Disconnect the 6-pin wiring harness connector.
3. Place the ohmmeter leads on pin 3 of the female connector and the test terminal *T* on the fuse-resistor terminal board (far-left terminal with blue wire).
4. Open and close the door to an eighth-inch. The ohmmeter *must* indicate zero.
5. Slowly open the door to 0.3125 inches; do not go beyond this. The ohmmeter *must* show an infinite reading.

RR-7 & RR-9 Multiprotector-Fuse Testing

1. Zero the ohmmeter on the lowest scale.
2. Disconnect the wiring from terminal *C* (Fig. 4-14).
3. Across the fuse leads, the meter should read under 1Ω when the thermal fuse is intact.
4. The meter will show an infinite reading if the thermal fuse is open.
5. If the fuse has opened, the door-switch malfunction *must* be corrected before the fuse is replaced.

RR-7 & RR-9 Multiprotector-Resistor Testing

1. Zero the ohmmeter on the lowest scale.
2. Disconnect the wiring from terminal C (Fig. 4-14).
3. The ohmmeter should read 145–155Ω across the resistor. If it does not, replace it.

Transformer Fuse Resistor

1. Remove all resistor leads from their terminals.
2. Zero the ohmmeter on the lowest scale.
3. Place the meter leads across thermal fuse (330°F). Meter should indicate 9.5–10.5Ω.

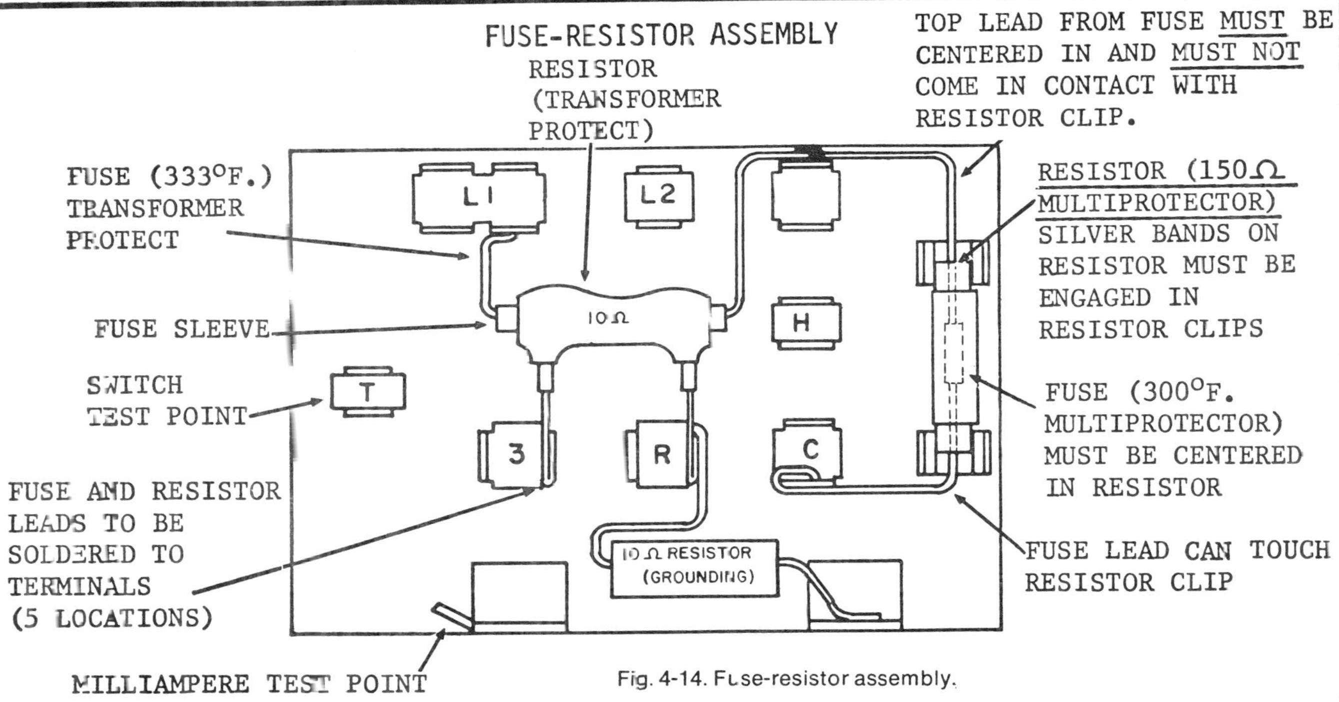

Fig. 4-14. Fuse-resistor assembly.

Note: If the thermal fuse is open, check all HV components for a shorted or grounded condition. A shorted or grounded condition in the 115V supply will not normally cause this fuse to open.

RR-6 Multiprotector Thermal & Transformer Fuse

Turn on the oven light. If the oven light operates, both fuses are intact. This test also shows that the tube and oven-cavity thermal cutouts are not open because of an overtemperature condition.

Note: If the multiprotector thermal fuse has opened, the left-door interlock-switch malfunction must be corrected before the thermal fuse is replaced.

RR-6 Tube & Oven-Cavity Thermal Protectors

Those are self-resetting and open when an overtemperature condition is reached on the tube housing or in the oven cavity.

Turn on the oven light. If the oven light operates, both protectors are operable. This test also shows that the multiprotector and transformer thermal fuses are intact.

RR-6 Oven Light Doesn't Operate on Above Tests

1. Disconnect the power.
2. Remove the side-access panel.
3. Zero the ohmmeter on the lowest scale.
4. Remove the leads from each component to be tested.
5. Place the ohmmeter leads on the component terminals.
6. The ohmmeter should read under 1Ω; if it does not, replace component. In case of the tube or cavity thermal-switch protector, wait for reset.

RR-4D & RR-7 5/30-Minute Timer

Timer terminals *L*, *TM*, and *TM1* are multiterminal connectors only; they are not connected internally to the timer or to each other, Refer to Fig. 4-15A.

The *T2* terminals are connected inside the timer to a switch arm. Terminals *T2*, *A*, and *B* are connected together when the timer is set to a *minutes* selection. These terminals are disconnected from each other when the timer is at zero.

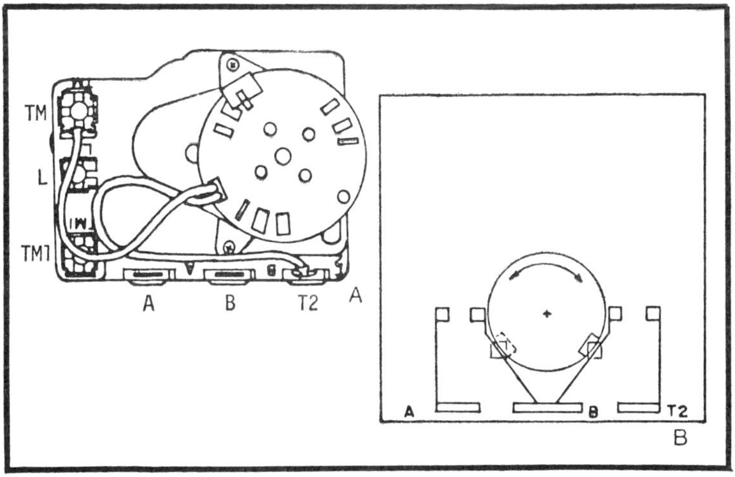

Fig. 4-15. Testing 5- or 30-minute timer.

The *T2*-to-*B* set of contacts control the timer motor. The *B*-to-*A* set of contacts control all other components.

RR-4D & RR-7 Timer Contact Testing

1. Disconnect the power from the unit; remove the leads from terminals *A*, *B*, and *T2*.
2. Zero the ohmmeter on the lowest scale.
3. When the timer is in the OFF position (click may be heard), the ohmmeter should show an open circuit (infinite reading between *T2* and *B*, *T2* and *A*, and *A* and *B*.
4. When timer is set to a *minutes* position (ON) a zero reading should be shown between the *T2*, *A*, and *B* terminals.
5. If the timer does not check as shown, replace it.
6. If the timer drags or stops during operation, unplug the unit to check for loose timer-motor lead connections at *TM* and *T2*. Also check that the timer shaft is centered in the control-panel clearance hole and that the knob is not dragging or broken. If these problems are not present, replace the timer.

Testing Oil-Filled Capacitor

Disconnect all wiring to the capacitor (0.85 μF). The oil-filled capacitor is to be checked for opens and shorts with

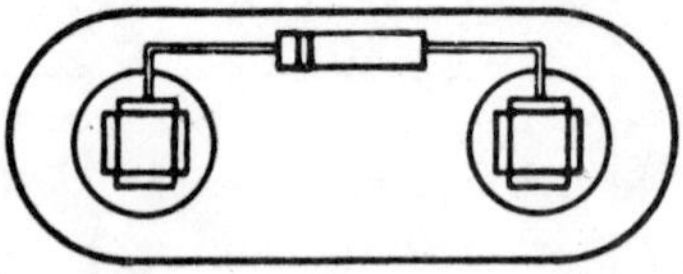

Fig. 4-16. Oil-filled capacitor.

an ohmmeter, using the R × 10K scale. The test can be made with or without the bleed resistor attached (see Fig. 4-16).

With Resistor	Without Resistor
10M	∞

The test should be conducted with meter set on the R × 10K scale by touching the leads to both sets of connections; then reversing the leads and conducting the test again. With the leads in one direction an open condition(∞)should be noted; a slow-discharge-to-open reading (10M) should be noted when the leads are reversed.

Test the capacitor for shorts between the terminals and the can. No continuity should be shown between the terminals and the can.

On reinstallation, observe the color dot to determine correct transformer tap (4 or 4A).

Blower Motor

The blower-motor coil is internally fused so the unit may be used in built-in installations. The fuse is embedded in the motor winding. If the fuse opens, the motor must be replaced. To check the motor, proceed as follows:

1. Zero the ohmmeter on the lowest scale.
2. Remove the leads from the blower motor.
3. Place the ohmmeter leads on the motor terminals.
4. The ohmmeter will show continuity (approximately 5–8Ω; value varies with motor brands).

COOKMATIC CONTROL MODULE OPERATION & TESTING

The control module (CM) is designed to govern operation by the activation of two internal solid-state switches and other circuitry.

First Solid—State Switch

This device conducts power between CM terminals 7 and 8, which supply power to the timers, timer pilot, latch solenoid, blower motor, and filament-control transformer. If this switch does not operate, none of the above items will operate, and the unit will not cook.

Second Solid-State Switch

This device conducts power between CM terminals 9 and 10, which supply variable power to the plate transformer. This switch, in conjunction with the slide control and other CM internal circuits, supplies variable power from 10% to 100%.

Caution: The following tests are conducted using live circuits. For these tests, the oven should be disconnected before starting and between tests. A voltmeter with a 120V scale is required.

Test leads with clamp-type clips should be used. The control panel may be tipped out for easier access to test terminals. A water load must be placed in the oven during these tests. The door must be closed and the timer set. Conduct the following tests if the blower, timer, and tube filament are not operating. Refer to Fig. 4-17. Proceed as follows:

1. Place the meter leads on CM terminal 9 and the START switch terminal with the black CM lead on it.
2. Depress the START switch.
3. The meter should indicate approximate line voltage (115–120V) with the switch depressed.
4. If no voltage is indicated, a component in the supply wiring is not functional; the CM is not at fault.

If the above testing is acceptable:

5. Place the meter leads on CM terminals 7 and 8.
6. The meter should indicate approximate line voltage (115–120V) before the START switch is depressed.
7. With the START switch depressed and released, the meter reading should be less than 5V. This indicates that the first solid-state switch is operational and so are the items controlled by it (blower motor, etc.). If they are not operational, check the wiring to these devices.

Conduct the following tests if the items in step 7 above (blower, etc.) operate but the unit does not cook.

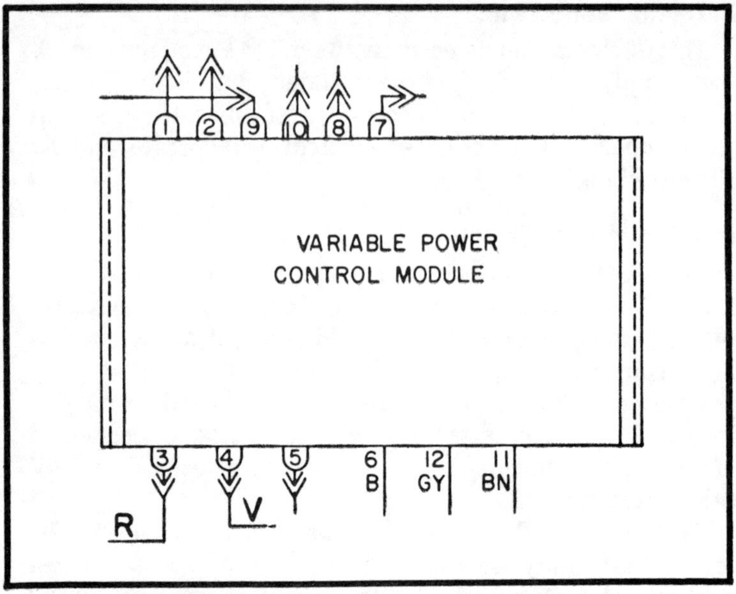

Fig. 4-17. Variable-power control module.

1. Place the meter leads on CM terminals 1 and 2.
2. The meter should indicate approximately 24V AC when the START switch is depressed and released, which indicates the CM is being supplied with voltage from the filament transformer.
3. If the meter does not indicate the correct voltage, test the filament control transformer.

If the CM tests above are acceptable:

1. Place the slide control to far-right position.
2. Place the meter leads on CM terminals 9 and 10.
3. The meter should indicate approximate line voltage (115–120V) before the START switch is depressed.
4. With the START switch depressed and released, the meter reading should be less than 5V. This indicates that the second solid-state switch is operational and the plate transformer should be in operation.
5. If the meter reading does not change when the START switch is depressed and released, the wiring or the plate-transformer primary winding may be open. Also test the slide control and slide-control wiring. If the above items test acceptably, replace the CM.

Conduct the following test if all components are operating but the Cookmatic function does not vary cooking time.

1. Check the slide-control wiring to the CM.
2. Place the voltmeter leads across the plate-transformer primary terminals (red and yellow).
3. Keeping your hands clear; start the unit. The voltmeter should indicate line voltage (115–120V) with slide control in far-right position. As the slide control is moved to the left, the voltage will decrease. *Note*: The meter needle will pulse as the second solid-state switch turns on and off. If it does not, and if the slide control and wiring have been tested, replace the CM.

RR-6 TOUCHMATIC TIMER OPERATION

The oven control will provide four separate and distinct modes of operation: 1 microwave, 2 timer, 3 time of day, 4 termination functions. The end-of-cycle (EOC) tone may sound when the unit is first connected to the power supply.

Microwave Mode

This mode consists of three separate submodes and two combinations of submodes.

Cook Submode. In this mode of operation the control will energize the magnetron at full power for the full amount of time entered. Entry into this mode is as follows:

1. Enter the amount of cook time in minutes and seconds.
2. Touch the DEFROST pad. The DEFROST indicator light will illuminate; the time entered will disappear.
3. Depress the START switch and release. The indicator light will remain on; the time entered will reappear and will count down second by second.

After the amount of time entered in step 1 has elapsed, the control will annunciate an EOC by means of an audible tone, and the indicator light will go out.

Defrost Submode. In this mode of operation the control will energize the magnetron at full power for 20 seconds of every 60 seconds of time entered. Entry into this mode is as follows:

1. Enter amount of defrost time in minutes and seconds.
2. Touch DEFROST pad. The DEFROST indicator light will illuminate; the time entered will disappear.
3. Depress the START switch and release. The indicator light will remain on; the time entered will reappear and count down second by second.

After the amount of time entered in step 1 has elapsed, the control will signal by means of an audible tone, and the indicator light will go out.

Slow-Cook Submode. In this mode of operation, the control will energize the magnetron at full power for a period of 20 seconds of every 40 seconds of time entered. Entry into this mode is as follows:

1. Enter the amount of cook time in minutes and seconds.
2. Touch the SLOW-COOK pad.
3. Depress the START switch and release; the SLOW-COOK indicator light will illuminate.

After the amount of time in step 1 has elapsed, the indicator light will go out and the control will annuciate an EOC by means of an audible tone.

Defrost/Cook Submodes. The control will energize the magnetron as described in *Defrost Submode* for the amount of defrost time entered, de-energize the magnetron for the amount of cook time entered, and energize the magnetron as described in *Cook Submode* for the amount of cook time entered. Entry into this mode is as follows:

1. Enter the amount of defrost time in minutes and seconds.
2. Touch the DEFROST pad. DEFROST light will illuminate; defrost time display will go off.
3. Enter the amount of cook time in minutes and seconds. DEFROST indicator will remain on.
4. Depress the START switch and release. DEFROST light will remain on; defrost time will appear and begin to count down.

After the amount of time entered in step 1 has elapsed, the DEFROST indicator light will go out. The control will automatically enter a hold cycle for the amount of time entered in step 3.

During the hold cycle, the HOLD indicator light will illuminate, and the magnetron will be disabled. After the completion of the hold cycle, the HOLD indicator light will go out and the control will automatically enter a cook submode; the appropriate indicator light will illuminate. After the amount of time entered in step 3 has elapsed, the control will annunciate an EOC by means of an audible tone, and the COOK indicator light will go out.

Defrost/Slow-Cook Submodes. In this type of operation, the control will energize the magnetron for the amount of defrost time entered, de-energize the magnetron for the amount of cook time entered, and energize the magnetron as described in *Slow-Cook Submode* for the amount of cook time entered. Entry into this mode is as follows:

1. Enter amount of defrost time in minutes and seconds.
2. Touch the DEFROST pad. The DEFROST indicator light will illuminate; the display will be blank.
3. Enter the amount of slow-cook time in minutes and seconds.
4. Touch the SLOW-COOK pad. The SLOW-COOK and DEFROST indicator lights will be illuminated. The cook time will be shown on unit.
5. Depress the START switch and release. The SLOW-COOK light will go out; DEFROST light will remain on.

After the amount of time entered in step 1 has elapsed, the control will automatically enter a hold cycle for the amount of time entered in step 3. During the hold cycle the magnetron will be disabled, and the HOLD indicator light will illuminate; the DEFROST indicator light will go out. After completion of the hold cycle, the control will automatically enter a slow-cook submode; the SLOW-COOK indicator light will illuminate; the HOLD indicator light will go out. After the amount of cook time entered in step 3 has elapsed, the control will annunciate an EOC by means of an audible tone.

Timer Mode

In this mode of operation, the control will completely disable the microwave oven. Entry into this mode will be as follows (this mode can be entered only if the control is not in a microwave mode):

1. Enter time in minutes and seconds.
2. Touch the TIMER pad. The TIMER indicator light will illuminate; the timer will commence counting down.

After the time entered has elapsed, the control will annunciate an EOC by means of an audible tone.

Time-Of-Day Mode

The TOD counter will operate whenever power is applied to the control. The control will be in this mode of operation any time that it is not in either of the other modes. The time of day is set by touching the CLOCK SET pad, entering the time in hours and minutes, and again touching the CLOCK SET pad. If the CLOCK SET pad is not touched the second time, the clock will not advance, and all mode indicator lights will be illuminated. If no time is entered, the reset must be touched to extinguish the mode lights. The timer automatically resets after EOC.

Termination Functions Mode

The control has the capability of terminating any mode of operation. Termination of any of the microwave modes is accomplished as follows:

Prior to Start. Any microwave mode may be terminated by touching the RESET pad and prior to depressing the START switch. This action will clear all counters except TOD; this will be displayed when RESET pad is touched.

After Start. Termination of any microwave mode after the START switch has been depressed may be executed by depressing the STOP switch. This action will disable the microwave oven and hold whatever numbers were present in the counters at the time the STOP switch was depressed. Complete termination of the microwave mode after the START switch has been depressed may be executed by depressing the STOP switch, then touching the RESET pad. This action will result in the same response as described in *Prior to Start*.

Resumption of the Cycle. If the RESET pad was not touched, resumption of operation is obtained by depressing the START switch.

Timer Mode. Termination of the timer mode is executed by touching the RESET pad. This action clears the timer counter and displays TOD.

TOD Mode. In the slow-cook and defrost modes, the indicator light will remain illuminated during the full time selected and will not cycle with the magnetron operation.

If desired, by touching *0* the TOD display can be turned off. Touch the RESET to illuminate the TOD display.

RR-6 TOUCHMATIC TIMER TESTING

Conduct a programming test per the preceding section, *Touchmatic Timer Operation*. See the troubleshooting guides for possible cause of failure.

RR-6 Touchmatic Timer Bench Testing

Use of the schematic or wiring diagram will assist you in understanding the following test information. The multi-connector terminal board on the rear of the timer has no internal connections to the timer.

All timer control leads exit from the bottom and are color coded but not numbered. White lead No. 9 and the timer power-supply ground symbols signify a PC (printed circuit) board connection only. Lead No. 9 and the timer power supply *do not* connect to the Radarange or timer chassis.

Caution: All test leads should have insulated terminals to prevent the possibility of shock and prevent touching of incompatible wiring, which could damage the Touchmatic timer. The test cord ground lead should be connected to the timer, using one of the mounting holes. Proceed as follows:

1. Remove the timer from the unit.
2. Lay the timer assembly face up on an insulated surface.
3. Connect the black wire (hot lead) of a test cord to wire lead 8 (black) of the timer. Connect a short black lead to this junction for a later connection.
4. Connect the white wire (neutral lead—may have a ribbed insulation instead of white color) to wire leads 1 and 5 (white, orange) of the timer.
5. Connect a small bulb and socket or indicator light to timer leads 1 and 2 (blue and red).
6. Using two momentary contact switches—one normally open (N.O.) and one normally closed (N.C.)—proceed as follows:

a. Connect the common terminal of both switches to timer lead 6 (gray). This is shown as point *C* on Fig. 4-18.

b. Connect the N.O. terminal of the START switch to point *A* on Fig. 4-18. This is the short black lead mentioned in step 3. The

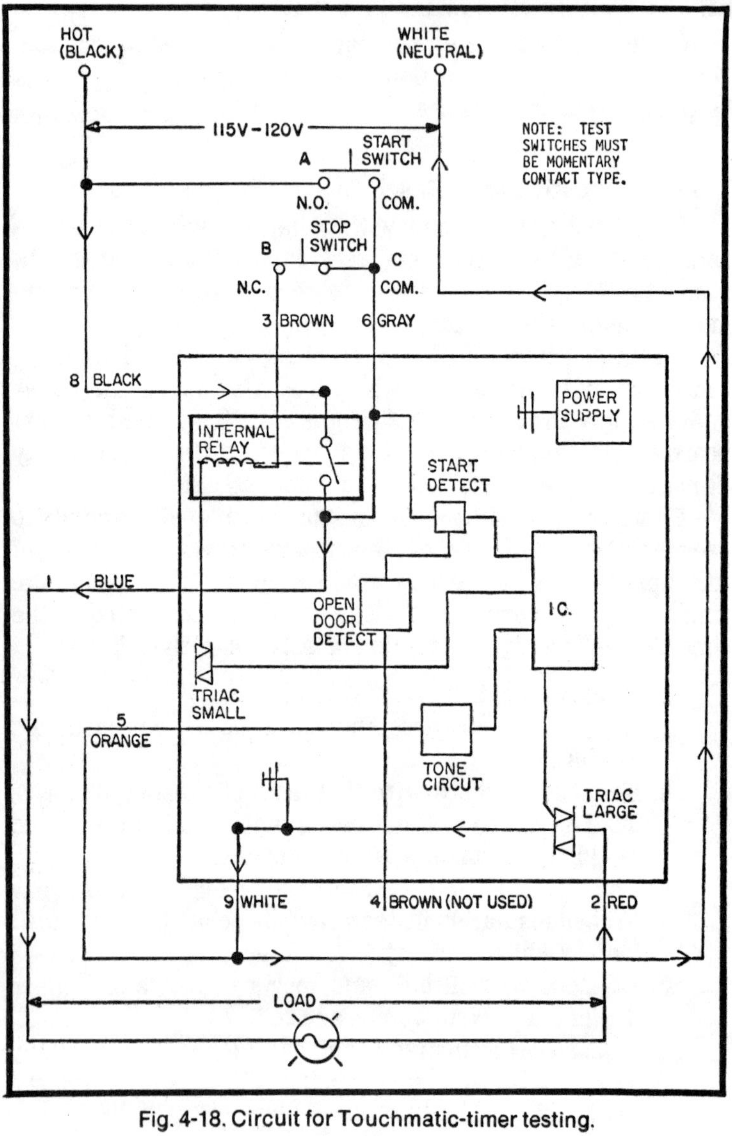

Fig. 4-18. Circuit for Touchmatic-timer testing.

START switch contacts must be closed when released.

c. Connect the N.C. terminal of the STOP switch to point *B* on Fig. 4-18. This is timer lead 3 (brown). The STOP switch must be closed when not depressed, and open when depressed.

7. Check all connections for correct hookup and for tightness. Plug the test cord into the outlet, keeping your hands clear of voltage areas.
8. Follow the Touchmatic programming-test procedure shown in *Touchmatic Timer Operation.*

RR-6 Internal Tests of Timer

An in-unit test on the large *triac* may be conducted as follows:

1. Place 1 cup of water in the unit.
2. Enter a 1-minute selection, touch the timer pad, and push the START button. The fan should operate.
3. If the unit heats water, the large internal triac is defective.

To conduct the tests shown below, the timer wiring must be disconnected from the Radarange components.

Test 1—Internal Timer Relay

1. Zero the ohmmeter on the R × 10K scale.
   Caution: Do not let your fingers touch probes or timer leads, as a false indication will be given.
2. Place the ohmmeter leads on timer wire terminals 8 (black) and 1 (blue. A zero reading should be indicated. If any reading (usually below 1Ω) is indicated, the internal-timer-relay contacts are not opening. This will cause the latch solenoid to close whenever the Radarange is connected to the 115–120V supply line.

Test 2—Large Triac

1. Place the ohmmeter leads on timer wire terminals 2 (red) and 9 (white) and note the reading.
2. Reverse the ohmmeter leads and note reading.

3. Infinite readings should be indicated in both tests. If any resistance (usually below 1Ω) is indicated, the internal timer's triac is shorted. This condition will not normally be noticed by the owner, but may cause failure of the interlock switches and failure of the multiprotector fuse.
4. Test or replace these components if the triac has shorted.

Test 3—Internal Wiring & Relay

1. Zero the ohmmeter on the lowest scale.
2. Place the ohmmeter leads on timer wire terminals 6 (gray) and 1 (blue).
3. A reading of under 1Ω should be indicated if the internal wiring from 6 to 1 is intact. A high reading usually indicates the internal-relay contacts are not closing; the unit may only operate when the door is closed and the START switch held in.
4. Check the START and STOP switches for proper operation if the relay contacts have failed.

RR-9 TOUCHMATIC TIMER OPERATION

The TM module is designed to control—by the actuation of an internal solid-state switch, an internal relay, and other circuitry—all functions of the oven components.

RR-9 Internal Relay

The internal relay contacts conduct power between TM terminals 3 and 4, which supply power to the latch solenoid, blower motor, and filament transformer. If this relay or its contacts do not operate, none of the items above will operate, and the unit will not cook.

A water load must be placed in the oven during TM tests. The door must be closed and the timer set to a time selection.

Internal Solid-State Triac Switch

This unit conducts power between TM terminals 1 and 2, which supply constant or variable power to the plate transformer, depending on the mode selected.

RR-9 TOUCHMATIC TIMER TESTING

Two methods of testing the timer module are shown in this chapter. 1) Touchmatic timer electrical testing, and 2) Touchmatic timer testing by programming instructions. In most cases, only *Method 2* will be used to determine that the TM requires replacement. In other instances, *Method 1* will need to be used.

Important: Do not replace a TM unless you have read the "Additional Comments" section of *Method 2*.

Touchmatic Timer Testing—Method 1

This is a method of electrically testing (using a voltmeter) the TM. (Also see *Method 2* for programming instructions.)

Caution: Some tests are conducted using live circuits. For these tests the oven should be disconnected before starting and between tests. A voltmeter with a 120V AC scale is required. Hands should be kept clear of all voltage areas during testing. Test leads with clamp-type clips should be used.

Conduct the following test if all components are not operating.

Test 1A—Testing for Input Voltage to TM. Plug in the power cord and close the door. If the TOD can be programmed, input power is available to the module. If the TOD cannot be programmed, check the power cord, tube and cavity thermal switches, thermal fuses, and wiring. If the oven light operates, look for a wiring problem only.

Test 1B—Testing for TM Output to Latch Solenoid, Blower Motor, and Filament Transformer. Proceed as follows:

1. Place a water load in the unit.
2. With the oven plugged in and door closed, program the TM for cook, defrost, or Cookmatic operation with the silde control in any position.
3. Push the START button.
4. The latch solenoid, blower motor, and filament transformer (look for magnetron glow) should operate.
5. Unplug the power cord.
6. If *none* of these components operated, look for loose wires and test the START and STOP switches.
7. If the items in step 6 test operational, replace the TM.
8. If only the latch solenoid operates, the TM is operational. Test the left- and latch-interlock switches and wiring.

9. If only blower motor or filament operates, check individual part per testing procedures.

Test 1C—Testing for TM Output to Power Transformer. If the items in *Tests 1A and 1B* are operational, proceed with the following steps:

1. Leave the water load in unit.
2. Make sure that the power cord is unplugged.
3. Connect the voltmeter (120V AC scale) leads to primary terminals (red and yellow wires).
   Caution: Be sure that the meter leads do not touch other transformer terminals because HV is present.
4. Plug in the power cord. Close the oven door.
5. Program TM for Cookmatic operation with the slide control in the far-right position (100% operation).
6. Push the START button.
7. Keeping hands clear, start the unit. The voltmeter should indicate line voltage (115–120V AC) with slide control in far-right position. As slide control is moved to the left, the voltage will decrease. The meter needle will pulse as the second solid-state switch turns on and off. If it doesn't with Tests *1A* and *1B* completed and the interconnecting wiring is okay, replace the TM.

Note: This test can also be conducted in the cook or defrost mode. In the cook mode, the meter should indicate a constant 115–120V (line voltage). In the defrost mode, the meter will indicate 115–120V for approximately one-third of the programmed time and no voltage for two-thirds of the programmed time.

Touchmatic Timer Testing—Method 2

Loss of power to the Radarange (as in a power outage) for a duration of approximately 0.1 second or more may cause timer to lose TOD. This duration of power outage will not normally be noticed by a customer, but in some instances it is indicated by a rolling TV picture.

When the unit is first connected to the power supply, or after a power interruption, approximately 30 seconds should be allowed before trying to program TOD.

Whenever the TOD is programmed into the timer module incorrectly, it can not be reprogrammed until approximately 30 seconds has elapsed. For example, suppose the correct TOD is 1:00 and the customer misprograms the timer to indicate 11:00. The timer will not accept the 1:00 program for approximately 30 seconds.

When the unit is first connected to the power supply, or after a power outage, the first programmed cook, Cookmatic, or defrost operation may not be accurately timed. For example, if customer programmed it for 2-0-0 of cook time and unit ran more or less than 2 minutes.

Any static charge applied to the timer panel may upset the operation of the panel. For example, a customer walking across a carpet, touching the panel, may cause loss of the correct TOD. This static charge will not normally cause the module to become inoperative; the TOD or cooking mode may be reprogrammed after 30 seconds. The Radarange should be installed on a separate grounded circuit to prevent voltage conditions which may cause loss of correct TOD or time selected for cooking modes.

In some instances, persons having a dry-skin condition cannot program the modules. Normal moistening of the fingers will correct this problem. The timer module cannot always be programmed by persons having long fingernails if the nails interfere with good finger contact to the glass sensing plate. The timer module cannot be programmed while wearing any type of gloves.

If the time module will not reset after touching the RESET pad, disconnect power to the unit; then reconnect the power and test module. If the timer module will not reset, it must be replaced.

Sometimes a part failure in the HV section (capacitor, magnetron, diode, etc.) can cause the timer module to misprogram shortly after entering a cook, Cookmatic, or defrost mode. If this condition is found, disconnect the unit from its power source. Then disconnect and tape apart the leads on the primary of the transformer. Test the timer module. If its operation is normal, check all HV components for failure.

Occasionally the timer module will not program unless the metal trim is touched while touching the pads. This can be traced to one of the following problems.

- Power plug isn't properly grounded or polarized
- Dry-skin condition exists
- Timer module has lost sensitivity

The TOD shown in the timer module is more accurate than time kept by any other household clock. In case of any customer complaint as to TOD accuracy, be sure that the customer is informed as how to correctly set the TOD and that only one member of the household is programming the TOD. In many instances, children have programmed the TOD without the parent's knowledge. A piece of clear tape may be placed over the clock pad to eliminate this possibility.

Have the customer monitor the TOD for 3 days. At the end of each day, the customer should check the correct TOD by using the local TOD telephone number. If the module is incorrect, have the customer record the error in minutes and reset the module to the correct TOD for the 3-day period. If the module is still incorrect, it should be replaced. Test the timer as follows:

1. If the timer module displays 8-8-8-8 and cannot be programmed in any mode to change the 8-8-8-8 display, replace module and do not proceed with following steps.
2. If the timer module will not reset when the RESET pad is touched (the STOP switch must be pushed if unit is running), replace module and do not proceed with following steps.
3. If the unit is opening a house fuse, check all primary (120V) wiring and components (except timer module) for short or ground. If all wiring and components test acceptably, replace timer module and check operation of replacement module. This condition may appear in some modes and not others (e.g., defrost or Cookmatic mode may open a fuse, while cook mode may operate normally).
4. Touch the RESET pad and program timer to read 8-8-8-8. Make sure all numbers are complete with no segments missing.
5. Touch the RESET pad and program timer with numbers 1-2-3-4-5-6-7-8-9-0.

   Note: After number 4 is programmed, the timer will indicate 1-2-3-4. As the numbers run past the left indicator, the final numbers shown will be 7-8-9-0.
6. With the timer still indicating 7-8-9-0, program the timer with 0-0-0-0. The third zero should extinguish the display.

7. Touch the RESET pad; then touch the CLOCK pad. At this point, the COOK, HOLD, DEFROST, COOKMATIC, and TIMER indicator lights should come on. Program timer with 1-2-5-9, touch the CLOCK pad, and all indicator lights should go out. The timer is now programmed for TOD (12:59 was used as an example for time of day).
8. Program the timer with 1-0 and touch the TIMER pad. The TIMER indicator should light; the timer should count down to zero, sound audible tone, and return to TOD.
9. Place a water load in the oven (3–4 cups).
10. Program the timer with 1-0-0, touch the DEFROST pad, and start the unit. The DEFROST indicator should light. The timer should count down to zero, sound audible tone, and return to TOD.
11. Program the timer with 3-0, touch the COOKMATIC pad (selector control any position), and start the unit. The COOKMATIC indicator should count down to zero, sound audible tone, and return to TOD.
12. Program the timer with 3-0; start the unit. The COOK indicator should light; timer should count down to zero, sound audible tone, and return to TOD.
13. Program the timer with 5, touch the DEFROST pad, program the timer with 5, touch COOKMATIC pad, and start the unit. The unit should defrost for 5 seconds (observe the indicator lights). The timer should sound the audible tone on complettion of this 15-second cycle and return to TOD.
14. Program the timer with 5, touch the DEFROST pad, program the timer with 5, and start the unit. The unit should defrost for 5 seconds, hold for 5 seconds, and cook for 5 seconds (observe indicator lights), The timer should sound the audible tone on completion of this 15-second cycle and return to TOD.
15. Program timer with 1-0. Start and stop unit three times within the 10-second period.
16. Touch the RESET pad. TOD should appear and read the 1-2-5-9 as programmed in step 7 plus the testing time. For example, if it took you 2 minutes to conduct the tests after step 7, the time would indicate 1-0-1.

17. Record the temperature of a water load and place it in the unit. Program the timer with 1-0-1, touch the TIMER pad, and push the START button. In this condition the fan should operate, the magnetron filament should glow, and the latch solenoid should operate; but the oven should *not* heat the water. If the unit heats the water, the solid-state switch (triac) is shorted and the timer must be replaced. A shorted triac may cause the failure of the START-STOP switch contacts. Test these contacts electrically.
18. Conduct an oven-performance test in cook, Cookmatic, and defrost modes.
19. If all the preceding tests fail to operate as described, check the following:
    a. Are all interlock switches and the START and STOP switches closing to supply power to the timer? If these switches do not operate correctly, the module will not operate per design specifications; only the switch should be adjusted or replaced.
    b. Is all wiring connected to the correct terminals?
    c. Are there any loose wires?
20. If steps 4–18 do not check as described and step 19 is okay, replace the Touchmatic timer module; test replacement module for correct operation.

Note: Although not normally seen, it is possible that during transportation of a unit or a replacement timer module, a static charge may be built up on the module which will cause the unit to operate improperly upon installation. On a properly grounded circuit, a 5–10 minute standing time must be allowed before the module will perform as designed. On ungrounded circuits, the standing time will be longer to bleed off the accumulated static charge. It will be somewhat dependent upon relative humidity of the air.

Chapter 5

Servicing Frigidaire Microwave Ovens

This chapter covers three of the newest models in the Frigidaire[1] line of microwave ovens—Models RCM-2, RCM-4, and RCM-6.

The RCM-2 is a new economy model with 500W of cooking power and a 14-minute timer. One of its advantages is the ease of operation: Just set the timer for the microwave cooking time required and push the START button.

Model RCM-4 offers more cooking power—675W—plus an automatic-defrost cycle and a fast-cook cycle. The control panel incorporates COOK and DEFROST signal lights and a 20-minute timer. This model is similar to the RCM-37S, but with a defrost cycle added.

The Model RCM-6 has all the advantages of the RCM-4 plus a few of its own. It offers a slow-cook cycle that varies the power to roast quality meats and cook delicate sauces and cheese and egg dishes. There is also a 35-minute timer with the first 5 minutes expanded for accurate setting.

HOW THE RCM-2 WORKS

The circuitry of a microwave oven can be considered in two sections—the 115V control section and the HV circuit.

---

1. All of the illustrations in this chapter are courtesy of Frigidaire Division, General Motors Corporation.

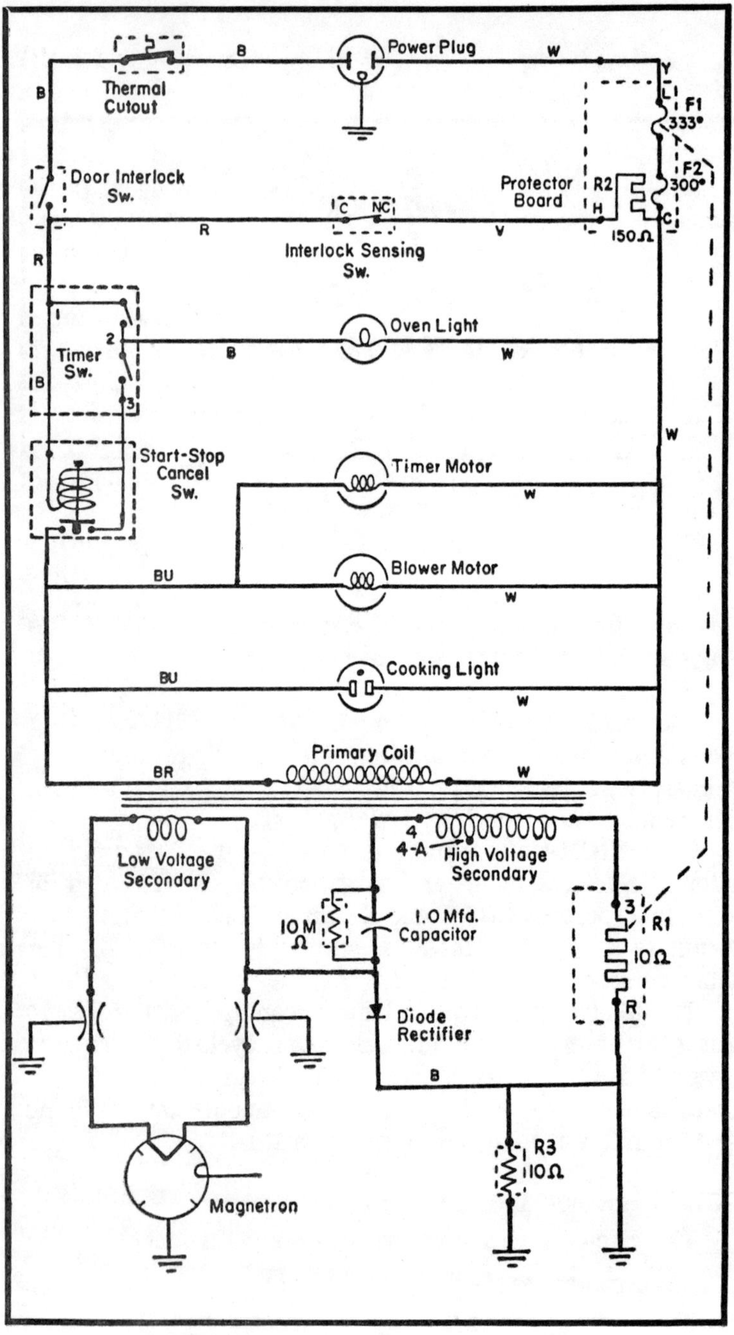

Fig. 5-1. Schematic of Frigidaire RCM-2.

RCM-2 Control Circuit

Figure 5-1 shows the circuit of the RCM-2. The black (B) wire from the line cord is connected to the thermal cutout switch, which will open in case of an oven fire or an overheated magnetron to cut all power to the rest of the circuit.

The door-interlock switch is the next component in the line. This switch can only be closed by shutting the door, and therefore prevents operation of the microwave oven with the door open.

If the door-interlock switch contacts should remain closed when the door is opened, a special protector circuit will open to interrupt power to the oven. This circuit consists of the interlock-sensing switch, which is closed when the door is open; heating resistor R2 on the protector board; and thermal fuse F2, which is located inside resistor R2. The mounting of these resistors and fuses is shown in Fig. 5-2.

If the interlock switch remains closed when the door is opened, current will flow through the interlock-sensing switch and resistor R2, heating thermal fuse F2 until it melts, opening the line. This fuse opens at approximately 300°F.

The other fuse and resistor on the protector board are operated by a high current in the HV circuit and will be discussed later.

Start-Stop Cancel Switch

This switch, located above the pilot light on the front panel, is operated by the START and STOP buttons, the timer contacts, and the door latch.

When the door is shut and a cooking time set on the timer, pressing the START button will lock the door and will close the circuit to the timer and blower motors, the cooking light, and the primary of the HV transformer. If the door is open, the button can be pushed but will not close the switch.

When time runs out, both sets of timer contacts open. With these contacts open, all of the transformer current passes through and energizes the cancel solenoid on the start-stop switch. This solenoid unlocks the door and returns the start switch to its open position. Pressing the STOP button also will open the start switch and unlock the door.

The oven light is on only when a cooking time has been set on the timer.

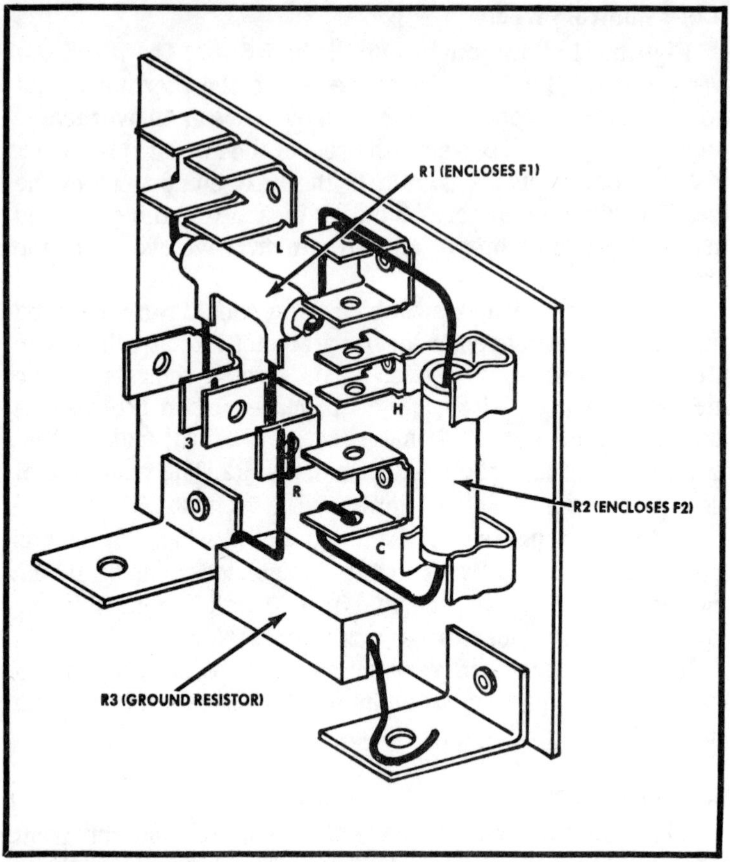

Fig. 5-2. Protector board with heating resistors and fuses.

RCM-2 High-Voltage Circuit

Figure 5-1 also shows the HV section of the RCM-2 circuit. As shown, the transformer secondary has two windings. One is a low-voltage (LV) winding to supply the current for heating the magnetron cathode. The other is a HV winding, which, with the diode and capacitor, supplies approximately 3600V DC to the magnetron. The circuit used to develop the 3600V is a half-wave voltage doubler, which operates as follows:

When the voltage in the transformer HV secondary winding is negative on the grounded end (see Fig. 5-3), electron flow through the rectifier will be in the direction shown by the curved arrow, charging the capacitor to 1800V with the polarity shown.

On the next half of the cycle, when the secondary voltage is positive on the grounded end, the rectifier will not conduct, and therefore can be considered an open in the circuit. (See Fig. 5-4.) The capacitor is still charged, and the polarity of the voltage on the transformer now is added directly to the capacitor voltage, giving 3600V total, which is applied between the cathode and anode of the magnetron.

The magnetron voltage therefore is that shown in Fig. 5-5. It will be 1800V on one half of the AC cycle and will rise to 3600V on the other half-cycle.

The 3600V in this circuit cannot be measured without special equipment. Do not attempt to read voltages in the secondary circuit. No useful purpose would be served by doing so, and you and your equipment will be subjected to an unnecessary hazard.

The 10M (10 mega ohms) resistor, R4, connected across the capacitor, will discharge it eventually; but since the capacitor can retain its charge for some time, it is important to *discharge the capacitor before making tests or replacements of any components in the HV circuit.* If this is not done, you could receive a serious electrical shock. The capacitor can be discharged by placing a screwdriver with an

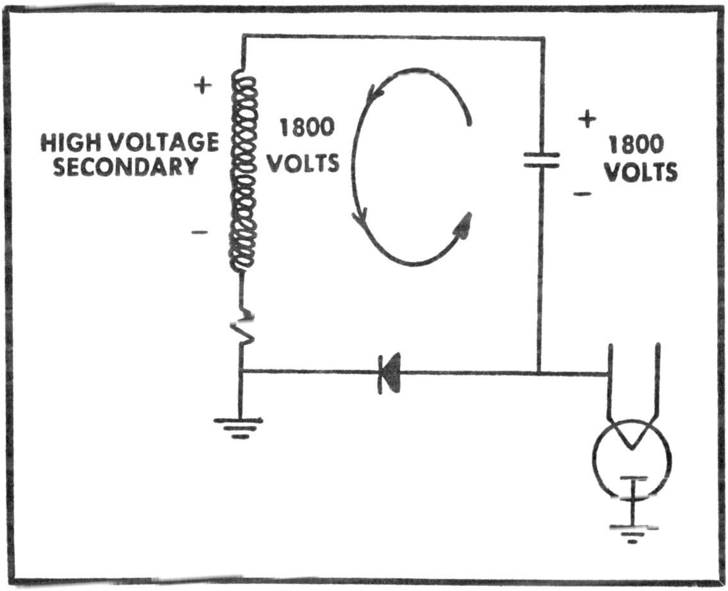

Fig. 5-3. High-voltage power supply, showing capacitor charging.

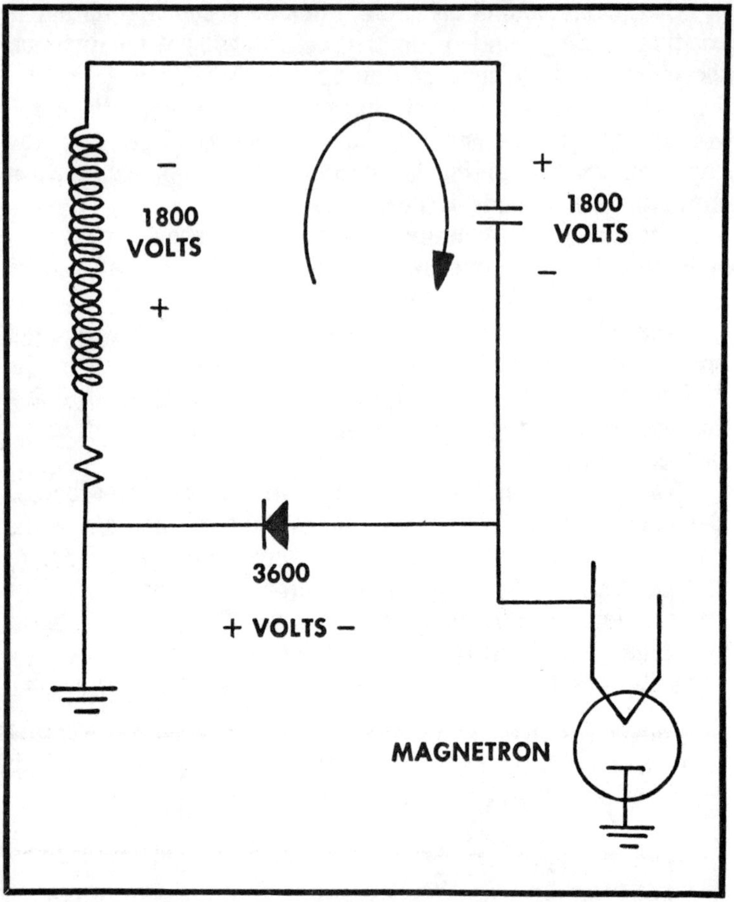

Fig. 5-4. Power supply, showing 3000V applied to magnetron.

insulated handle across its terminals or between one of the feedthrough capacitor's terminals and ground.

Thermal Fuse Protects Against High Current

As shown in the schematic (Fig. 5-1), the current from the HV secondary of the transformer passes through resistor R1, mounted on the protector board. This is a tubular resistor; inside it is thermal fuse F1, which opens when it is heated to 333°F.

Whenever the current in the secondary circuit becomes excessively high because of a short or any other malfunction, the resistor will heat, causing the fuse to open. Since the fuse is in the control circuit, power to the entire unit is interrupted when it opens. A sleeve around the fuse protects it from HV.

Rectifier Must Withstand 3600V

As shown in Fig. 5-3, the diode rectifier will have 3600V applied to it on one half of every AC cycle. For this reason the rectifier should be checked whenever thermal fuse F1 is blown.

RCM-2 TROUBLESHOOTING

Figure 5-1 is a complete schematic of the RCM-2, while Fig. 5-6 is its wiring diagram.

By observing what components are energized, you can often tell approximately where a trouble is. For instance, if the oven light and cooking light come on and the timer and blower motors both operate, but the oven will not heat the food, you know that the trouble is in the HV circuit—an open rectifier or transformer, a defective magnetron, or an open connection anywhere in the HV section.

If the oven light, cooking light, timer, and blower motor remain energized, the LV part of the circuit is open. One of the thermal fuses may be open, the door-interlock switch or oven-cavity thermal switch may be open, or the timer contacts may not be closing. If only the oven light can be energized, the contacts in the start switch may not be closing.

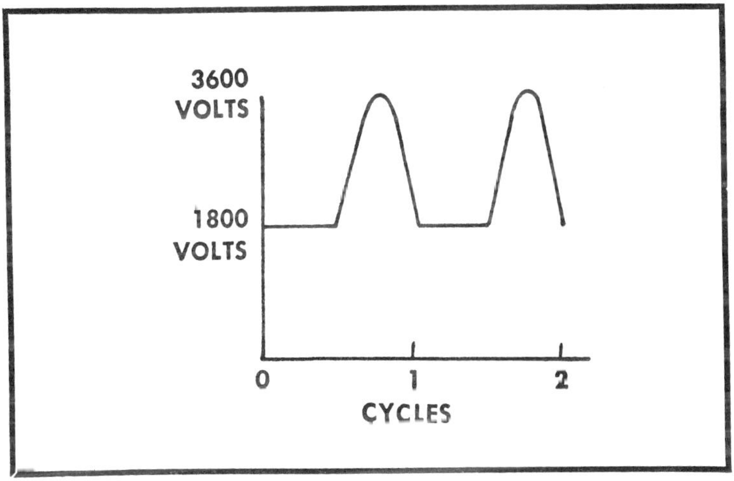

Fig. 5-5. Magnetron voltage waveform.

Fig. 5-6. Wiring diagram of Frigidaire RCM-2.

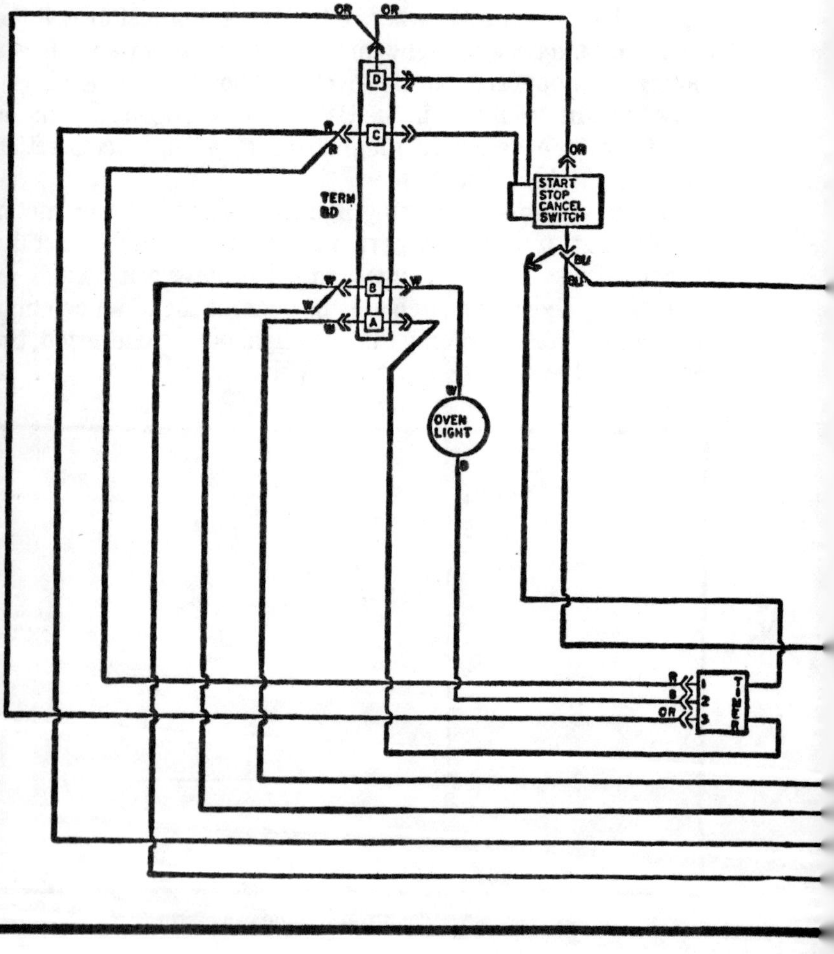

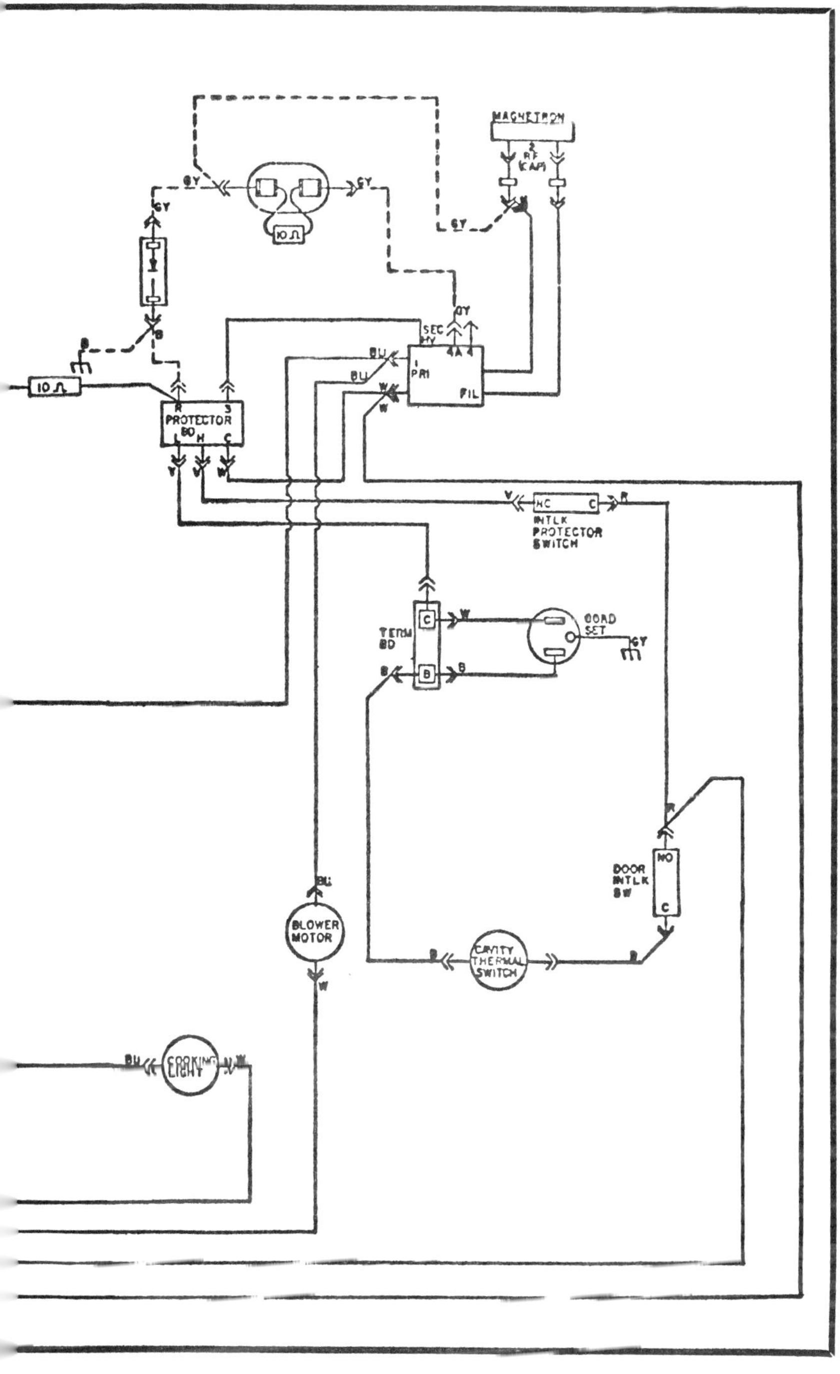

MAGNETRON
2
RF
(CAP)
GY
GY
GY
GY
10 Ω
GY
B
B
10 Ω
R
3
PROTECTOR
BD
L
H
C
Y
V
W
SEC
HV
GY
4A
4
BU
1
PRI
BU
W
W
FIL
V
NC
C
R
INTLK
PROTECTOR
SWITCH
C
W
TERM
BD
B
B
B
CORD
SET
GY
R
NO
DOOR
INTLK
SW
C
BU
BLOWER
MOTOR
W
R
CAVITY
THERMAL
SWITCH
R
BU
COOKING
LIGHT
W

RCM-2 TESTING PROCEDURES

Because of the voltages used in the microwave-generating circuit, access to the components has been made relatively difficult by using pop rivets to mount the access panels to the cabinet. These rivets must be drilled to remove the panel. Over one of the rivets is a seal that must be destroyed to remove the rivet.

After you have serviced the unit, you should install new rivets and a new seal. Extreme care should be exercised if a unit is encountered that has the seal or rivets disturbed because safety devices may have been bypassed. Because of the basic design of the unit, it is very unlikely that there will ever be any problem of microwave radiation unless the door seals have been damaged.

Access to the operating components at the front is gained by removing the shell. To do this, it is necessary to remove the three screws along the top, two screws on each side, and four screws and a rivet in the back. The shell may then be removed.

If nothing is operating, the thermal fuses are most likely open. Remove the yellow lead (Y) to terminal *L* on the protector board. Then check both fuses with an ohmmeter.

If F1 (in the horizontally mounted resistor at the top of the board) is open, check for a short in the HV circuit—possibly in the rectifier, the large capacitor, or one of the feedthrough capacitors. It is possible, though it does not often happen, that the transformer or the magnetron may have become shorted.

If thermal fuse F2 is open, a failure of the door-interlock switch is indicated. It is mandatory that this switch be replaced.

Do not forget to replace the yellow lead on the protector board after correcting the trouble.

Checking the Rectifier

To test the diode rectifier, remove one of the leads to the diode. Use an ohmmeter set on R × 10K (or higher) scale. Measure the resistance between the diode terminals. Reverse the leads and measure the resistance again. One reading should be at least ten times the other. If it is not, replace the diode.

Checking the Capacitor

The oil-filled capacitor should be checked with an ohmmeter using the R × 10K (or higher) scale. It should be an open circuit if the parallel resistor is disconnected, and should be 10M with the resistor connected. It should be an open circuit between each terminal and the can of the capacitor.

Oven Power Test

This test is sensitive to changes in supply voltage. It should be run with the line 115–118V. Proceed as follows:

1. Use 1 quart of tap water in a Pyrex container for this test.
2. Measure the temperature of the water. Remove the thermometer from the water before heating it.
3. Set the oven timer for 2 or 3 minutes. Use a watch or clock with second hand (or a stopwatch) to time the actual heating of the water.
4. Heat the water in the microwave oven for 1 minute as timed by the watch.
5. Stir the water with the thermometer; check its temperature. The temperature rise should be at least 10°F for 1 minute of heating.

Measuring Magnetron Current

The magnetron current in this model should be 275 mA DC. To measure this current, a DC milliammeter with a scale of 300 mA or higher is necessary. If the meter has not been used for this purpose before, it is necessary to determine the current due to the meter impedance. This is done as follows:

1. Unplug the unit.
2. Remove the shell. Discharge the capacitor.
3. Remove the black ground lead (HV circuit) from the spade terminal and attach the lead to the red meter lead.
4. Connect the black meter lead to the grounding-spade connection.
5. Set the meter to a milliampere scale higher than 300.
6. Check all the meter lead connections to ensure that they are secure.
7. Place a water load in the oven.
8. Plug in and start the unit. Record the reading in milliamperes
9. Repeat the test with the grounding resistor R3 (Fig. 5-2) disconnected from the fuse-board mounting screw.

10. Record the new meter reading in milliamperes. Calculate the difference between the two readings. The difference between the two readings is the current due to the internal impedance of the meter. Record this figure on the face of the meter. When taking later readings, steps 9 and 10 can be omitted. Take the milliampere reading obtained in step 8 and add the figure recorded on the face of the meter.
Example:

Meter reading with R3 disconnected	275mA
Meter reading with R3 connected	250mA
Meter correction to be added	25mA

If the total magnetron current is lower than 275 mA, the gray wire from the oil-filled capacitor may be moved from tap 4A to 4 on the transformer. If the reading is high, the connection may be changed from 4 to 4A.

If the reading is still high or low, the capacitor may be changed to correct the problem.

If the current is zero or is very low, the magnetron may need to be changed.

RCM-2 COMPONENT REPLACEMENT PROCEDURES

The transformer, capacitor, rectifier, and fuse board can best be removed by first removing the transformer assembly. Disconnect the transformer secondary leads in the terminal box on the magnetron assembly. Hold the feedthrough capacitor terminals with a pair of pliers while doing this to prevent damaging the capacitors. Disconnect the yellow, white, and violet leads to the protector board, tagging them with masking tape if necessary to ensure that they can be replaced on the correct terminals. Disconnect the black lead from the ground terminal. Remove the mounting nut from the transformer base (see Fig. 5-7). The transformer assembly may then be slid to the rear and removed.

The defective component may then be removed and replaced. If the capacitor is replaced, notice the color of the dot on one side. If this dot is black or white, the gray lead to the transformer should be connected to terminal 4. If the dot is yellow or red, this lead should be connected to terminal 4A.

If the rectifier is replaced be sure that the positive (+) end is to the rear and is connected to the black ground wire.

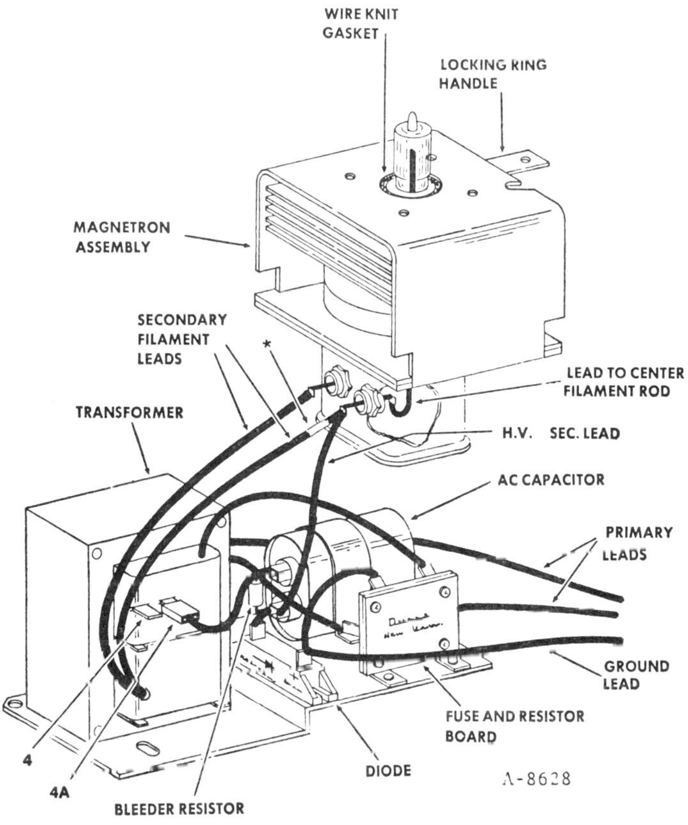

* Black tape is used to indicate phasing of transformer. It may be on either secondary lead. The lead marked with the tape must be connected to the right-hand feed-through capacitor.

Fig. 5-7. Transformer and magnetron connection.

When connecting the assembly after installing it in the unit, be sure to connect the transformer lead with the black tape on it to the rear feedthrough capacitor.

Magnetron Replacement

The magnetron can best be replaced by first removing the transformer assembly as previously described. Then remove the aluminum tape from the transformer side of the magnetron; remove the fish-paper air chute from the other

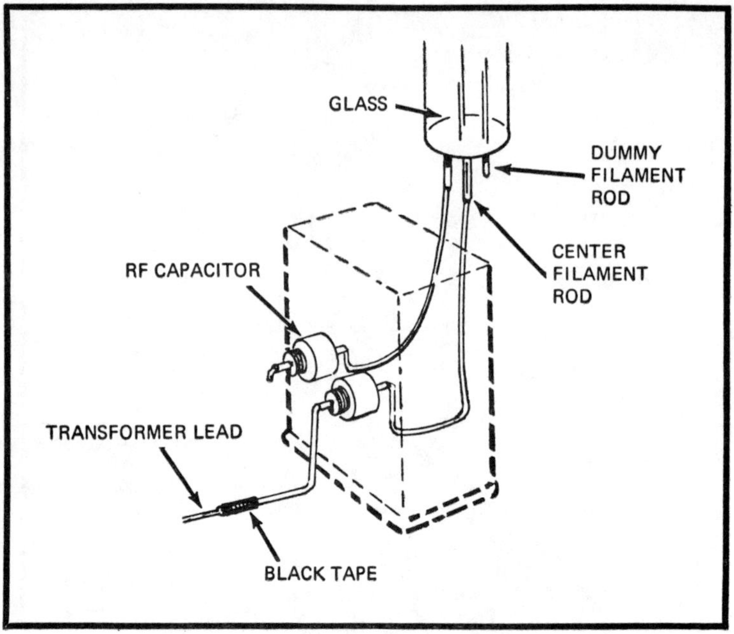

Fig. 5-8. Feedthrough capacitor connection.

side. Remove the two snap-in plugs in the upper edge of this part.

The magnetron-release lever will then be exposed. Push it toward the front of the oven while supporting the lower end of the magnetron assembly with your other hand. The magnetron then will drop down to be removed.

Note: When installing a new magnetron, *be careful not to touch the glass part of the tube.* If the glass is touched, oil from your fingers will be left on the glass. This oil heats when exposed to microwaves and can cause an early failure of the magnetron.

When connecting the magnetron, ensure that the right-hand feedthrough capacitor is connected to the center filament rod on the magnetron. Figures 5-7 and 5-8 show the connections. The black tape may be on either transformer lead.

Ensure that the tube leads and ferrite rings are well separated from all grounded parts. Return the defective magnetron to the factory in the same packing in which the replacement magnetron was received.

Blower & Housing

The blower housing can be removed after removing the housing screws. The blower wheel can then be removed after loosening the setscrew. With the wheel removed, the motor-mounting screws are accessible; the motor may be removed and replaced.

Timer, Start-Stop Switch, & Pilot Light

The pilot light can be replaced by disconnecting the leads and pushing the bulb forward through the escutcheon.

To remove the timer and the switch assembly, the escutcheon must first be removed. The escutcheon is mounted with four screws. Remove these four screws from the rear, remove the timer knob, and pull the escutcheon forward. The mounting screws for the switch assembly and the timer are now accessible.

The escutcheon must also be removed to replace the push buttons on the start-stop switch.

Grease Shield, Stirrer & Waveguide Filter

The grease shield is removed by taking out the three screws in front and the snap-in plugs in the rear. The mounting screw for the stirrer is then accessible. Do not overtighten the mounting screw when replacing the stirrer to prevent binding.

It should never be necessary to replace the waveguide filter, but if it must be removed for an oven-liner change, it can be taken out after removing the grease shield and stirrer. It is mounted with six screws.

The filter must be installed with the high ramp toward the magnetron, as shown in Fig. 5-9. The six mounting screws must hold the filter in good contact with the waveguide.

Door

The door of the RCM-2 must be replaced as a single unit. The only part replaceable separately is the handle insert. Proceed as follows:

1. Remove the outer shell.
2. Release the interlock-switch lever by prying the door-stop pin up to remove the door stop (two nylon pieces).
3. Remove the six door-mounting screws and remove the door.

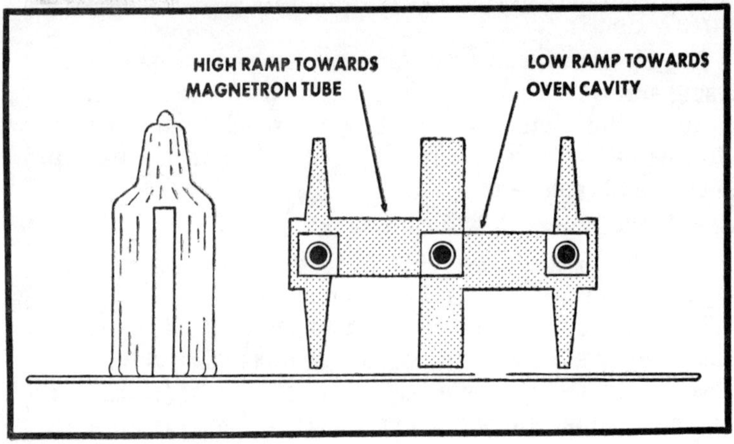

Fig. 5-9. Correct placement of waveguide filter.

Oven Liner

The oven liner of the RCM-2 is an integral part of the welded framework on which the entire unit is built. Replacing the liner involves transferring all other parts to the replacement liner.

HOW THE RCM-4 WORKS

The circuit of the RCM-4 is shown in Fig. 5-10. The black line (B) from the plug leads to the magnetron thermal cutout, which opens when the magnetron temperature rises too high.

The left interlock switch is the next component in the black line. Connected to it is the interlock-sensing switch. The sensing switch is mounted on top of the left interlock switch and is operated by the same lever. It is linked, however, so that when the door is opened it will not close until the left interlock switch has opened. When the door is closed, the sensing switch will open before the interlock switch closes.

The sensing switch is connected to resistor R2 on the protector board. If the left interlock switch does not open when the door is opened, current through the sensing switch and R2 will melt thermal fuse F2.

The start switch is the next component in the black line. It is a double-pole, single-throw (DPST) switch and interrupts both lines when it is opened.

The right interlock switch and the timer contacts are both in the white line (W). The right interlock switch is a

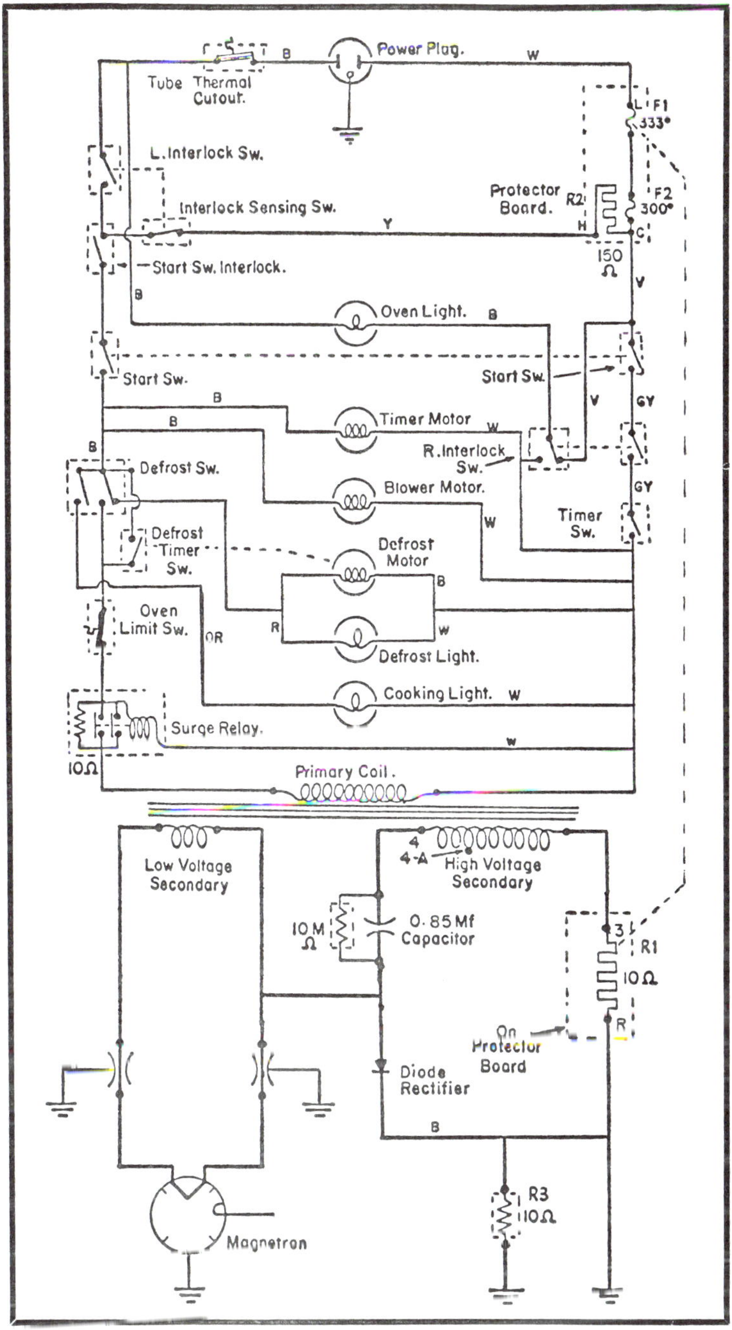

Fig. 5-10. Schematic of Frigidaire RCM-4.

double-pole, double-throw (DPDT) switch, with one pole used to turn on the oven light when the door is opened. When the door is closed and the timer running, the oven light will be turned on by the timer and the start switch contacts.

The defrost switch also is a DPDT switch, one pole of which turns on the defrost motor and defrost light. The other pole of the switch turns on the cooking light for fast cooking.

The defrost timer switch is in parallel with the defrost switch. It is operated by cam driven by the defrost timer motor. It is closed 24 seconds and open 36 seconds of each minute. The defrost timer motor is a bidirectional motor, which means that it may run either direction. The on-off cycle is the same either way.

The oven limit switch is the next component in the black line, followed by the surge relay. A 10Ω resistor across the contacts of this relay limits the surge of currrent in the transformer primary when the current is turned on.

The operation of the HV circuit of the RCM-4 is similar to that of the RCM-2, but larger components are used to generate more microwave power for the oven.

The power supply voltage of the RCM-4 is 4000V DC. A larger magnetron is used. This magnetron's current should be 300 mA instead of 275 mA. As shown in the schematic, the power-supply capacitor is a 0.85 μF (microfarad) capacitor.

ACCESS TO RCM-4 COMPONENTS

Access to the transformer, capacitor, magnetron, and other parts in the rear is gained by removing the back panel. The interlock switches, timer, and other parts in the front are made accessible by removing three screws at the bottom rear of the oven liner and pulling the unit out of the shell a few inches.

To remove the shell completely from the cabinet, the transformer leads must be disconnected from the feedthrough capacitors. The capacitor terminals should be supported with a pair of pliers when disconnecting the leads to prevent damage to the capacitor.

RCM-4 TROUBLESHOOTING

Figure 5-10 is the schematic of the entire RCM-4 circuit, while Fig. 5-11 is the wiring diagram.

Troubleshooting in the HV section of this model will be the same as in the RCM-2, except for the higher current and power of the RCM-4. The magnetron current is measured in the same way, but it should be 300 mA instead of 275 mA. Other parts in the HV circuit are checked in the same way.

To check the control circuit, open the oven door; if the oven light comes on, you will know the thermal fuses and the thermal cutout are not open. Put a water load in the oven. Close the door, turn on the timer, and press the START button. Ensure that the blower motor and timer motor are running. Set the defrost switch on DEFROST; ensure that the defrost timer motor is running and the defrost light is on. With the defrost switch off, the cooking light should be on. If it is, ensure that the surge relay is closing. If it is not, the oven limit switch must be open.

RCM-4 Oven Power Test

Check the supply voltage before running this test. It should be as near as possible to 115–118V.

1. Draw 1 quart of tap water in a Pyrex container.
2. Measure and record the temperature of the water. Remove the thermometer from the water before heating it.
3. Set the microwave-oven timer for 2 or 3 minutes. Use a watch or clock with a second hand (or a stopwatch) to actually time the heating of the water.
4. Heat the water in the microwave oven for 1 minute as timed by the watch.
5. Remove the container of water from the oven, stir with the thermometer, and check its temperature. The temperature rise should be approximately 15°F per minute.

RCM-4 COMPONENT REPLACEMENT PROCEDURES

Removing the transformer assembly is the same on the RCM-4 as on the RCM-2, except that two nuts must be removed on the RCM-4 instead of one. The capacitor, rectifier, and fuses are also checked and replaced in the same way as on the RCM-2.

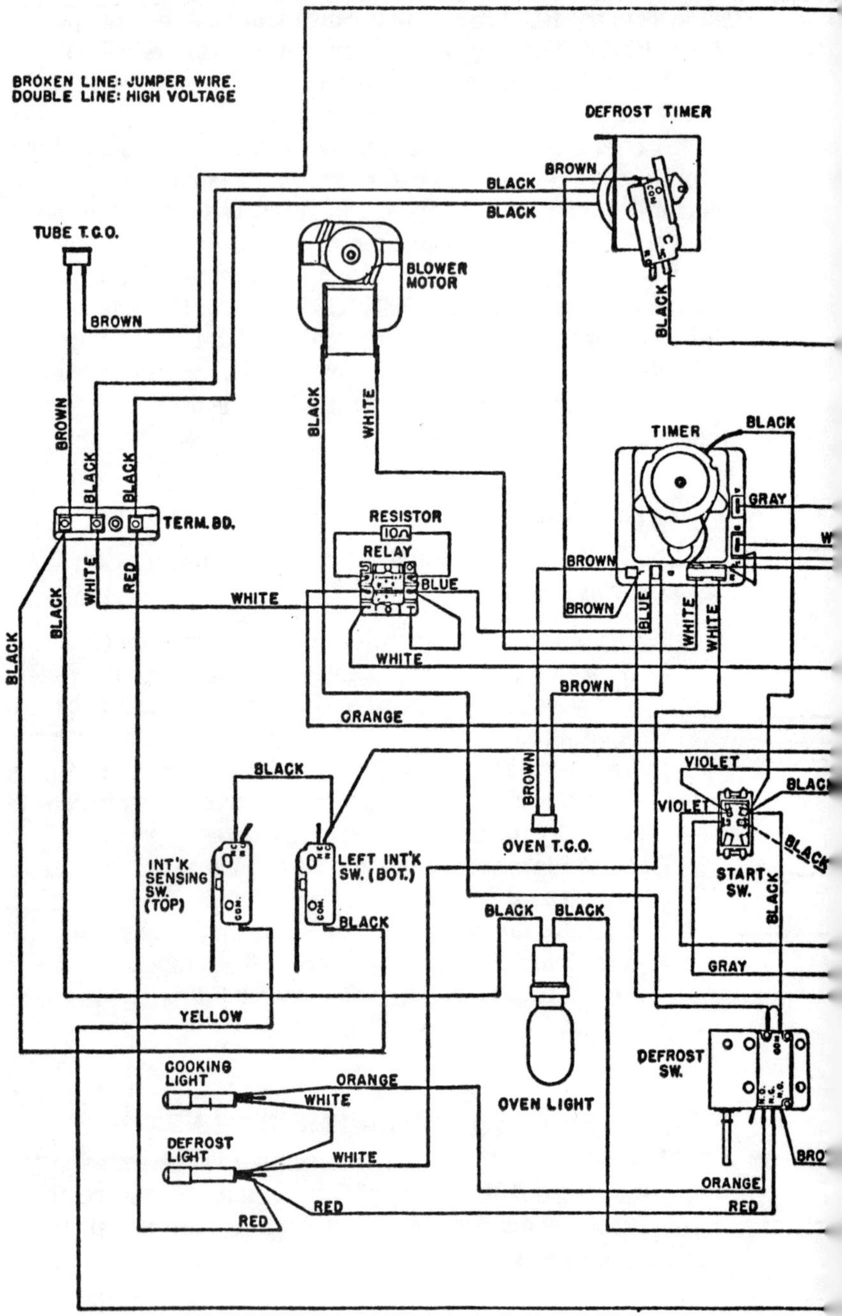
BROKEN LINE: JUMPER WIRE.
DOUBLE LINE: HIGH VOLTAGE
DEFROST TIMER
BROWN
BLACK
BLACK
TUBE T.C.O.
BLOWER MOTOR
BROWN
BLACK
WHITE
BROWN
BLACK
BLACK
TIMER
BLACK
GRAY
TERM. BD.
RESISTOR
RELAY
BLUE
WHITE
BROWN
BROWN
BLUE
WHITE
WHITE
WHITE
RED
BLACK
BLACK
BROWN
ORANGE
BLACK
BROWN
VIOLET
VIOLET
OVEN T.C.O.
INT'K SENSING SW. (TOP)
LEFT INT'K SW. (BOT.)
BLACK
START SW.
BLACK
BLACK
BLACK
GRAY
YELLOW
DEFROST SW.
COOKING LIGHT
ORANGE
WHITE
OVEN LIGHT
DEFROST LIGHT
WHITE
ORANGE
RED
RED
RED

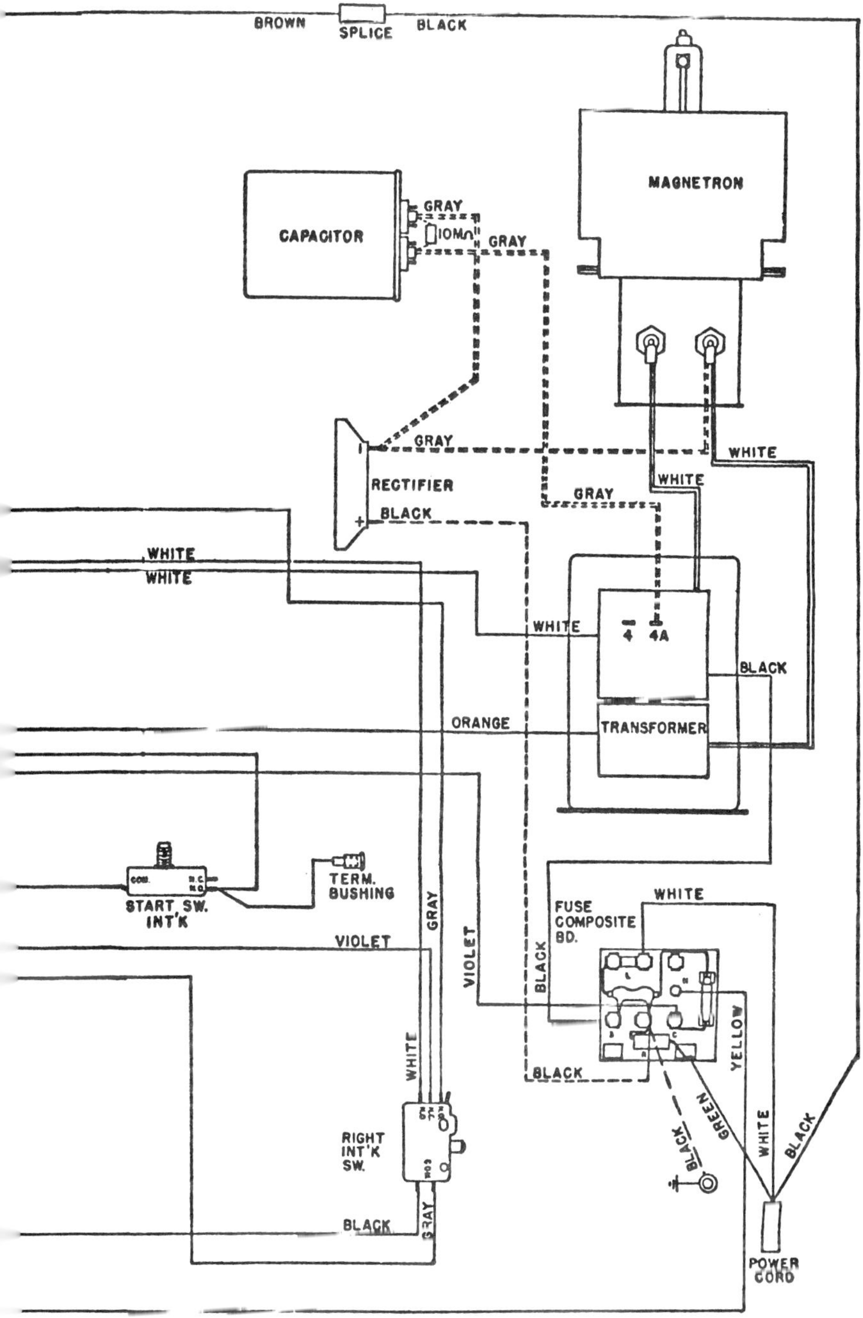

Fig. 5-11. Wiring diagram of Frigidaire RCM-4.

Fig. 5-12. Magnetron in RCM-4.

Magnetron Replacement

The same types of magnetron failures may be experienced with this model as with the RCM-2. To replace the magnetron:

1. Discharge the capacitor by shorting its terminals with a screwdriver.
2. Remove the HV cover from the capacitor box. Remove the heater leads from the feedthrough capacitor

terminals. Support the capacitor terminals with a pair of pliers when removing the leads to prevent damage to the capacitors.

3. Release the magnetron by pushing in with one hand on the release lever (see Fig. 5-12). Hold the magnetron assembly with the other hand to prevent it from falling. Pushing on the release lever rotates the locking ring and releases the magnetron from the mounting posts.
4. Lower the tube assembly carefully to remove it from the oven.
5. If the feedthrough capacitors are in good condition they may be removed and installed on the replacement magnetron. The right-hand capacitor must be connected to the center filament rod on the magnetron. (See Fig. 5-8.)
6. Check the ferrite rings on the magnetron leads to ensure that neither rings nor leads are near any grounded parts.
7. To install the new magnetron, insert the antenna glass carefully into the waveguide. Hold the tube assembly firmly against the waveguide with one hand and pull the release lever to the rear with the other. *Do not touch the glass parts of the magnetron with your fingers,* as doing so leaves oil on the glass, which heats when exposed to microwaves and can cause early failure of the magnetron.
8. Ensure that the magnetron assembly is properly locked in place on all four mounting posts. Improper installation will result in incorrect functioning of the magnetron.
9. Connect the transformer lead marked with black tape to the right-hand feedthrough capacitor. (See Fig. 5-7.)
10. Connect the other leads as they were originally.
11. Return the defective magnetron to the factory in the same packing in which the replacement magnetron was received.

Defroster Timer Replacement

1. Remove the mounting screws of the defrost timer. A ratchet screwdriver for Phillips-head screws is needed for removing these screws.

2. Transfer the leads to the replacement timer. Install the new timer in place.

Surge-Relay Replacement

1. Remove the cabinet shell as instructed previously.
2. Remove the mounting screws for the relay.
3. Transfer the leads to the replacement. Install the relay.

Blower & Housing

1. Disconnect the leads from the blower motor.
2. Remove the motor bracket mounting screws.
3. Remove the aluminum tape on the blower housing.
4. Remove the motor cooling fan.
5. Slide the motor, mounting bracket, and blower housing to the rear to remove from the cabinet.
6. Remove the blower housing mounting screws to lift off the housing.
7. Remove the blower from the shaft.
8. Remove the motor mounting screws. Lift the motor from the bracket.
9. Install replacement parts in reverse order.
10. When installing blower wheel, the hub of the wheel must be flush with the end of the motor shaft.

Testing the Left Door Interlock Switch

1. With the oven unplugged, remove the back panel and pull the oven forward a few inches from the cabinet shell.
2. Connect an ohmmeter to the terminals of the left interlock switch, which is the bottom one of the two switches mounted on the common bracket.
3. Close the oven door. The meter should show continuity.
4. Open the oven door. The switch should open before the door is opened 0.3125 inches.
5. The left door interlock and the interlock-sensing switches are mounted on a common mounting bracket. The interlock-sensing switch is on top of the left door interlock. Adjustment is made by loosening the screw securing the mounting bracket. A slotted hole is provided in the bracket for adjustment.

Testing the Interlock-Sensing Switch and Fuse

1. With the oven unplugged, connect an ohmmeter to the prongs of the line cord.
2. With the start switch open, the door open, and the timer off, the ohmmeter should read approximately 60Ω (resistance of the oven light bulb). This resistance will increase rapidly if ohmmeter is left connected.
3. If the ohmmeter shows no continuity, the light bulb is open, the tube thermal protector is open, or one of the thermal fuses open.
4. If F2 is open, it is mandatory that the left interlock switch be replaced.
5. If the ohmmeter reads approximately 60Ω, close the oven door. The ohmmeter reading should increase to infinity when the right-hand interlock switch is actuated.
6. Open the door slowly. The resistance reading will be about 60Ω. If the resistance momentarily drops further, the interlock-sensing switch is not opening until after the left interlock switch closes. Adjust the left interlock switch to open slightly later.

Door Parts Replacement

1. Move the oven cavity far enough out of the cabinet to reach the top-hinge mounting screws.
2. Remove the top hinge, lift the door assembly off the bottom hinge, and disengage the door stop and interlock-switch actuating arm. Note the 6599501 shim washers used to adjust the hinge side of the door.
3. Peel back the carbon-loaded vinyl seal. Drive out the roll pins.
4. Lift the outer plastic panel off, sliding it to the left slightly to clear the handle.
   a. To replace the door handle, peel back the vinyl door seal and remove the two handle-mounting screws.
   b. To replace the door strike, squeeze the locking tabs together and push it through to the other side.
   c. To remove the inner-door panel or transparent insert, remove the speed nuts.

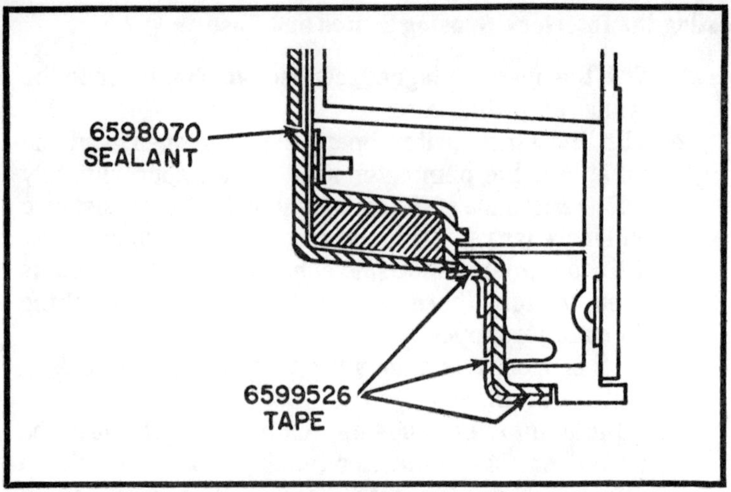

Fig. 5-13. Door and seal construction—side view.

Note: The transparent inner-door insert must be sealed to the inner-door panel with 6598070 sealant.

d. To replace the carbon-loaded seal, peel off the old seal. Lay three strips of 6599526 double-sided tape in the locations shown in Fig. 5-13 to repair or replace the tape originally used.

5. When remounting the door, assemble it loosely and close it so that the strike engages. Adjust the hinge plates and add or remove the 6599501 spacers to align on left side of the door.

Grease Shield Removal

1. Using a small instrument such as a screwdriver in the slots in the front of the grease shield, bend the shield back to release the pins.
2. Remove the shield and replace.

Stirrer Replacement

1. Remove the grease shield.
2. Remove the mounting screw from the stirrer shaft. *Note:* This is a left-handed screw.
3. Remove the stirrer and bushing assembly.

4. Separate the stirrer bushings.
5. Replace it, taking care not to invert the stirrer.

Stirrer Support Replacement

If the left-handed threads on the stirrer-support shaft are damaged, it will necessary to unsolder the old shaft and solder in a new one. The dimension shown in Fig. 5-14 is quite important.

Oven Liner Replacement

1. Remove the rear panel.
2. Remove the three screws securing the timer to the cabinet flange.
3. Disconnect the leads at the rear.
4. Pull the liner out of the outer shell.
5. Remove parts necessary to get at the nuts mounting the liner to the front bezel.
6. Transfer all of these parts to the new oven liner.
7. Assemble it in reverse order.

Oven Light Bulb Replacement

1. Remove the timer knob and escutcheon.
2. Raise the oven light barrier, pressing lightly on end of bayonet-type bulb, turning the bulb a quarter-turn counterclockwise to remove it.
3. Obtain a new bulb locally. The oven light is a 115V, 25W 25T8DC bulb. If the bulb is not available locally, it may

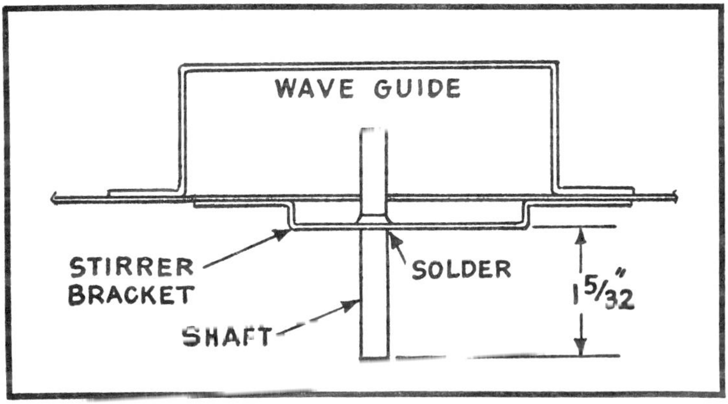

Fig. 5-14. Replacement of stirrer support.

be ordered from Frigidaire. The part number is 9948178.

Oven Light Socket Replacement

1. Slide the oven cavity out.
2. Remove the air baffle.
3. Remove the clips securing the light shield to the oven cavity.
4. Spread the light shield toward the outside of the oven and remove.
5. Remove the pop rivets. Replace the socket assembly.
6. Upon assembly, the 6595973 aluminum tape *must* be replaced.

HOW THE RCM-6 WORKS

The circuit of the RCM-6 is given in Fig. 5-15. As shown, the oven light may be turned on by the oven light switch or by the right-interlock switch (when the door is opened).

When the door is closed and the timer is turned on, a circuit is completed from the black line (B) through the right interlock switch, the timer switch, and the stop switch to the power relay. The start switch connects the white line (W) to the other side of the power-relay coil, energizing it. With this relay energized, its own contacts maintain the connection to the white line after the start switch has been released so that the relay remains energized.

In parallel with the power relay is the defrost motor, the cooking light, and the latch solenoid. The latch solenoid keeps the door locked shut and the latch switch closed.

The latch switch completes the circuit from the white line to the selector switch and energizes the blower and timer motors, which are connected to the black line through the left interlock switch and the oven tube thermal-cutout switches.

The selector switch connects the white line to the surge relay. It is operated by the user to select defrost, slow cook, or fast cook. On defrost and slow cook, it connects the white line to the surge relay through the contacts of the defrost-timer switch. Contact is made (applying power to the unit) 24 seconds out of 60 on defrost, and 36 seconds out of 60 on slow cook, On fast cook, the defrost timer switch is bypassed. The surge-relay operation is the same as in the RCM-4.

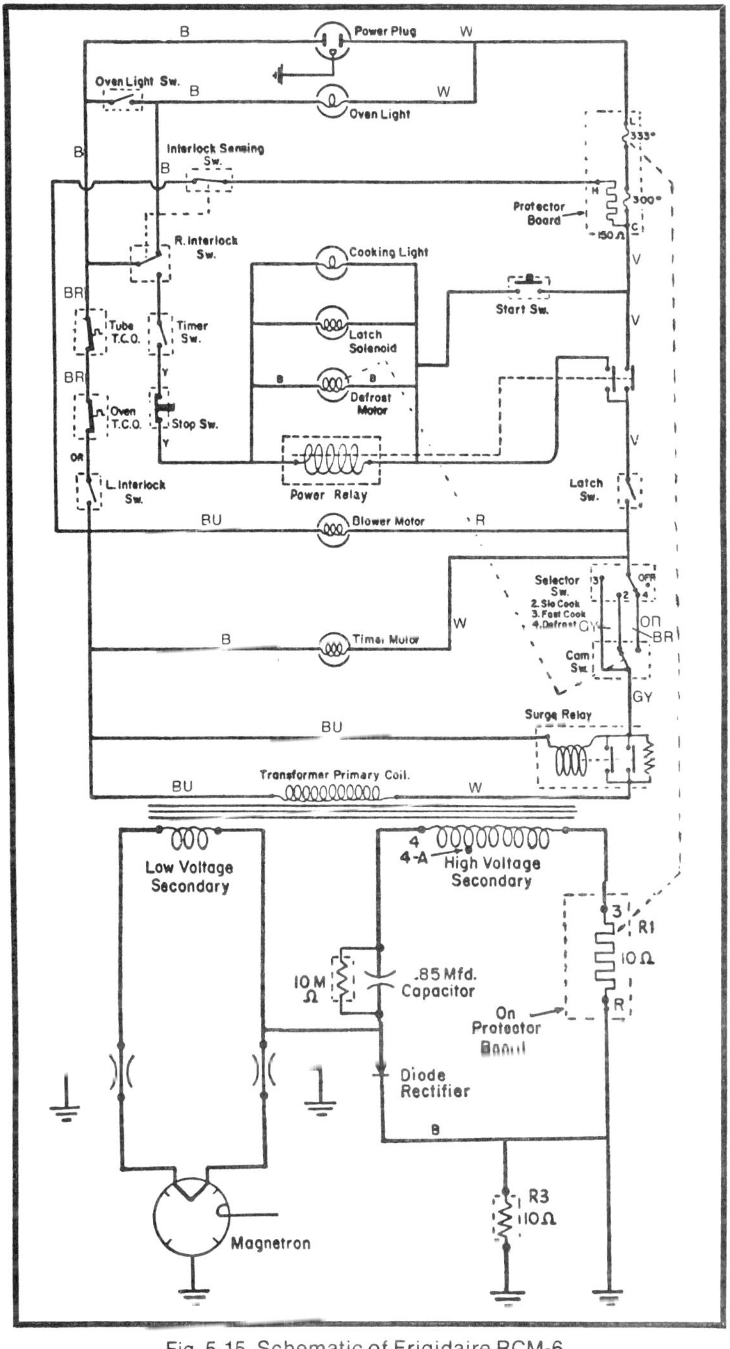

Fig. 5-15. Schematic of Frigidaire RCM-6.

The right interlock switch on the RCM-6 is a DPDT switch. The second pole is used as the interlock-sensing switch. If the left interlock switch fails to open when the door is opened, the sensing switch connects line current to resistor R2 on the protector board. In a few seconds, this will melt thermal fuse F2.

Operation of the HV circuit of the RCM-6 is exactly the same as that of the RCM-4.

RCM-6 TROUBLESHOOTING

The complete schematic of the RCM-6 is shown in Fig. 5-15, while the wiring diagram is in Fig. 5-16.

RCM-6 Left Interlock Switch Test

1. Unplug the unit and remove the cabinet.
2. Remove one lead from the left interlock switch. Connect an ohmmeter across the switch.
3. With the door closed, the ohmmeter should indicate a short circuit.
4. Open the door slowly.
5. The switch should open before the door is open 0.3215 inches at the top. If it does not, the switch may be adjusted by bending the arm slightly.

RCM-6 Interlock Switch Sequencing

1. Turn the oven light on.
2. Unplug the microwave oven. Connect the ohmmeter across the prongs of power plug.
3. Open the oven door. Set the selector switch to OFF and timer on OFF.
4. The ohmmeter should read approximately 60Ω (resistance of the oven light.) The resistance will increase rapidly while ohmmeter is connected.
5. Close the oven door slowly. The resistance reading on the ohmmeter should not change.
6. If the resistance drops momentarily while the door is being slowly closed, it means that the left interlock switch is closing before the interlock-sensing switch opens. Adjust the left interlock switch to close slightly later. If necessary, bend the switch arm slightly upward.

RCM-6 Control Circuit Check

1. Plug in the microwave oven. Close the door and press the START button, but leave the selector switch on OFF. Set a cooking time on the timer.
2. If the cooking light does not come on, check thermal fuses F1 and F2. If F2 is open, it is mandatory that the left interlock switch be replaced. If F1 is open, look for a short in the HV section.
3. Ensure that the door is locked shut. If it is not, either the latch solenoid is open, the wiring to it is open, or the latch mechanism has a mechanical malfunction.
4. Ensure that the blower motor and timer motor are both running. If neither is running, the trouble may be that the latch switch is not closing or the power relay is not closing.
5. Turn the selector switch to DEFROST. Listen for an opening and closing of the surge relay. The magnetron should be on 24 seconds and off 36 seconds.
6. If power is obtained only while the start switch is held, a connection to one of the contacts on the power relay probably is open.

Other Tests

The oven power test is exactly the same as it is for the RCM-4. The magnetron-current test is also exactly the same as it is for the RCM-2, except that the total current measured should be 300mA for the RCM-6 instead of 275 mA.

RCM-6 COMPONENT REPLACEMENT PROCEDURES

Most of the operating parts are accessible after removing the right side panel, but sometimes it will also be necessary to remove the outer shell.

Side Panel Removal

1. Remove the seal and three screws from the bottom-right side.
2. Drill out the pop rivet.
3. Lift off the side panel.
4. When installing the side panel, the screws, pop rivet, and seal must be replaced.

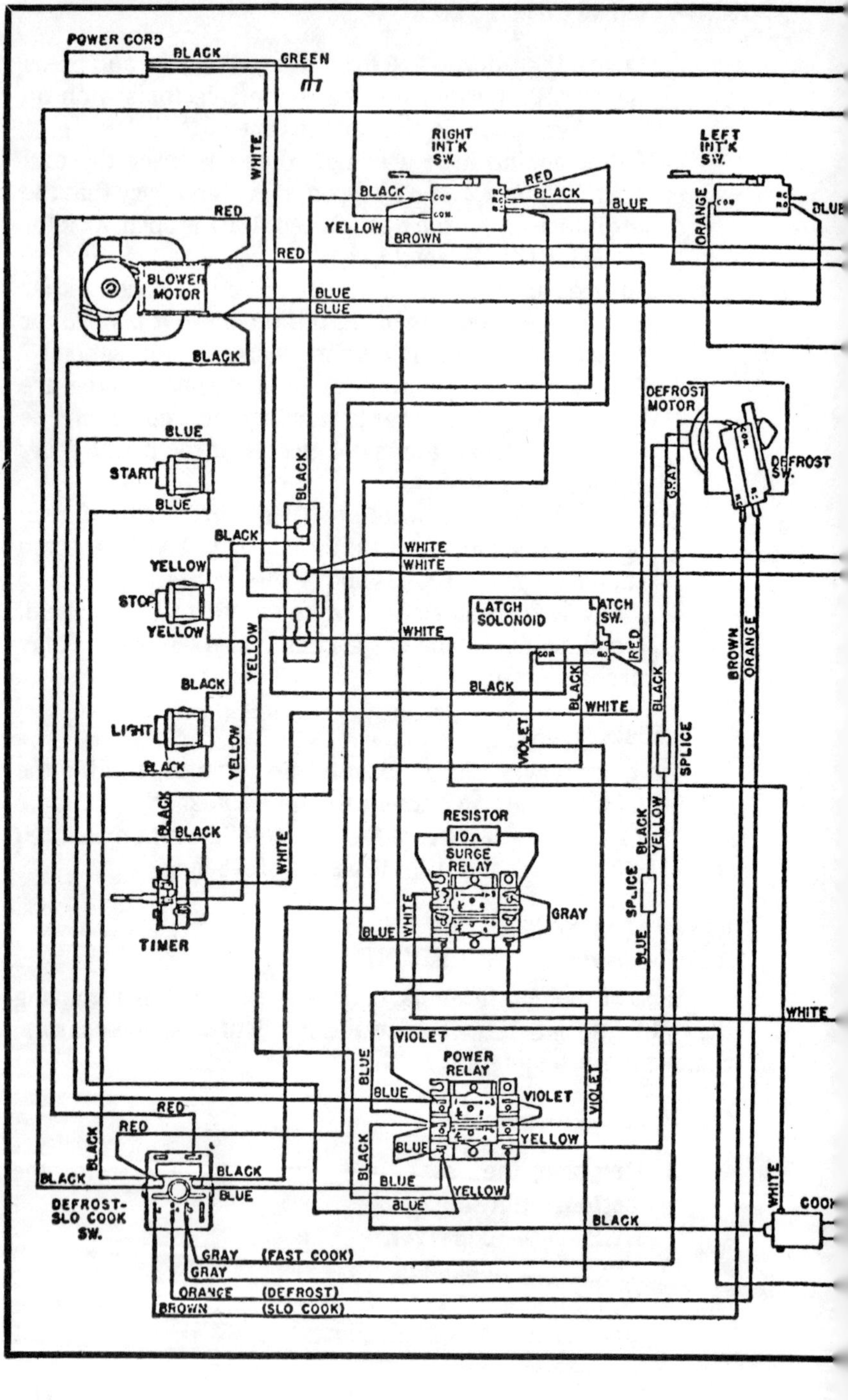
POWER CORD
BLACK
GREEN
WHITE
RIGHT INT'K SW.
LEFT INT'K SW.
RED
BLACK
BLUE
YELLOW
BROWN
ORANGE
BLOWER MOTOR
BLUE
BLUE
BLACK
DEFROST MOTOR
DEFROST SW.
START
STOP
LIGHT
YELLOW
WHITE
WHITE
LATCH SOLONOID
LATCH SW.
GRAY
BROWN
ORANGE
VIOLET
SPLICE
RESISTOR
10Ω
SURGE RELAY
TIMER
WHITE
POWER RELAY
VIOLET
DEFROST-SLO COOK SW.
GRAY (FAST COOK)
GRAY
ORANGE (DEFROST)
BROWN (SLO COOK)
COOK

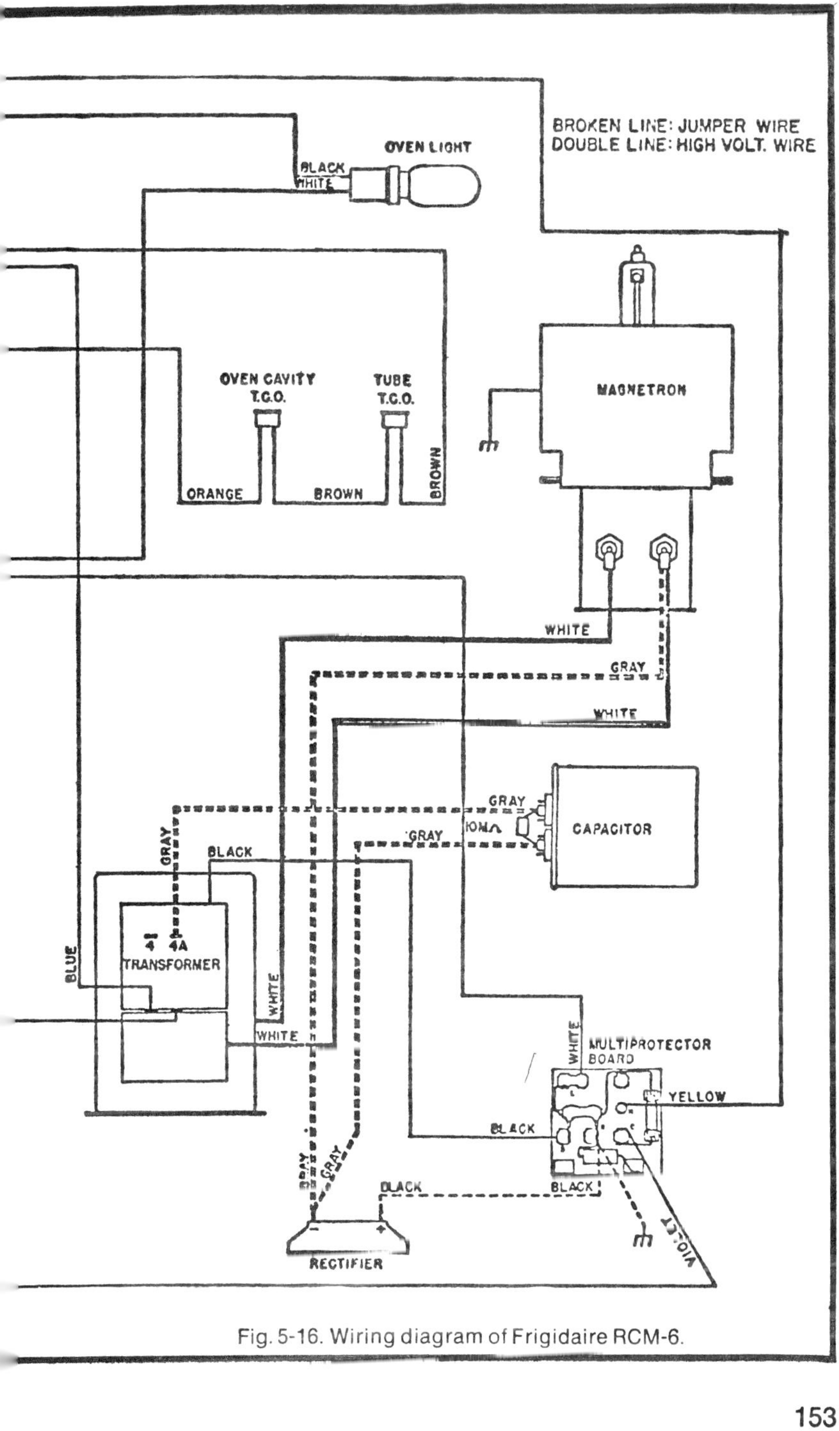

Fig. 5-16. Wiring diagram of Frigidaire RCM-6.

Exterior Shell Removal

1. Remove the right side panel.
2. Remove the screws from bottom-left side of cabinet.
3. Open the door and remove the screws from behind the flap of the door seal along the top and left side of the door.
4. Raise the shell and remove the hex screw in upper right-hand corner from inside.
5. Lift off the shell.

Control Panel Removal

1. Remove the side panel.
2. Remove the panel-mounting screws from the inside. To remove some of the screws, a quarter-inch socket with an extension will be necessary.
3. Pull the bottom end of the control panel outward and push down to release the top end.

Control Panel Trim Removal

1. Remove the control panel as previously described.
2. Remove the screw from the bottom-left corner of the trim.
3. Remove the six screws from the inside edge of the trim.
4. Remove the trim.

Control Panel Glass Removal

1. Remove the control panel trim as previously described.
2. Remove the control knobs.
3. Remove the glass.

Timer, Switch, or Cooking Light Removal

1. Remove the control panel glass as previously described.
2. Remove the dial.
3. Remove the mounting screws from timer or nut from selector switch.
4. Transfer the leads to replacement control and light.
5. Assemble it in reverse order.

Defrost Timer Removal

1. Remove the right side panel.
2. Remove the timer mounting screws (Fig. 5-17).
3. Transfer the leads to the replacement timer.
4. Assemble it in reverse order.

Air Duct Removal

1. Remove the side panel and the exterior shell.
2. Remove the three air-duct mounting screws from inside the oven.
3. Lift out the air duct.

Blower Housing Removal

1. Remove the air duct.
2. Remove the foil connecting the housing and the waveguide.

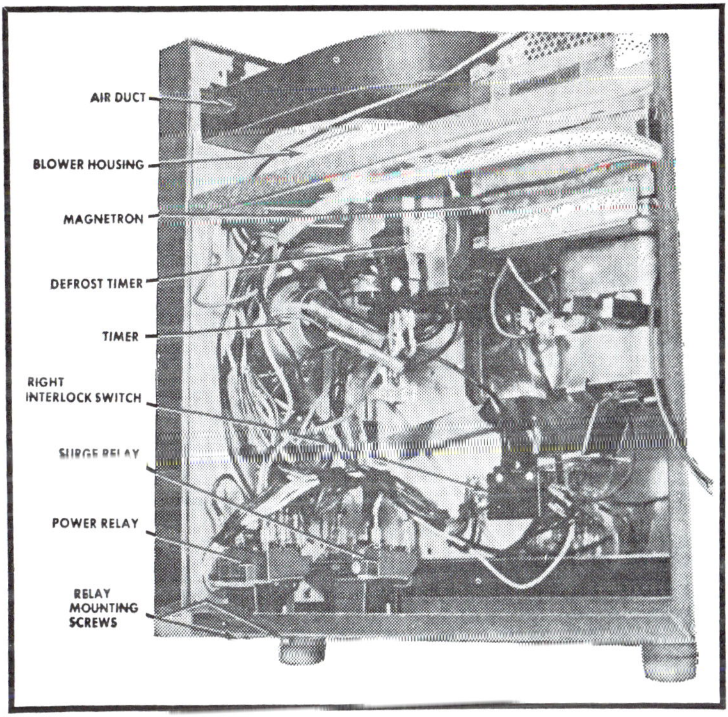

Fig. 5-17. Model RCM-6 with side panel and transformer assembly removed.

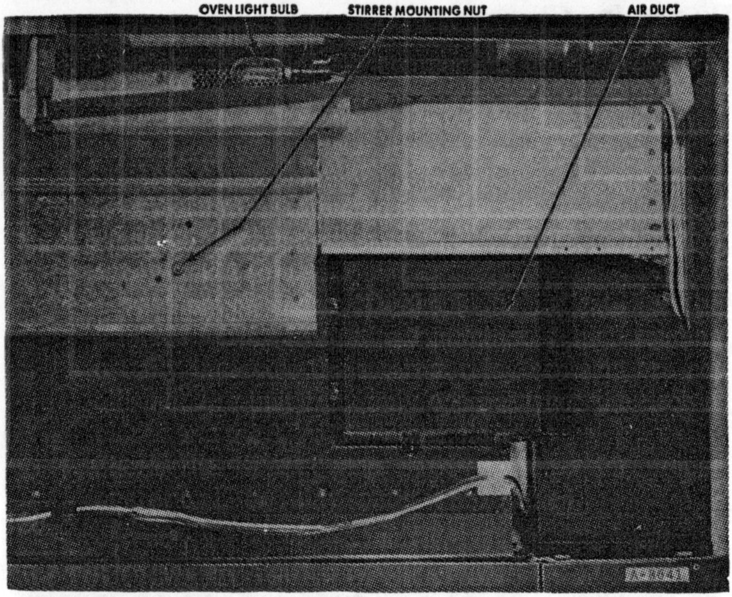

Fig. 5-18. Model RCM-6 with shell removed—top view.

3. Remove the mounting screws for the housing.
4. Lift the housing up, unhook the tab on the inside edge, and lift out.

Blower Wheel & Motor Removal

1. Remove the blower housing as previously described.
2. Remove the blower wheel by loosening the setscrew.
3. Remove the motor-mounting bracket and the motor leads.
4. Install the motor mount on the replacement motor and connect the leads.
5. Place the rubber grommets in the mount. Install the motor mount with the four shoulder screws.
6. Slide the blower wheel on the shaft. Tighten the setscrew. Test the blower wheel by spinning it. See that it does not rub. Adjust the wheel upward if necessary to prevent rubbing.
7. Replace the blower housing and air duct.

Grease Shield Removal

1. Remove the screws from the front edge of the grease shield.

2. Drop the front edge, unhook the rear edge, and remove the grease shield.

Stirrer Removal

1. Remove the grease shield.
2. Remove the right side panel and the cabinet shell.
3. Remove the mounting screw for the stirrer. Hold the mounting nut with a wrench. (See Fig. 5-18.)
4. When replacing the stirrer, be careful not to tighten the screw enough to cause binding of the stirrer.

Door Lock Mechanism Replacement

1. Remove the side panel and cabinet shell.
2. Remove the air duct.
3. Remove the leads from the switch and the solenoid.
4. For easier access, remove the control panel.
5. Remove the latch-bracket mounting screws.
6. Open the oven door and drill out the latch-bracket-mounting pop rivet. Remove the assembly. Parts of the door-lock mechanism are shown in Fig. 4-19.

High-Voltage Components Replacement

These parts are replaced in exactly the same procedures as on the RCM-4 magnetron replacement will be easier if the transformer assembly is removed first.

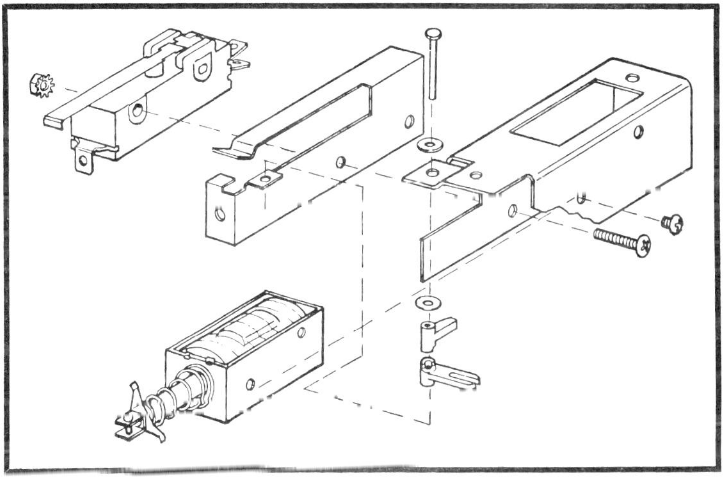

Fig. 5-19. Exploded view of door-lock mechanism—RCM-6.

Ceramic Tray Replacement

1. Turn the timer to OFF.
2. Open the oven door and cut seal along edges. Remove any broken pieces of ceramic.
3. Remove any remaining sealant.
4. Apply a bead of RTC-732 cement (6599908) around edge. Set in the new tray.
5. Leave the oven door open and allow the cement to stand for 1 hour before using.

Chapter 6

Servicing General Electric Microwave Ovens

This chapter discusses the servicing of a large number of General Electric[1] microwave ovens manufactured from 1971 to now. Some of the topics embraced by this chapter are; theory of circuit operation, troubleshooting, plate-current tests, and the testing and replacement of door-interlock switches.

HOW MODELS JET-70001, 78001 & 80002 WORK

The description and operation of the microwave oven circuit is as follows (refer to appropriate oven schematic, Fig. 6-1 or 6-2).

When the timer is set, the timer contacts close and establish a circuit to the oven light and blower motor. The remainder of the oven circuits are activated by the COOK switch.

When the COOK switch is pushed, it energizes the cook relay and timer motor through both sections of the latch switch, which are normally closed with the door closed. The cook relay operates two sets of contacts. One circuit is a latching or holding circuit around the cook switch (momentary type). The second circuit energized is the power transformer and stirrer motor.

1. All of the illustrations in this chapter are courtesy of General Electric Company.

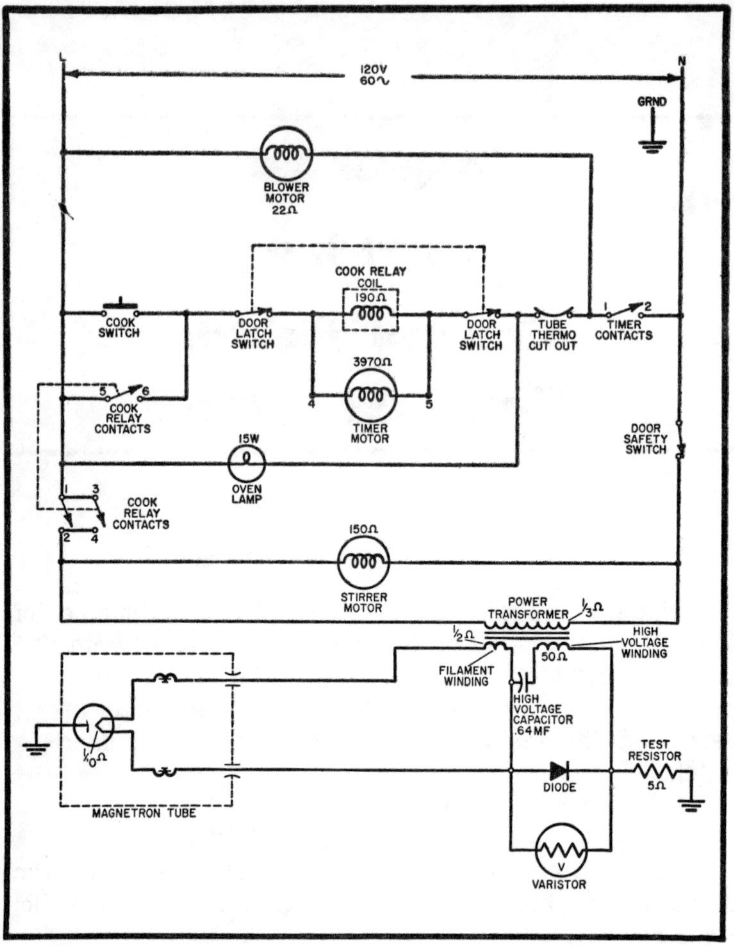

Fig. 6-1. Schematic diagram of General Electric Model JET-70001.

Voltage-Doubler Circuit

The magnetron requires a high DC voltage source for operation. This voltage is supplied by a half-wave voltage-doubler circuit consisting of a HV capacitor, a diode, and a varistor.

The 2300V secondary winding of the transformer is connected to the capacitor and diodes. The capacitor increases the voltage to about 3800V due to its peak-voltage-charging capabilities. The diodes rectify the voltage, and the resultant half-wave DC voltage is applied to the magnetron. The negative output from the DC circuit is connected to the filament of the magnetron. The positive output is connected to the plate of the magnetron, which is grounded to the chassis.

The diode is therefore connected with opposite polarity from the magnetron in order that the capacitor will charge

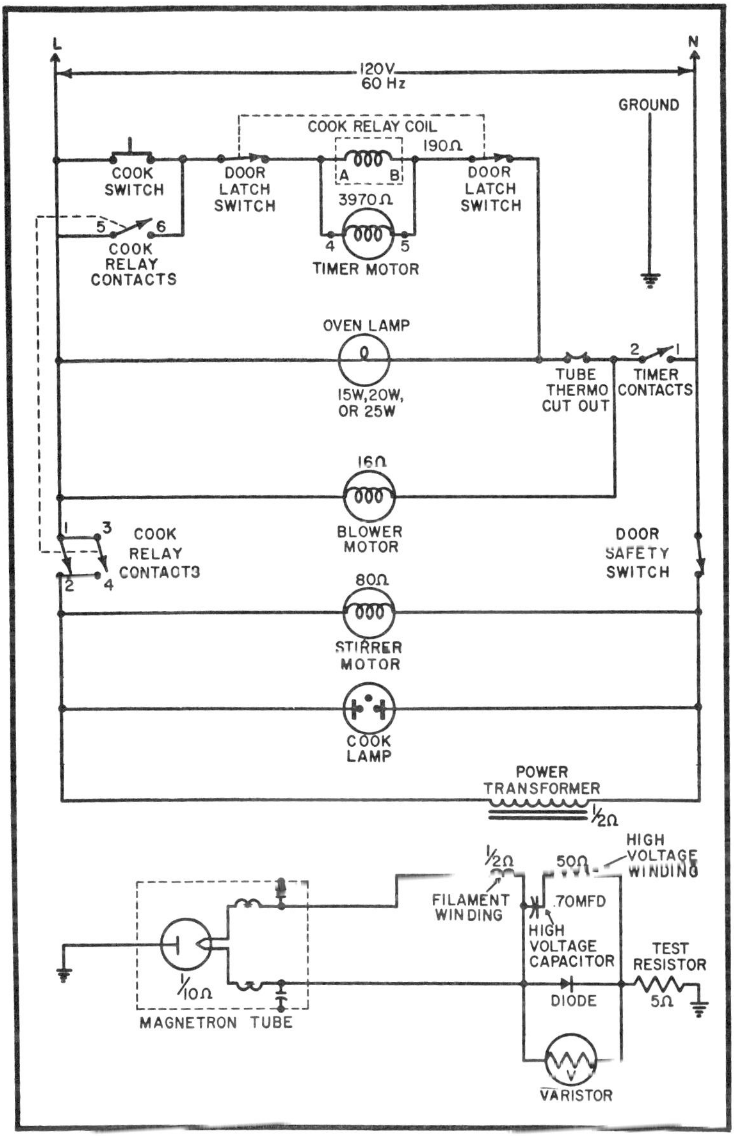

Fig. C-2. Schematic diagram of General Electric Models JET-78001 and JET-80002.

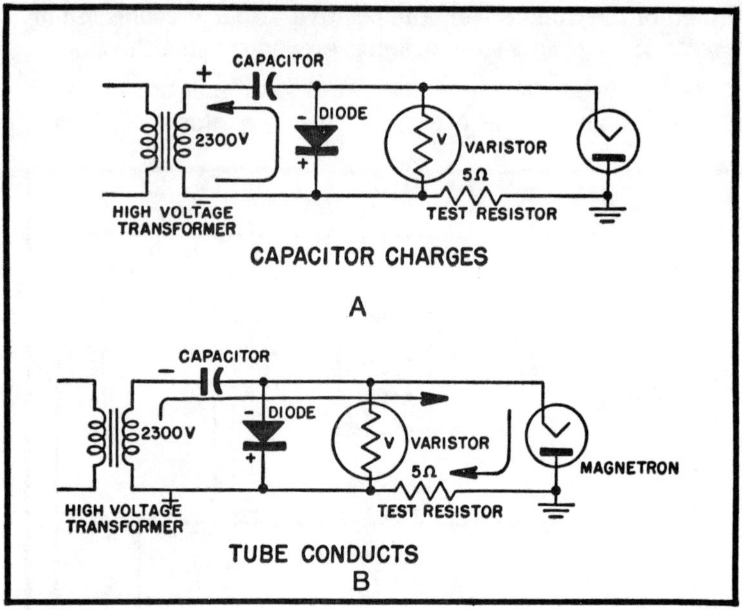

Fig. 6-3. (A) Capacitor charges to peak voltage. (B) Capacitor discharges to aid transformer.

through the diode during one half-cycle (Fig. 6-3 top) and will discharge through the magnetron on the alternate half-cycle (Fig. 6-3 bottom). On the discharge cycle, the capacitor aids the transformer voltage, thereby providing the voltage-doubling action. While the circuit is called a *voltage doubler*, it does not actually provide *double* the applied voltage.

The varistor serves as a protection to the diode in case of an unusually high surge of voltage. Normally the varistor has a very high resistance and serves no function at normal operating voltages. At unusually HV levels, the resistance of the varistor reduces rapidly to a low level and shunts the resultant current around the diode, thus protecting the diode from failure.

HOW THE JET-80001 WORKS

The description and operation of the microwave oven circuit is as follows (refer to oven schematic, Fig. 6-4).

When the ON-OFF switch is pushed to ON, the switch contacts close and complete a circuit to the blower motor, magnetron-filament (heater) transformer, and the time-delay relay. After about 7 seconds, the time-delay relay operates and its contacts close. The time delay is necessary to allow the magnetron filament to warm up before applying voltage.

When the relay contacts close, it completes a circuit through the magnetron thermal protector (N.C.) and turns on the oven lights. This is the indicator to the user that the magnetron is ready to use.

When the timer is set, the contacts close completing a circuit through half of the dual latch switch to the COOK

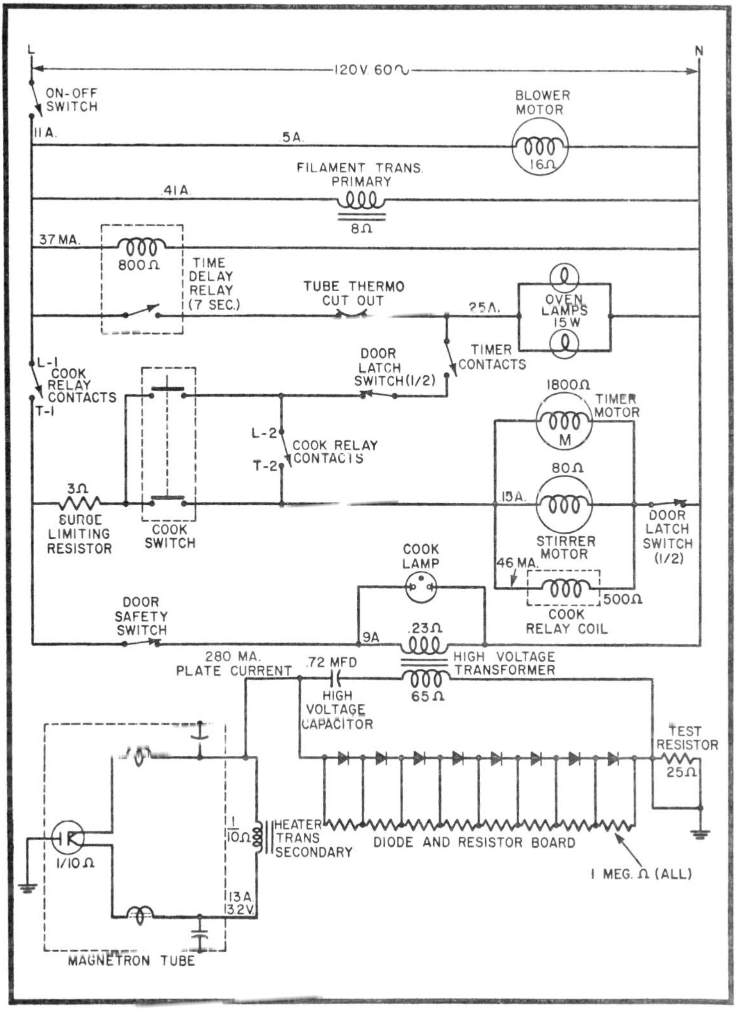

Fig. 6-4. Schematic diagram of General Electric Model JET-8000 I.

switch. With the oven lights on, when the COOK switch is pushed to ON, it completes two circuits:

1. One circuit is through both sets of contacts in the COOK switch, which establishes a circuit for, and energizes, the cook relay, timer motor, and stirrer motor through the other half of the latch switch.
2. The second circuit energized is the primary winding of the HV transformer, through the 3Ω surge-limiting resistor and the door-safety switch. This also turns on the COOK lamp which indicates that the HV is on and the food should be cooking.

Note: Just as soon as the cook relay is energized, it closes two sets of contacts which latch in the above two circuits. This latching effect is necessary to maintain the circuits since the COOK switch is only a momentary switch.

The HV transformer's secondary winding provides approximately 2300V AC to the voltage-doubler circuit.

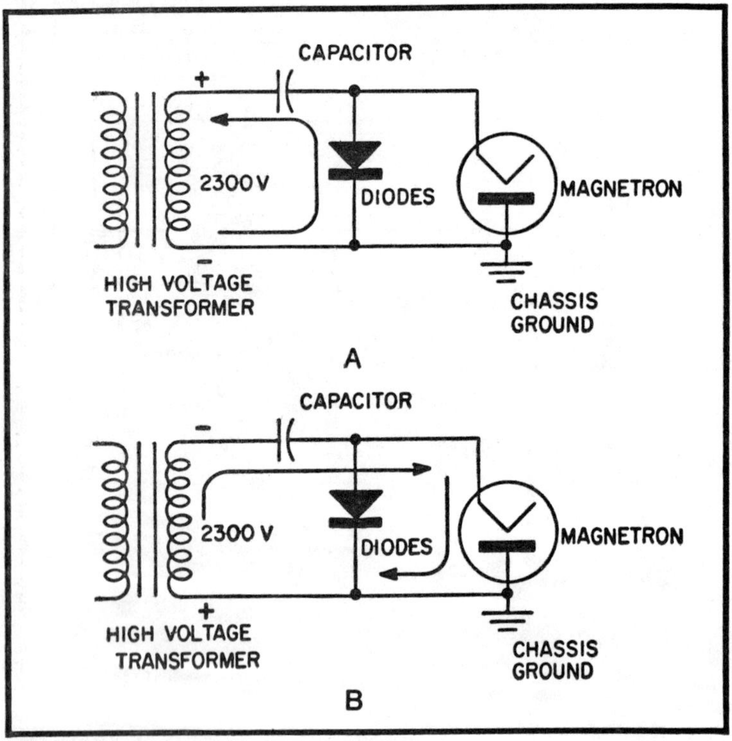

Fig. 6-5. Voltage-doubler circuit—JET-80001, -82001, -83001.

JET-80001, 82001 & 83001 Voltage Doubler

The magnetron requires a high DC voltage source for operation. This voltage is supplied by a half-wave voltage-doubler circuit consisting of an HV capacitor and a diode board.

The 2300V secondary winding of the transformer is connected to the capacitor and diodes. The capacitor increases the voltage to about 3800V due to its peak voltage-charging capabilities. The diodes rectify the voltage, and the resultant half-wave DC voltage is applied to the magnetron. The negative output from the DC circuit is connected to the filament of the magnetron. The positive output is connected to the plate of the magnetron, which is grounded to the chassis.

The diodes are therefore connected with opposite polarity from the magnetron in order that the capacitor will charge through the diodes during one half-cycle (Fig. 6-5A) and will discharge through the magnetron on the alternate half-cycle (Fig. 6-5B). On the discharge cycle, the capacitor aids the transformer voltage, thereby providing the voltage-doubling

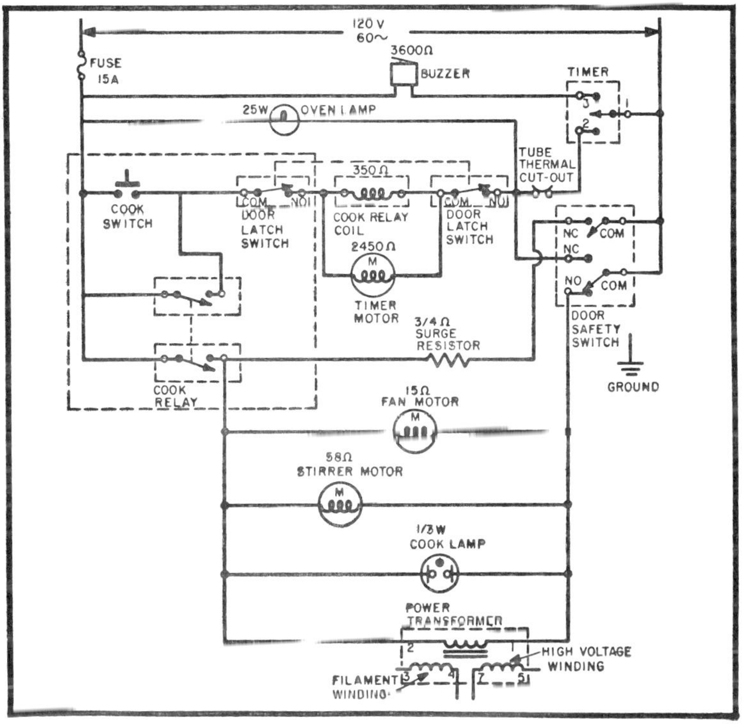

Fig. 6-6. Schematic diagram of General Electric Model JET-02001.

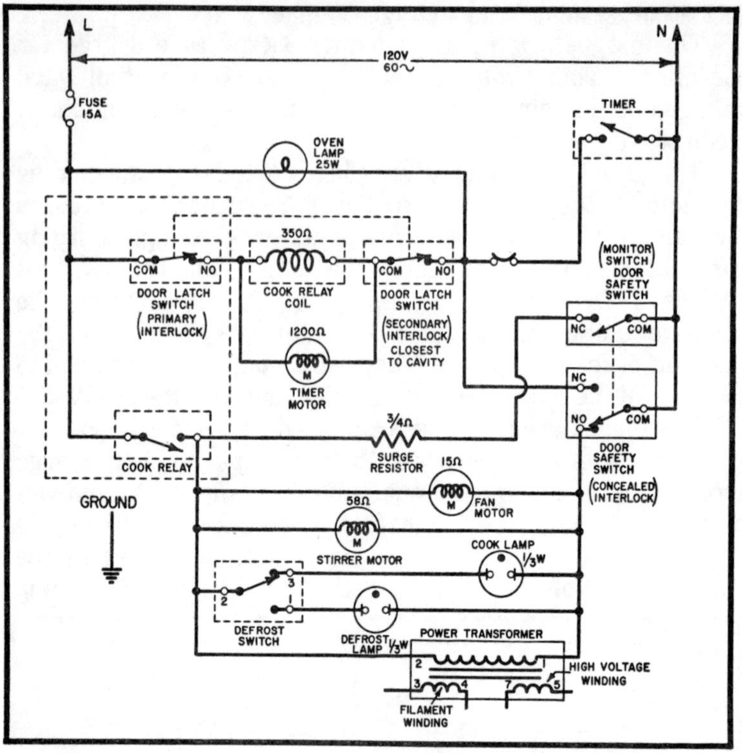

Fig. 6-7. Schematic diagram of General Electric Model JET-83001.

action. The circuit does not provide exactly *double* the applied voltage.

HOW THE JET-82001 & 83001 WORK

The description and operation of the microwave oven circuit is as follows. (Refer to oven schematic, Fig. 6-6 and 6-7.)

When the timer is set, the timer contacts close and establish a circuit to the cook-relay coil through both sections of the latch switch and magnetron thermal switch. The cook-relay contacts then close and energize the power transformer, blower, stirrer, and COOK light. This is therefore a one-control, instant-on system.

Defrost Circuit

The defrost feature is provided by adding a 1200Ω resistor in series with the diode. A set of contacts in the

COOK-DEFROST switch parallel the resistor (Fig. 6-8). During cooking, the contacts are closed to bypass the resistor. During defrosting, the contacts are open and the resistor becomes part of the circuit. The resistor reduces the charge on the HV capacitor and thereby reduces the power output of the magnetron to approximately 60% full power. (See the previous section for an explanation of the voltage doubler.)

HOW THE JET-82003 WORKS

The description and operation of the microwave oven circuit is as follows. (Refer to oven schematics Figs. 6-9 and 6-10.)

When the timer is set, the timer contacts close and complete the circuit to the series-connected interlocks (both latches and the door-safety switch), which are electrically closed at any time the door is closed. This series-connected circuit then energizes the power transformer, timer motor, stirrer motor, and COOK lamp. The magnetron thermal cutout is the only switch in the neutral (N) leg, and it interrupts power to all components except the fan and oven light at any time the magnetron overheats.

TROUBLESHOOTING AND TESTING

Since the outer case must be removed to service any component, the following safety precautions should be observed in order to prevent injury due to hazardous electrical shock.

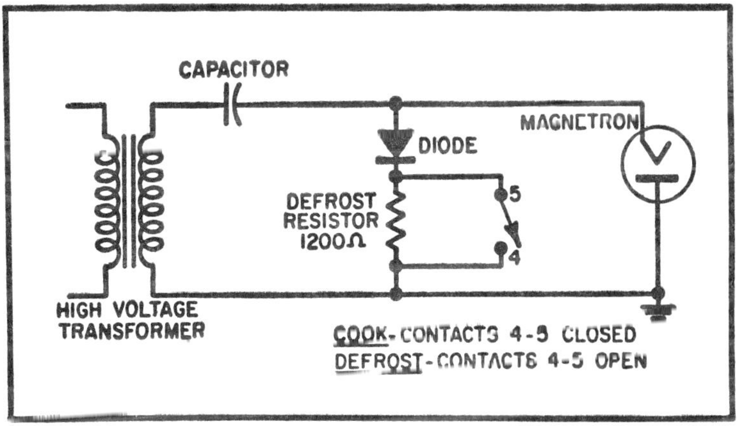

Fig. 6-8. Defrost circuit.

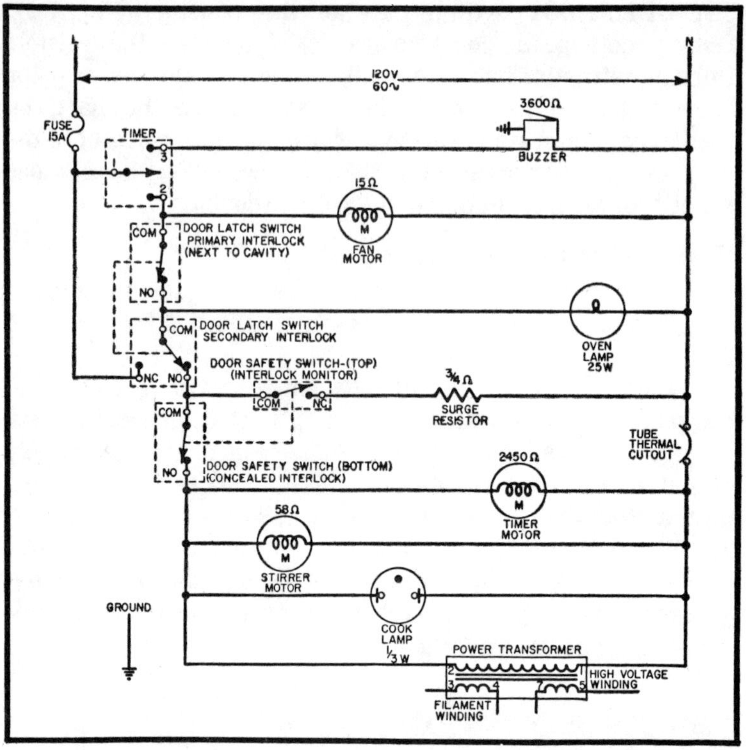

Fig. 6-9. Schematic diagram of General Electric Model JET-82003.

1. Unplug the oven.
2. Remove the screws at the sides and rear of case.
3. Push the outer case to the rear slightly to free the retaining clips along the front edge. Lift off the case.
4. Discharge the HV capacitor by shorting across its terminals with a shorting bar. Two plastic-handled screwdrivers can be used by simultaneously touching each capacitor terminal and the screwdrivers together.

Magnetron Diagnosis

When the COOK light and blower operate, but little or no heat is produced, the trouble is either in the HV or magnetron circuit. To diagnose the problem, the following simple test procedure should be followed:

1. Unplug the oven and remove the outer case.
2. Discharge the HV capacitor.

3. Install the voltmeter on the filament leads, or clamp on an ammeter around one filament lead to determine if the filament circuit is energized. Plug in the power cord and turn the oven on. The voltmeter should read 3–4V AC. Or an ammeter should read 13A.
   a. If the filament circuit is normal, proceed with step 4.
   b. If the filament circuit isn't normal, check the transformer winding and connections.
4. Turn off the oven.
5. Discharge the HV capacitor:
   a. If a discharge flashes, the problem is an open diode or magnetron.
   b. If there is no flash, proceed to step 6.
6. With the oven off, remove both of the filament leads from magnetron to isloate it. Turn the oven on and off again, and repeat the capacitor discharge test.

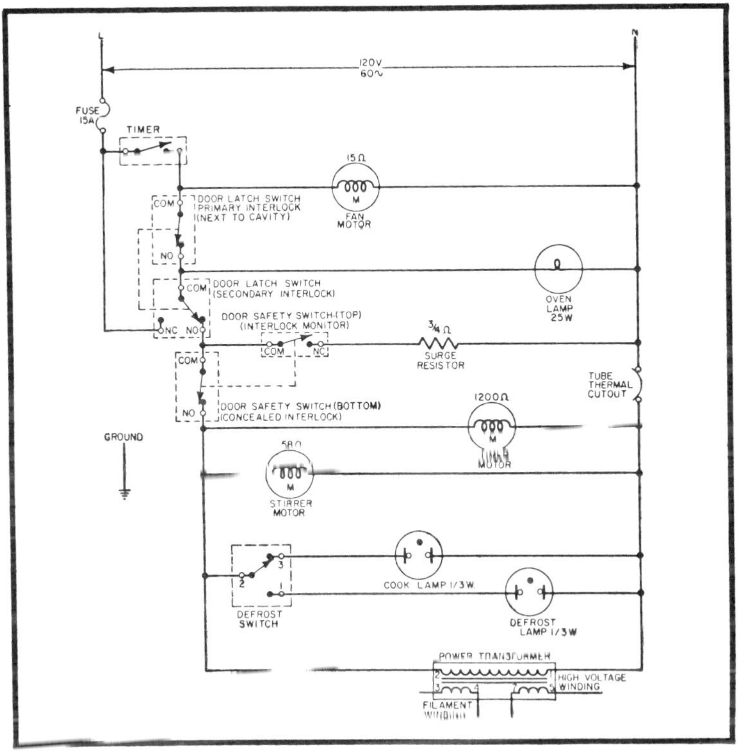

Fig. 6-10. Schematic diagram of General Electric Model JET-83002.

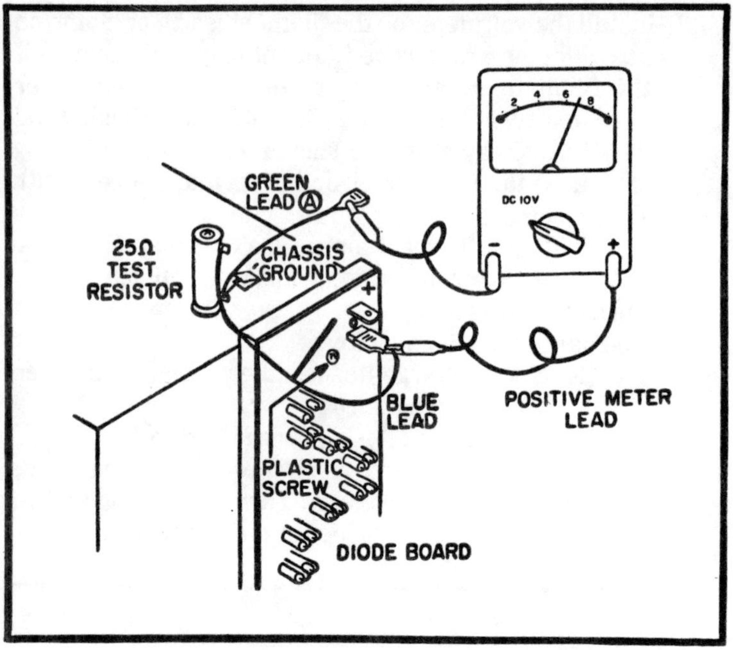

Fig. 6-11. First plate-current test—01 model.

a. If a discharge flashes now, the magnetron is the problem
b. If there is still no flash, the problem is in the power supply. Check the transformer, capacitor, and diode.

Model 01 Plate Current Tests

WHEN THE COOK light turns on, but little or no heat is produced, the trouble is either *in* the HV or magnetron circuit. To diagnose the problem, *one* or *two* plate-current tests have to be performed. The result of each test will indicate the source of the problem. For the first plate-current test, proceed as follows:

1. Unplug the oven and remove outer case.
2. Discharge the HV capacitor.
3. Disconnect the grounded test-resistor green lead (*A*, Fig. 6-11) from the positive terminal of diode board. Connect the voltmeter across the test resistor by connecting the test leads between the disconnected lead and the positive (+) terminal on the diode board

(observe polarity). Set the voltmeter to the 10–12V DC scale.

4. Plug in the oven, use a water load, and set the oven for cooking.
5. Observe the voltmeter reading to follow the first plate-current diagnosis chart.

First Plate-Current Diagnosis

DC Test Voltage	Diagnosis & Procedure
7.0V (approx.)	Indicates normal plate current (280 mA). Run performance test.
5.5V (approx.)	1. Short circuit to ground between voltage-doubler circuit and magnetron. Check wiring.
	2. Shorted magnetron tube.
0V	Perform second plate-current test.

For the second plate-current test (use this test *only* if indicated by first plate-current test), proceed as follows:

1. Unplug the oven cord.
2. Discharge the HV capacitor.
3. Using a jumper wire with alligator clips, connect one end to the negative terminal on the diode board. Connect other end to chassis ground (Fig. 6-12).
4. Disconnect the grounded test-resistor green lead (A) from the positive terminal of the diode board. Connect the voltmeter between the disconnected lead and the positive terminal on the diode board (observe polarity). Set the voltmeter to the 10–12V DC scale.
5. Plug in the oven, use a water load, and set the oven for cooking.
6. Observe the voltmeter reading to follow the second plate-current diagnosis chart.

Second Plate-Current Diagnosis

DC Test Voltage	Diagnosis & Procedure
5.5V	1. Magnetron filament open. 2. Filament-transformer open. 3. Open wiring to magnetron or filament transformer.
0V	1. HV transformer open or shorted. 2. Diode board open or shorted. 3. HV capacitor open or shorted.

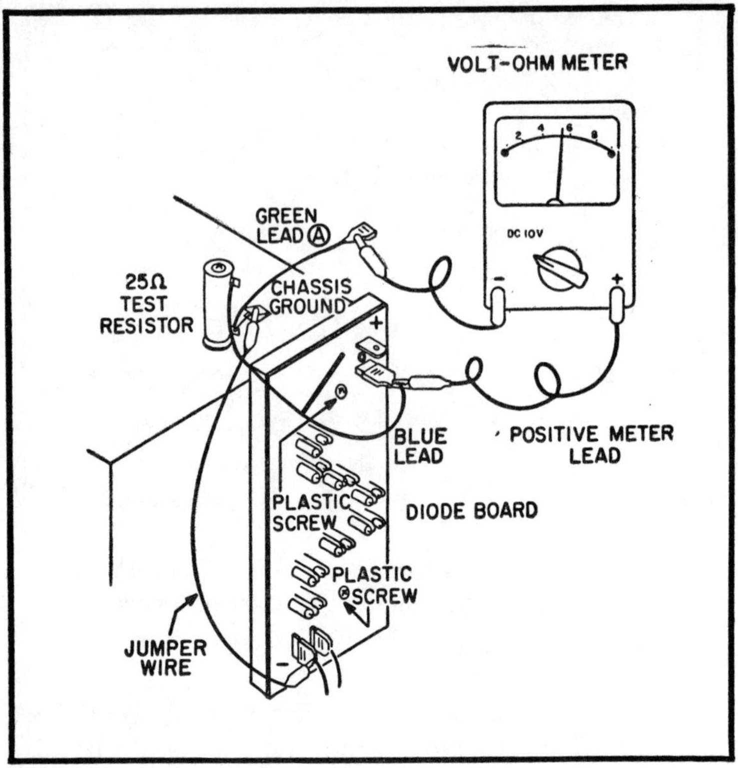

Fig. 6-12. Second plate-current test—01 model.

Model 02 Plate Current Tests

When the COOK light turns on but little or no heat is produced, the trouble is either in the HV or magnetron circuit. To diagnose the problem, one or two plate-current tests have to be performed. The result of each test will indicate the source of the problem. For the first plate-current test, proceed as follows:

1. Unplug the oven and remove outer case.
2. Discharge the HV capacitor.
3. Connect the voltmeter across the test resistor by connecting the test leads between chassis ground and the positive terminal, observing polarity. Set the voltmeter to the 10–12V DC scale (Fig. 6-13).
4. Plug in the oven, use a water load, and set the oven for cooking.

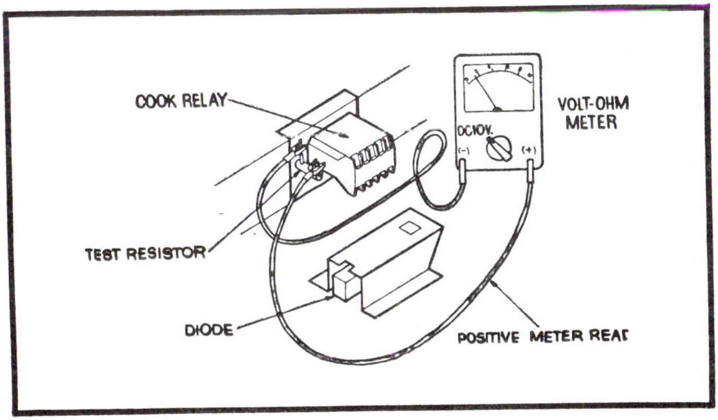

Fig. 6-13. First plate-current test—02 model.

5. Observe the voltmeter reading to follow the first plate-current diagnosis chart below:

First Plate-Current Diagnosis

DC Test Voltage	Diagnosis & Procedure
1.4V (approx.)	Indicates normal plate current (280 mA). Run performancc test.
0V (approx.)	1. Short circuit to ground someplace between voltage-doubler circuit and magnetron. Check wiring. 2. Shorted magnetron tube.
0V	Perform second plate-current test.

When voltmeter is set to the AC voltage scale, reading will be approximately 3.2V AC.

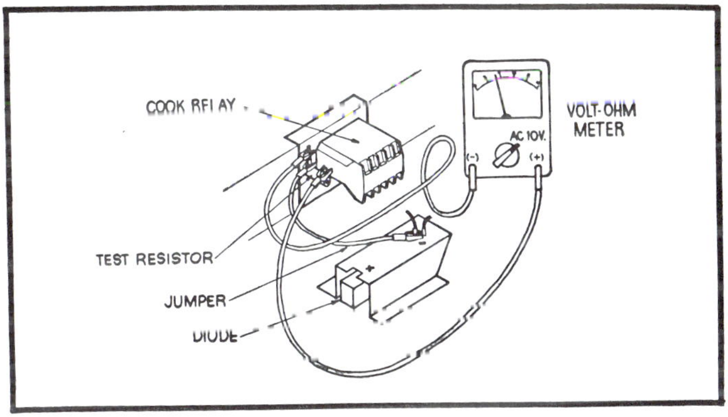

Fig. 6-14. Second plate-current test—02 model.

For the second plate-current test (use this test only if indicated by first plate-current test), proceed as follows:

1. Unplug the oven cord.
2. Discharge the HV capacitor.
3. Using a jumper wire with alligator clips, connect one end to negative terminal on diode. Connect other end to chassis ground (Fig. 6-14).
4. Connect the voltmeter across the test resistor. Set it to the 10–12V AC scale.
5. Plug in oven, use water load, and set oven for cooking.
6. Observe the voltmeter reading to follow the second plate-current diagnosis chart below:

Second Plate-Current Diagnosis

AC Test Voltage	Diagnosis & Procedure
3.2V AC (approx.)	1. Magnetron filament open—test tube for open circuit. 2. Filament transformer open—test transformer. 3. Open wiring to magnetron or filament transformer.
0V	1. High-voltage transformer open or shorted—test transformer. 2. Diode or varistor shorted—test diode and varistor. 3. High-voltage capacitor open—test capacitor.

Jet-70001 Plate-Current Tests

When the COOK light turns on, but little or no heat is produced, the trouble is either in the HV or magnetron circuit. To diagnose the problem, one or two plate-current tests have to be performed. The result of each test will indicate the source of the problem. For the first plate-current test, proceed as follows:

1. Unplug the oven and remove outer case.
2. Discharge the HV capacitor.
3. Connect the voltmeter across the test resistor by connecting the test leads between chassis ground and the positive terminal observing polarity. Set the voltmeter to the 10–12V DC scale (Fig. 6-15).
4. Plug in the oven, use a water load, and set the oven for cooking.

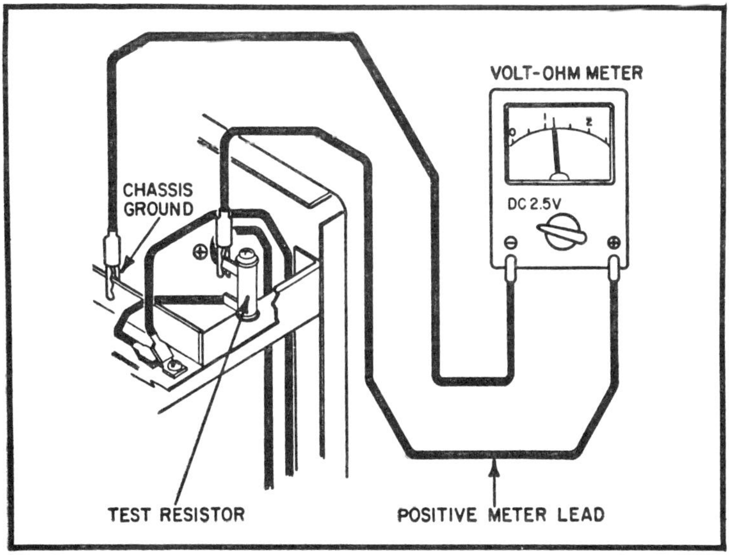

Fig. 6-15. First plate-current test—JET-70001.

5. Observe the voltmeter reading to follow the first plate-current diagnosis chart below.

First Plate-Current Diagnosis

DC Test Voltage	Diagnosis & Procedure
1.2V (approx.)	Indicates normal plate current (240 mA). Run performance test.
0V (approx.)	1. Short circuit to ground someplace between voltage-doubler circuit and magnetron. Check wiring. 2. Shorted magnetron tube.
0V	Perform second plate-current test. When voltmeter is set to the AC voltage scale, reading will be approximately 3.0V AC

For second plate-current test (use this test only if indicated by first plate-current test), proceed as follows:

1. Unplug the oven cord.
2. Discharge the HV capacitor.
3. Using a jumper wire with alligator clips, connect one end to negative terminal on diode. Connect other end to chassis ground (Fig. 6-16).
4. Connect the voltmeter across the test resistor. Set it to the 10–12V AC scale (refer to Fig. 6-16).

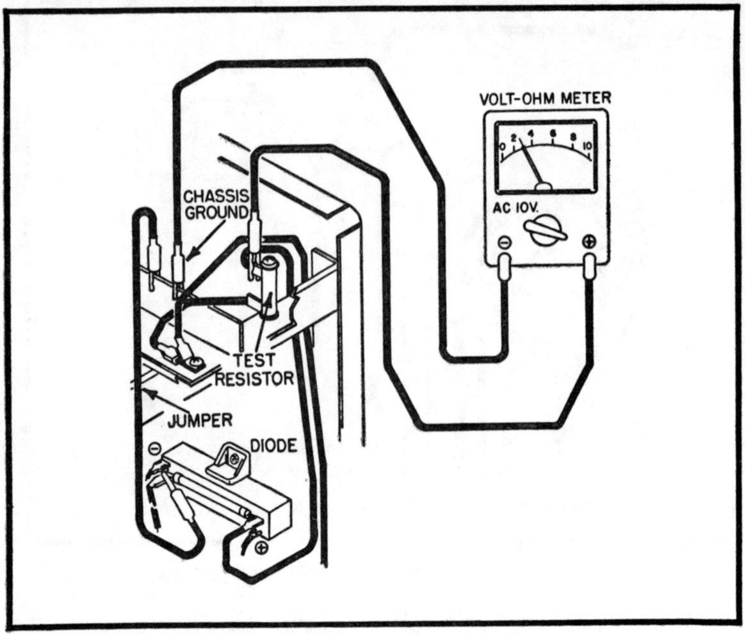

Fig. 6-16. Second plate-current test—JET-70001.

5. Plug in the oven, use a water load, and set the oven for cooking.
6. Observe the voltmeter reading to follow the second plate-current diagnosis chart below:

Second Plate-Current Diagnosis

AC Test Voltage	Diagnosis & Procedure
3V AC (approx.)	1. Magnetron filament open. 2. Filament transformer open. 3. Open wiring to magnetron or filament transformer.
0V	1. HV transformer open or shorted. 2. Diode or varistor shorted. 3. HV capacitor open.

SERVICING DOOR INTERLOCKS

Most GE microwave ovens have three interlocks and a monitor switch to ensure that the magnetron cannot be energized with the door open. The monitor switch blows a fuse and interrupts power to the oven if the other switches are maintained in the closed position because of a defect or tampering. Some early models do not have a monitor circuit, because they were produced before the monitor circuit was required by federal regulation.

In each category of interlock, there are several different types of switches used, depending on the model. The types are indicated by letters in the following paragraphs and may be identified by the accompanying photographs and drawings.

Type A Door-Latch Switch

This is a dual latch switch, operated by a mechanical-finger lever, which is located at the top center of the oven. The lever is mounted inside the door and operates the latch switch as the lever is moved.

The switch assembly is actually two separate miniature pushbutton switches mounted together with screws (Fig. 6-17). Each switch is actuated by a metal lever or actuator operated by the motion of the mechanical lever in the door. The lever engages a strike which prevents opening the door without operating the lever.

When the door is in the closed position, the lever raises the actuator and presses the switch button. This closes the switch

Fig. 6-17 Door-latch switch—type A.

contacts. When the lever is raised, the actuator releases the button and opens the contacts.

The two switches operate simultaneously but control two different circuits. One switch controls the circuit to the timer motor, stirrer motor, and cook-relay coil. The other switch controls the circuit to the COOK switch and the latching contacts of the cook relay. To test the door-latch switch:

1. Unplug the oven and remove outer case.
2. Discharge the HV capacitor.
3. Raise the finger lever on the door. An audible click should be heard as the lever operates the switch actuator. If necessary, loosen the switch-mounting screws to move the switch to ensure contact with the actuator.
4. Make a continuity check of each switch with door open and closed. A normal reading should indicate closed contacts when door is shut. The contacts should open as soon as finger lever is raised.
5. Check the engagement of the lever with the strike in the switch assembly. The strike should be adjusted for minimum play between the strike and trigger when the door is closed and latched. Adjust the strike, if necessary, by loosening the mounting screws and moving stike in the correct direction.

To replace the door-latch switch:

1. Unplug the oven and remove the outer case.
2. Discharge the HV capacitor.
3. Open the door and remove the two nuts from the switch bracket on back of the front frame.
4. Transfer the leads to the new switch assembly and install it.
5. Check the operation according to the previous tests.

Type A Door Safety Switch

The door safety switch is a backup safety device for the latch switch (primary interlock). It is a 125V, 15A actuator-operated pushbutton switch, located on the left side of the oven, and is operated by the door-hinge lever (Fig. 6-18).

The door switch is connected in series with the HV transformer primary winding and will break the circuit when the door is opened. This provides secondary protection in case

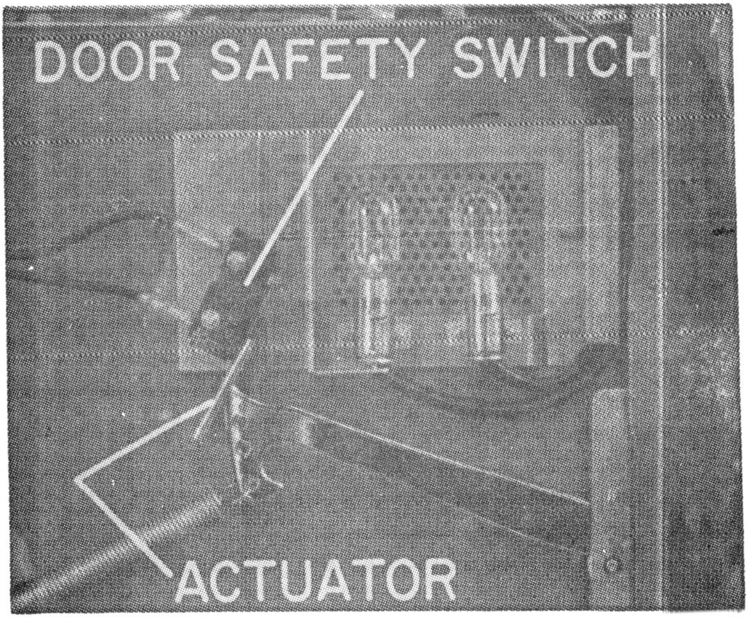

Fig. 6-18. Door safety switch—type A.

of failure of the primary interlock, the latch switch. To test the door safety switch:

1. Unplug the oven and remove the outer case.
2. Check the continuity of switch. Normal operation will show the switch closed when the door is closed. The contacts should open when the door is opened.
3. This door switch must open the circuit by the time the oven door is opened 1 inch as measured between the front frame and the top of the door.
4. If necessary, loosen the switch-mounting screws to adjust the position of switch for proper operation.

Type B Door-Latch Switch

This is a dual latch switch, operated by a mechanical lever, which is located between the oven and the control panel. The lever is mounted inside the door and operates the latch switch as the handle thumb button is pushed.

The switch assembly is actually two separate miniature pushbutton switches mounted together with screws (Fig. 6-19). Each switch is actuated by a metal lever or actuator operated by the motion of the mechanical lever in the door.

Fig. 6-19. Door-latch switch—type B.

The lever engages a strike which prevents opening the door without operating the lever.

When the door is in the closed position, the lever presses the actuator and operates the switch button. This closes the switch contacts. When the thumb button is pressed to open the door, the actuator releases the switch and opens the contacts.

The two switches operate simultaneously and control both sides of the line to the cook relay coil. To test the door-latch switch:

1. Unplug the oven and remove the outer case.
2. Discharge the HV capacitor.
3. Push the door-handle thumb button. An audible click should be heard as the lever operates the switch

actuator. If necessary, loosen the switch-bracket mounting screws to move the switch up or down to ensure contact with the actuator. (Refer to Fig. 6-19.) The holes arc slotted for vertical adjustment.

4. Make a continuity check of each switch with door open and closed. Normal reading should indicate closed contacts when door is closed. Contacts should open as soon as the thumb button is pushed. To replace the door-latch switch:

1. Unplug the oven. Remove the outer case and control panel.
2. Discharge the HV capacitor.
3. Open the door and remove the two switch-bracket mounting screws from front frame.
4. Remove the switch from mounting bracket.
5. Transfer the leads to new switch assembly and install it.
6. Check the operation according to the above tests.

Type B Door Safety Switch

The door safety switch is a backup safety device for the latch switch (primary interlock). It is a 125V, 15A actuator-operated push-button switch, located on the left side of the oven and operated by the door-hinge lever (Fig. 6-20).

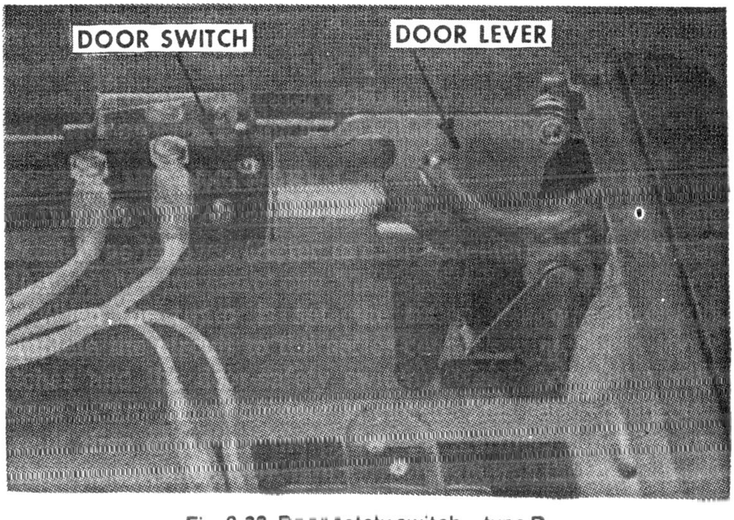

Fig. 6-20. Door safety switch—type B.

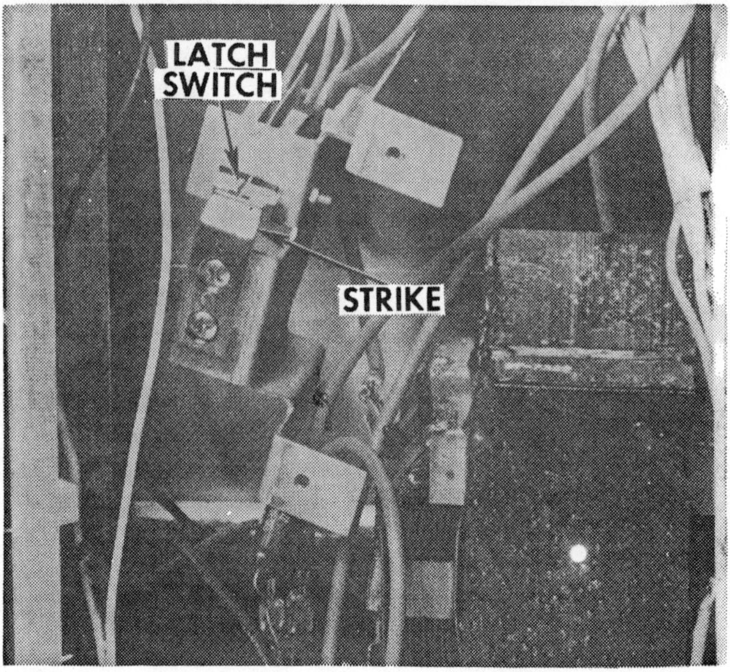

Fig. 6-21. Door-latch switch—types C and D.

The door switch is connected in series with the HV transformer primary winding and will break the circuit when the door is opened. This provides secondary protection in case of failure of the primary interlock, the latch switch. To test the door safety switch:

1. Unplug the oven and remove the outer case.
2. Discharge the HV capacitor.
3. Check the continuity of the switch. Normal operation will show the switch closed when the door is shut. The contacts should open when the door is opened.
4. This door switch must open the circuit by the time the oven door is opened a maximum of 1.0625 inches as measured between the front frame and inside edge of the top of the door.

Type C Door-Latch Switch

The primary and secondary interlocks are the two door-latch switches, which control the relay contacts for the magnetron power transformer and are located behind the oven front frame, beside the oven cavity (Fig. 6-21).

The primary interlock is the switch on the right or outside position. The secondary interlock is the switch closest to the oven cavity.

The door-latch switches are operated by the door-latching catch after the door has reached the closed position. The door-latch circuit is broken before the door makes any movement when one is opening the door. To test the door-latch switch:

1. Unplug the oven and remove the outer case.
2. Discharge the HV capacitor.
3. Push the door-handle thumb button. An audible click should be heard as the lever operates the switch actuator. If necessary, loosen the switch-bracket mounting screws to move the switch up or down to ensure contact with the actuator. (Refer to Fig. 6-21). The holes are slotted for vertical adjustment.
4. Make a continuity check of each switch with door open and closed. Normal reading should indicate closed contacts when door is closed. Contacts should open as soon as the thumb button is pushed.

To replace the door-latch switch:

1. Unplug the oven. Remove the outer case and control panel.
2. Discharge the HV capacitor.
3. Remove the control panel and remove two switch-bracket mounting screws from front frame.
4. Remove the switch from the mounting bracket.
5. Transfer the leads to the new switch assembly and install it.
6. Check the operation according to the previous tests.

Type C Door Safety Switch

The concealed door safety switch is a double-pole switch with one pole in series with the cook-relay contacts for the magnetron transformer; the relay contacts are controlled by the primary- and secondary-interlock switches. If the primary- and secondary-interlock (door latch) switches fail in the closed position or the power-transformer cook relay fails in the on position, the unit does not present a safety hazard because the concealed interlock (door-safety interlock) senses the door position by a hinge lever connected to the door and

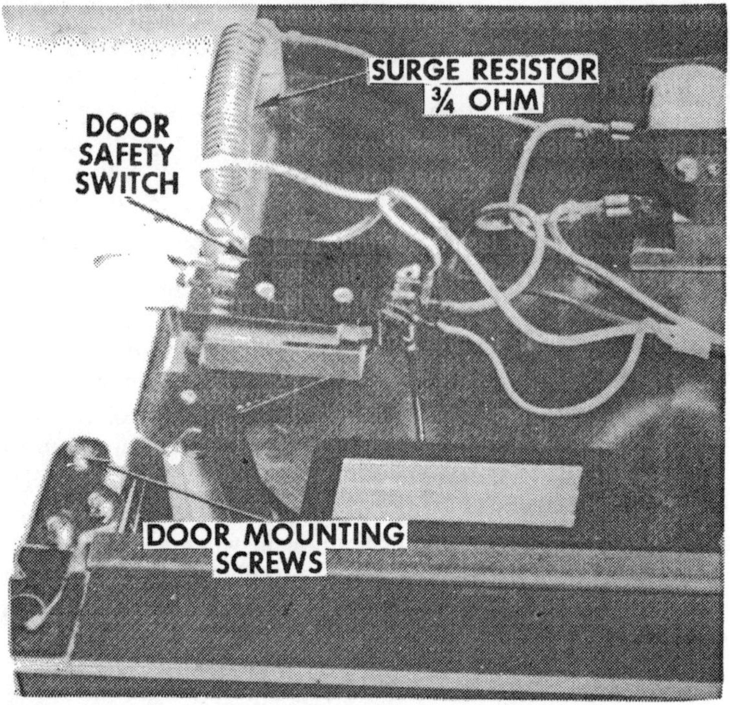

Fig. 6-22. Monitor-types A, B, and C.

extending through the oven front frame to operate the concealed interlock (Fig. 6-22). If the door is moved toward the open position with the primary and secondary interlock or cook relay failed, the concealed interlock de-energizes the unit by opening the transformer circuit.

As the door is opened, two audible clicks will be heard. The first click is the door safety switch; the second is the monitor.

Type A, B, C Monitors

A separate set of switch contacts located in the concealed door safety switch monitors the primary and secondary interlocks (latch switches). In case of a primary and secondary failure, the monitor switch closes and connects a 0.75Ω surge resistor across the 120V line, which blows the 15A oven fuse.

The complete monitor circuit must be tested according to the following procedure after each service call on the appliance. (Refer to Figs. 6-22 and 6-23.)

1. Unplug the oven and discharge the HV capacitor.
2. Check the 15A fuse for proper size. Do not use any other size except a 15A fuse (GE catalog number WB27X7).
3. Check the adjustment of the door safety switch. Two audible clicks should be heard both when opening and closing the oven door. This should occur during a 1.5-inch travel from fully closed position. Loosen the mounting screws to adjust the switch bracket.
4. The exact setting should then be made so that the second click (monitor) occurs when the door has opened 0.75–1.5 inches from the face of the frame. Tighten the switch-mounting screws.
5. Remove the door-safety-switch wiring and check the continuity of the switch while operating door as in step 4. Continuity must show as follows:

Door Position	Contacts Closed	Contacts Open
Closed	Common, N.O.	Common, N.C.
Open	Common, N.C.	Common, N.O.

Replace the wiring.

6. To test the operation of the circuit, proceed as follows:
   a. Connect an ohmmeter across the two line terminals of the appliance cord. Use the R × 1 scale.
   b. Remove the gray primary leads from the No. 1 terminal on the power transformer.
   c. *For types A and B only:* Place a jumper across the cook-relay contacts or hold the relay closed manually while opening and closing door.
   Note: The ohmmeter should read 0.75Ω when the monitor operates. This occurs with the second click of the door safety switch as the door moves 0.75–1.5 inches from the face of the frame.
   d. Confirm the test by removing the 15A fuse. The circuit should open.
   e. Replace the transformer leads and then remove jumper, if used, from the relay. Replacement of any part of the monitor circuit necessitates repeating this entire test procedure.

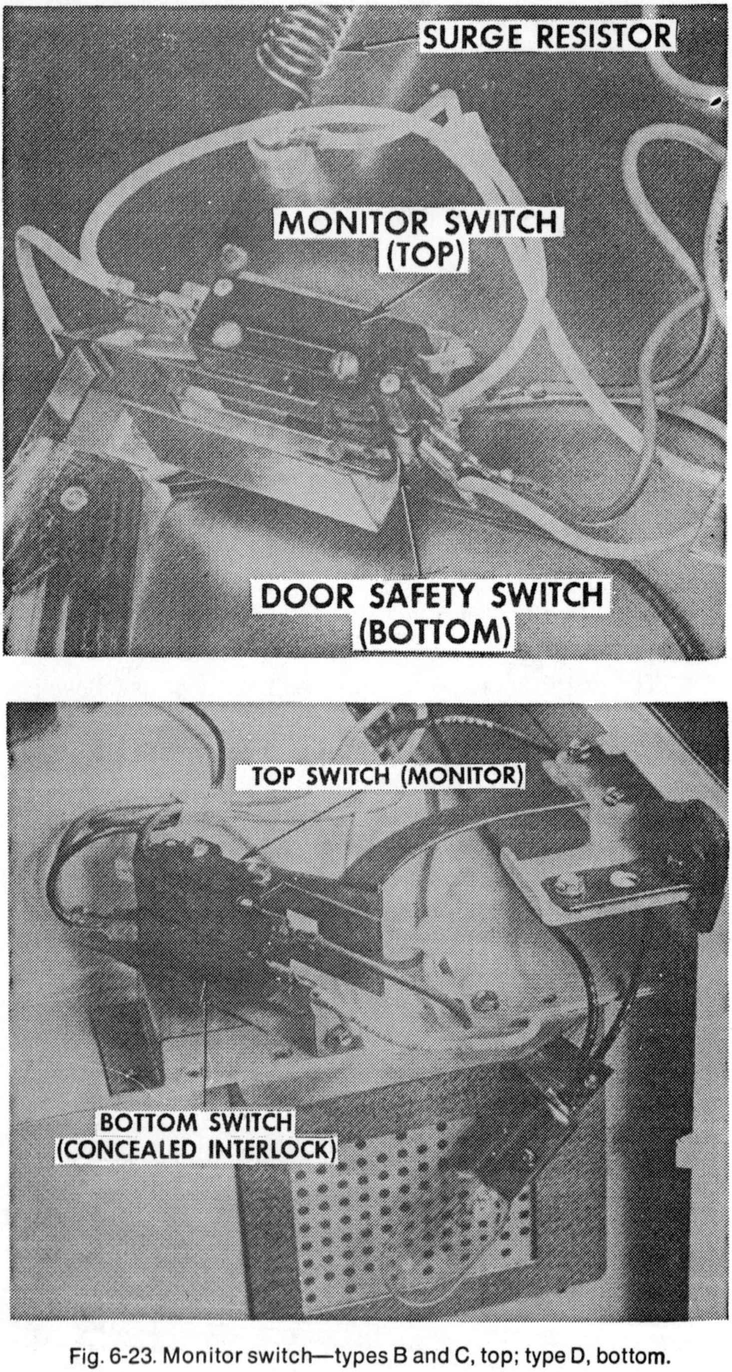

Fig. 6-23. Monitor switch—types B and C, top; type D, bottom.

Type D, E, and F Door Safety Switch

The door safety switch (concealed interlock) is the bottom switch of a dual switch assembly located on the hinge side of the top-front corner of the oven cavity (refer to Fig. 6-23). (The top switch is for the monitor circuit.) The two switch actuators are mechanically connected to each other and are operated by the same door-hinge lever.

The function of the concealed interlock is to serve as a backup for the primary and secondary interlocks by sensing door position. The switch is connected in the neutral line and breaks the circuit to the power transformer when the door is open. To test it, proceed as follows:

1. Unplug the oven and remove the outer case.
2. Discharge the HV capacitor.
3. Disconnect the leads from the switch. Check the continuity of the switch. Continuity must show as follows:

Door Position	Contacts Closed	Contacts Open
Closed	Common, N.O.	Common, N.C.
Open	Common, N.C.	Common, N.O.

To replace it, proceed as follows:

1. Unplug the oven and remove the outer case.
2. Discharge the HV capacitor.
3. The concealed interlock and the monitor switch must be replaced as an assembly. Transfer all leads to the new switch assembly. Then remove the old switch by taking out one mounting screw.
4. Mount the new switch assembly.
5. Test the monitor circuit according to the following instructions.

Type D Door-Latch Switch

The primary and secondary interlocks are the two door-latch switches, located behind the oven front frame on the right side of the oven cavity. (Refer to Fig. 6-21.) The primary interlock is the switch closest to the cavity wall. The secondary interlock is the switch at the outside position.

The door-latch switches are operated by the door latch. When the door is fully closed, the door latch depresses both interlock switches and closes their contacts to complete the electrical circuit of the two interlocks, which are connected in series. (Refer to proper oven schematic.) The circuit is broken, by depressing the latch button on the door handle, before the door can move. To test it, proceed as follows:

1. Unplug the oven and remove the outer case.
2. Discharge the HV capacitor.
3. Push the door-handle thumb button and observe the operation of both switches. An audible click should be heard as the latch operates the switches.
4. Make a continuity check between the front and rear bottom terminals of each switch with the door open and closed. Normal reading should indicate closed contacts when the door is closed. Contacts should open as soon as the latch button is pressed and before the door can move. The bottom rear lead must be removed during test.

Note: No adjustment of these switches is available. To replace them, proceed as follows:

1. Unplug the oven and remove the outer case.
2. Discharge the HV capacitor.
3. Remove the control panel and timer-bracket assembly (two screws).
4. Remove two switch-bracket mounting screws from the front frame.
5. Remove the switch from the mounting brackets. Install a new switch.
6. Transfer the leads and install the assembly.
7. Check the operation according to the previous tests.

Type E & F Door-Latch Switches

The primary and secondary interlocks are the two door-latch switches located at the top and bottom of the front frame. The switches are operated by separate latch levers at the top and bottom of the oven door.

When the door is fully closed, the door-latch levers depress both interlock switches and close their contacts to complete the circuit of the two interlocks, which are connected in series. The circuit is broken (by depressing the latch button on the door handle) before the door can move. To test them, proceed as follows:

1. Unplug the oven, remove the outer case, and discharge the HV capacitor.

2. Push the door-handle thumb button and observe operation of latch switches. An audible click should be heard as the switches operate.
3. Make a continuity check between the COM and N.O. terminals of each switch with door open and door closed. Normal reading should indicate closed contacts when the door is closed. Contacts should open as soon as latch button is pressed and before the door can move.

   Note: No adjustment of the switches is available.

To replace these switches, proceed as follows:

1. Unplug the oven, remove the outer case, and the HV discharge capacitor.
2. Top latch switch:
   a. Remove the two switch-mounting screws (Fig. 6-24).
   b. Transfer the leads to the new switch.
   c. Transfer the switch cover and mount the switch.
   d. Check its operation.

Fig. 6-24. Door-latch switch—types E and F.

3. Bottom switch:
    a. Remove the complete control-panel and instrument-bracket assembly. This assembly is fastened by two screws at the bottom, one screw in top brace, and two screws in flange of front frame.
    b. Carefully move the assembly to gain access to the bottom latch switch.
    c. Change the switch in the same manner as the top switch.
    d. Check its operation.

Type D Monitor

The monitor switch is the top switch of the dual switch assembly located on the hinge side of the top-front corner of the oven cavity. (The bottom switch is the door safety switch.) The two switch actuators are mechanically connected to each other and are operated by the same door-hinge lever. (See bottom of Fig. 6-23.) The complete monitor circuit should be tested according to the following procedure after each service call in the appliance:

1. Unplug the oven and discharge the HV capacitor.
2. Check the 15A fuse for the proper fuse size. Do not use any other size except a 15A fuse (GE catalog number WB27X7).
3. Check the adjustment of the door safety switch. Two audible clicks should be heard, both when opening and closing the oven door. (This should occur during a 1-inch travel from the fully closed position.) Loosen the mounting screws to adjust switch bracket.
4. The exact setting should then be made so that the second click (monitor) occurs when the door has opened 0.25–1.00 inch from the face of the frame. Measure it at the latch side of the door. Tighten the switch-mounting screws.
5. Repeat the test while closing the door. Both switches should operate before the door is closer than a quarter-inch to the front frame.
6. Remove the switch wiring and check the continuity of switch while operating door as in step 4. Continuity must show as follows:

Door Position	Contacts Closed	Contacts Open
Closed	Common, N.O.	Common, N.C.
Open	Common, N.C.	Common, N.O.

Replace the wiring.

7. To test the operation of the circuit:

    a. Connect an ohmmeter across the two line terminals of the appliance cord (use the R × 1 scale).
    b. Connect a temporary jumper lead between COM of the top latch switch and COM of the monitor switch.
    c. Slowly open the oven door.
    Note: This should read zero (short circuit) when the monitor operates. This occurs with the second click of the door safety switch as the door moves 0.75–1.00 inches from the face of the frame.
    d. Confirm the test by removing the 15A fuse. The circuit should open.
    e. Remove the temporary jumper lead. Replacement of any part of the monitor circuit necessitates repeating this entire test procedure.

OVEN LIGHTS

The oven lights are GE-types 15T7DC (15W) or 25T8DC (25W) and are conveniently replaced after removing a small panel on the oven case.

Type A

Two 15W, 120V incandescent lamps with bayonet-type bases are used for oven-cavity illumination. The GE equivalent for this lamp is 15T7DC (double-contact, bayonet-base), commonly called a *sewing machine lamp*. The lamps are accessible through a small panel on the left side of the oven case.

During normal operation, the lamps are also used as an indicator that the oven is ready for cooking. The lamps come on after the 7-second time-delay relay closes.

Type B

A single 15W, 120V incandescent lamp with a bayonet type base is used for cavity illumination. The GE equivalent for this lamp is 15T7DC (double-contact, bayonet-base), commonly called a *sewing machine lamp*. The lamp is accessible through a small panel on the left side of the oven case.

The oven lamp comes on as soon as the timer is set.

Type C

A single 25W, 120V incandescent lamp with a bayonet-type base is used for cavity illumination. The GE equivalent for this lamp is 25T8DC (double-contact, bayonet-base), commonly called a *sewing machine lamp*. The lamp is accessible through a small panel on the left side of the oven case. The panel is held with a single screw. The oven lamp comes on as soon as the timer is set or when the door is opened.

Chapter 7
Servicing Litton Microwave Ovens

This chapter describes the operation and servicing of three representative Litton[1] microwave ovens—two that are countertop models and one that is combined with a conventional electric range.

Unlike most microwaves ovens, which cook by time alone, Litton's Model 419 countertop oven cooks by temperature or time. This provides flexibility in cooking and may seem easier to many cooks who are used to cooking by temperature. The Vari-Temp control is set to the desired temperature, and the Vari-Temp sensor probe is inserted into the food, or load. When the probe senses that the load has reached the selected temperature, the Vari-Temp control causes the oven to stop cooking. The Model 419 also offers timed cooking with its 99-minute digital timer and variable-speed cooking with its solid-state Vari-cook oven control.

Remember the gas-range advertising slogan of some years back that puffed, "I'm a gas-top burner with a brain"? Well, the Litton Model 420 Memorymatic microwave oven really does have a brain. It's a solid-state computerlike memory control that programs time and cooking speeds automatically, and it's easy to use. For best results, some foods should be cooked at one cooking speed and finished at a

1. All of the illustrations in this chapter are courtesy of Litton Microwave Cooking Products.

different speed. With the Memorymatic control, you set both cooking speeds at one time—no need to return to the oven to set the second cooking speed. You just punch a few buttons to tell the Memorymatic control how you want your food cooked, and the oven remembers your instructions and carries them out automatically.

Like many microwave manufacturers, Litton offers combination microwave ranges. The servicing of conventional resistance-heating ovens is beyond the scope of this book, but this chapter includes extensive servicing information on the microwave section of Litton's Model 650 combination range. This model allows you to cook with a combination of microwave speed and conventional heat in one oven, with microwaves only, or with conventional baking and broiling.

HOW MODEL 650 WORKS

The following paragraphs provide insight into the operation of the microwave section of the Model 650, and by extrapolation, into the operation of other Litton microwave circuits.

Off Condition

When the oven-door handle is in the up position and the microwave timer is in the OFF position, no electrical power is allowed past the open contacts of the left interlock switch and microwave-timer switch, therefore no components operate (refer to Fig. 7-1).

Cook Condition

To put the oven into a cook condition, the door must be closed and the door handle pushed down. This activates three door-interlock switches. The left and right interlock switches are activated by the door-handle hooks to provide a circuit to the microwave-timer contacts. The interlock-monitor switch is activated by the right door handle guide. By selecting a cooking time with the microwave timer, the timer-switch contacts close to provide a circuit to the following components.

Microwave timer contacts L1-2 provide a circuit to the microwave timer and to the blower motor through the CLEAN switch. The blower motor forces cooling air across the microwave components, including the magnetron, and is exhausted through the back of the range. The timer motor begins timing the selected cooking time.

Microwave contacts 3-MW provide voltage for the following operation:

- The microwave indicator light illuminates to indicate the oven is in a cook cycle.
- The stirrer motor rotates the stirrer at the bottom of the oven cavity.
- The 120V AC input to the primary winding of the power transformer is provided through selector switch contacts 6-6A and is converted to approximately 4000V AC output from the HV secondary. At the same time, the filament-secondary winding of the power transformer provides approximately 3V AC to heat the magnetron filaments.
- The 4000V AC output from the power transformer's HV secondary is sent to a half-wave voltage-doubler circuit comprised of a single diode and one capacitor, where approximately −4000V DC (peak voltage) is provided to the magnetron-tube assembly.
- The −4000V DC applied to the cathode of the magnetron tube causes oscillation within the magnetron, to produce the 2.45 GHz cooking frequency.
- The RF energy produced by the magnetron antenna is channeled through the cavity feedbox, past the stirrer blade, and finally into the cavity, where the food is placed to be heated.
- Upon completion of the selected cooking time, the timer bell rings once as the timer switch is deactivated. Since no electrical power is allowed past the open timer-switch contacts, the oven reverts to the *off* condition.

Defrost Condition

The defrost condition occurs when the oven is in a cook cycle and the selector switch is turned to MICROWAVE DEFROST. The defrost timer motor rotates a cam, which activates and deactivates the defrost timer switch leading to the power transformer at 30-second intervals. While in the

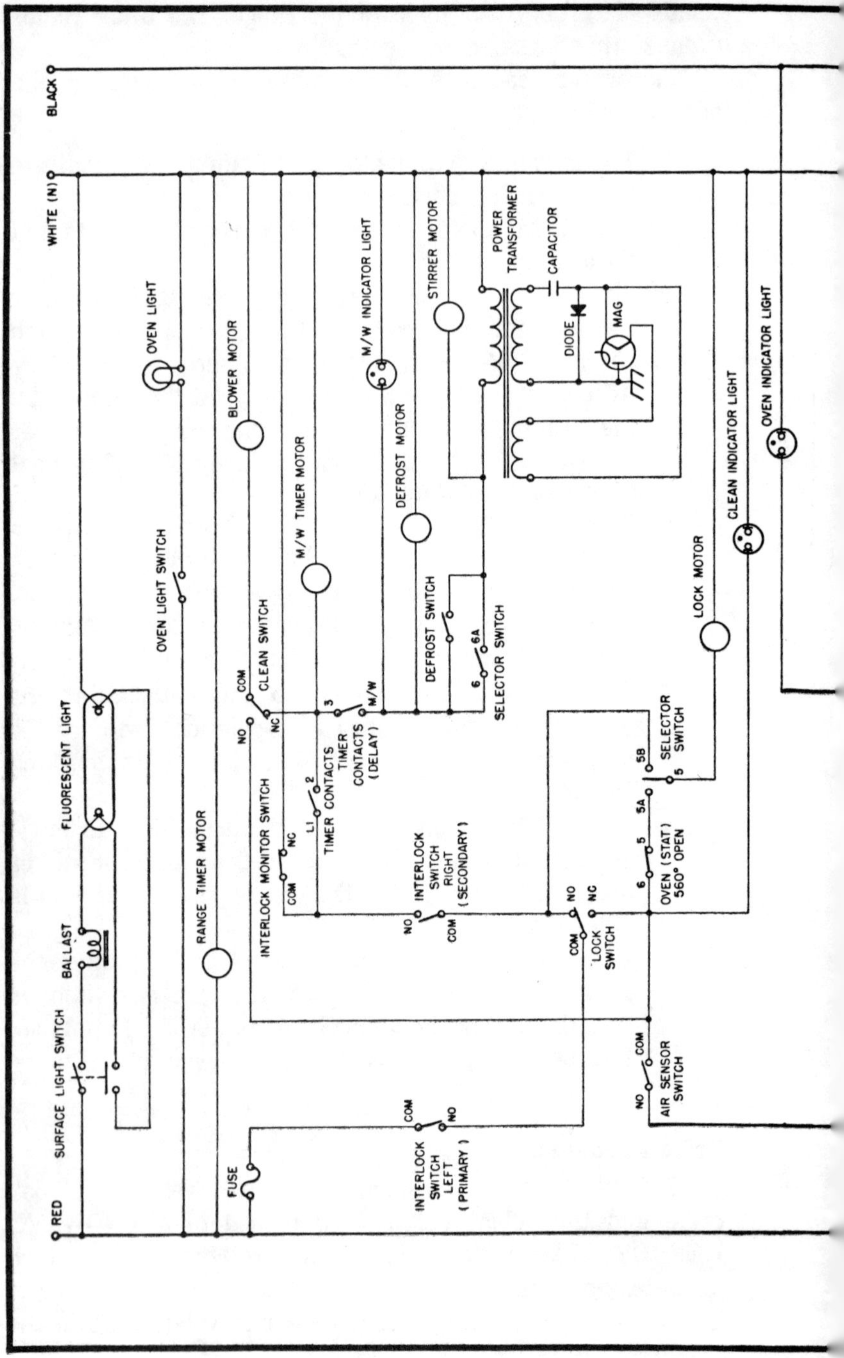
BLACK
WHITE (N)
RED
OVEN LIGHT
BLOWER MOTOR
M/W INDICATOR LIGHT
STIRRER MOTOR
POWER TRANSFORMER
CAPACITOR
DIODE
MAG
OVEN LIGHT SWITCH
M/W TIMER MOTOR
DEFROST MOTOR
DEFROST SWITCH
6A
6
SELECTOR SWITCH
LOCK MOTOR
CLEAN INDICATOR LIGHT
OVEN INDICATOR LIGHT
FLUORESCENT LIGHT
COM
CLEAN SWITCH
NO
NC
3
M/W
TIMER CONTACTS (DELAY)
2
L1
TIMER CONTACTS
RANGE TIMER MOTOR
INTERLOCK MONITOR SWITCH
COM
NC
INTERLOCK SWITCH RIGHT (SECONDARY)
NO
COM
NO
NC
COM
LOCK SWITCH
5B
5
SELECTOR SWITCH
5A
5
6
OVEN (STAT) 560° OPEN
BALLAST
SURFACE LIGHT SWITCH
COM
NO
AIR SENSOR SWITCH
FUSE
COM
NO
INTERLOCK SWITCH LEFT (PRIMARY)

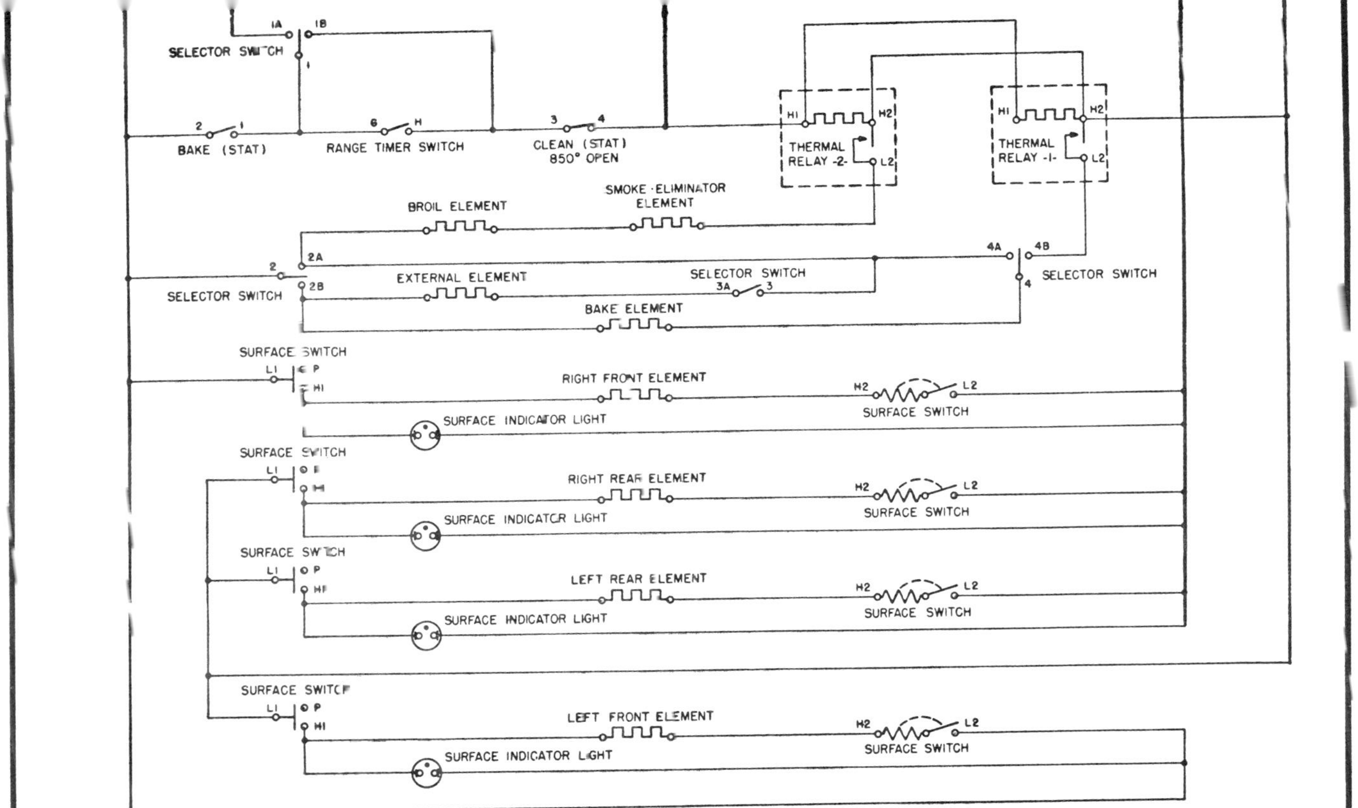

Fig. 7-1. Schematic diagram of Litton Model 650.

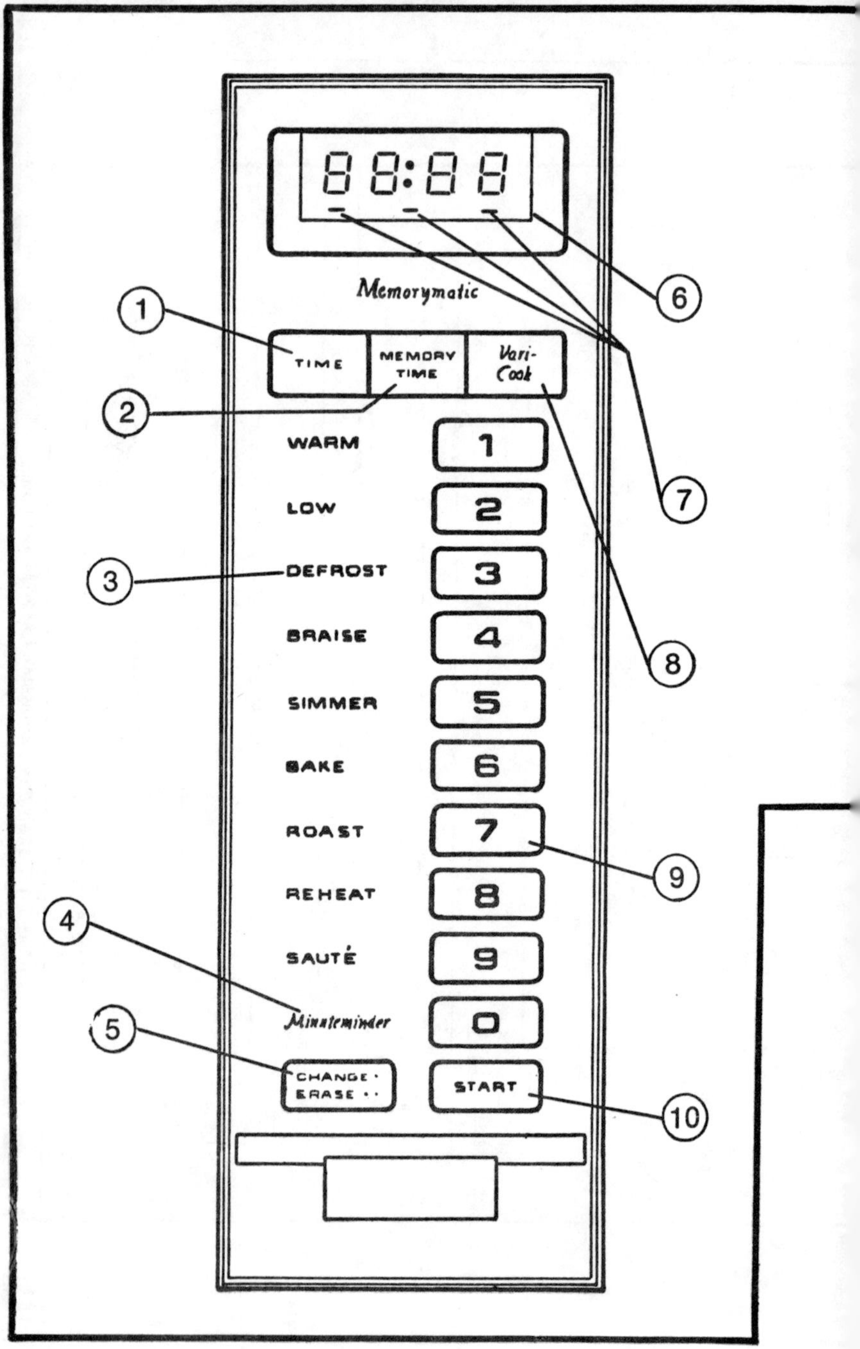

Memorymatic
1
TIME
MEMORY TIME
Vari-Cook
2
6
7
8
WARM
LOW
DEFROST
BRAISE
SIMMER
BAKE
ROAST
REHEAT
SAUTÉ
Minuteminder
3
4
5
9
10
1
2
3
4
5
6
7
8
9
0
CHANGE •
ERASE • •
START

1 **TIME BUTTON**
USED IN SETTING COOKING TIME AND ALSO IN OPERATION OF MINUTE MINDER.

2 **MEMORY TIME BUTTON**
SETS SECOND TIME SETTING FOR USE IN TWO-PHASE MICROWAVE RECIPES.

3 **VARI-COOK SETTINGS**
VARIABLE POWER LEVELS SELECTED BY TOUCHING NUMBERS. USED WITH VARI-COOL BUTTON TO SELECT WHETHER OVEN WILL OPERATE ON WARM, LOW, DE-FROST, BRAISE, SIMMER, BAKE, ROAST, REHEAT OR SAUTE SETTING.

4 **MINUTE MINDER**
OPERATES AS CONVENTIONAL TIMER, DELAY/COOK, OR COOK/REST TIMER.

5 **CHANGE/ERASE BUTTON**
CHANGES ENTRY WITH ONE TOUCH. ERASES ALL ENTRIES WHEN BUTTON IS TOUCHED TWICE

6 **DISPLAY WINDOW**
SHOWS TIME AND VARI-COOK POWER LEVEL. ACTS AS COUNTDOWN TIMER WHEN COOKING OR MINUTE MINDER IS IN OPERATION. NEW LONGER, 99-MINUTE, 99-SECOND TIMER.

7 **INDICATOR LIGHTS**
SHOW WHICH TIME OR VARI-COOK SETTING IS SHOWING IN DISPLAY WINDOW.

8 **VARI-COOK BUTTON**
USED WITH VARI-COOK VARIABLE POWER LEVELS TO SELECT VARI-COOK SETTINGS.

9 **NUMBERS**
TOUCHED TO ENTER COOKING TIMES AND ALSO USED TO SELECT VARI-COOK OR MINUTE MINDER SETTINGS.

10 **START BUTTON**

Fig. 7-2. Control panel of Litton Model 420.

defrost condition, the oven alternately cooks for 30 seconds and then rests for 30 seconds. The microwave timer continues to operate during the 30-second rest mode of the defrost condition.

HOW MODEL 420 WORKS

The microwave-generating circuitry is similar to that described for Model 650. A detailed knowledge of the internal workings of the additional electronic-control circuitry on the Model 420 is not necessary, since the control panel and interface controller are to be replaced when defective. To know when to replace these components, however, you must understand the normal user operation of the controls. The control panel is shown in Fig. 7-2, and the circuitry is shown in Fig. 7-3 and 7-4.

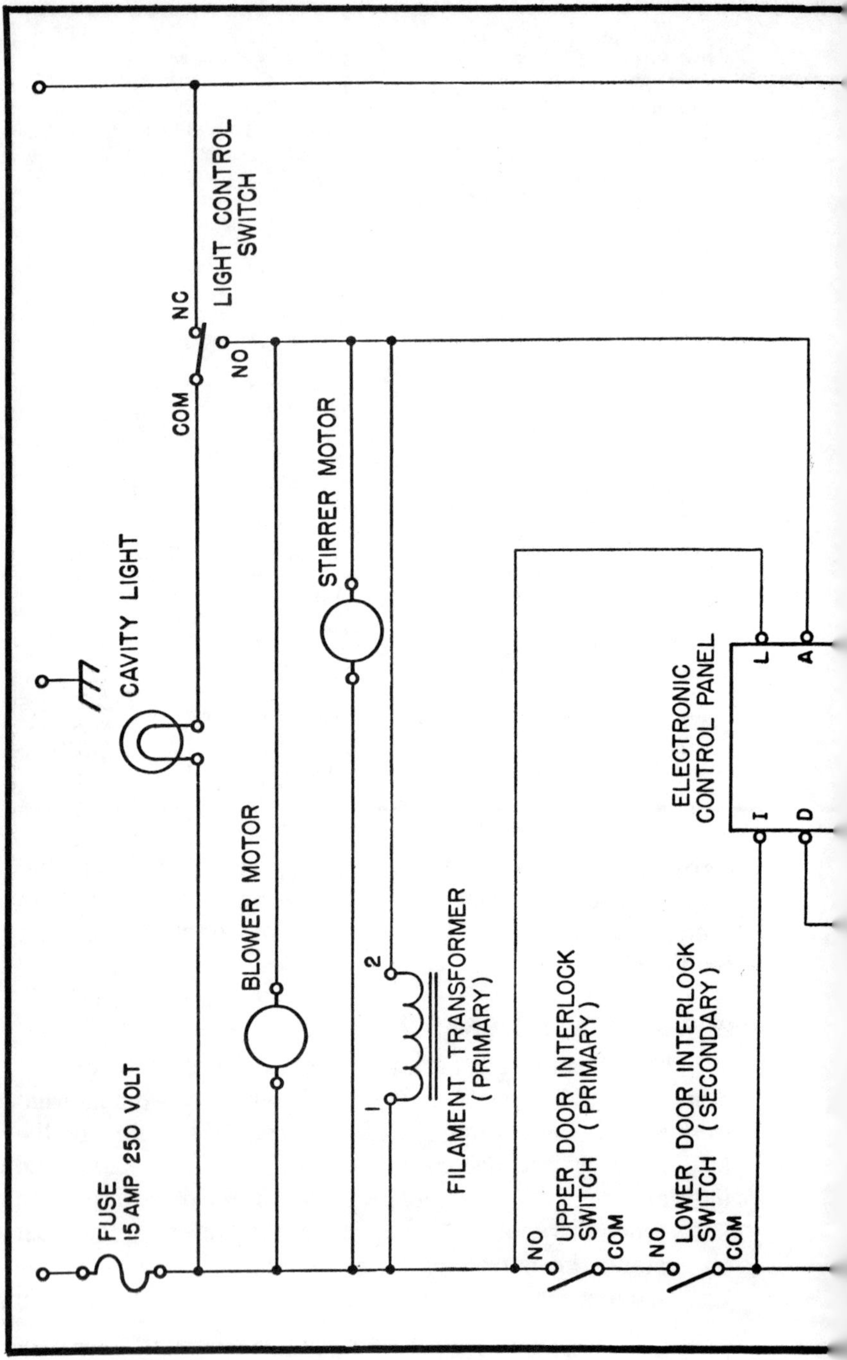
LIGHT CONTROL SWITCH
NC
NO
COM
STIRRER MOTOR
CAVITY LIGHT
ELECTRONIC CONTROL PANEL
L
A
I
D
BLOWER MOTOR
2
1
FILAMENT TRANSFORMER (PRIMARY)
FUSE 15 AMP 250 VOLT
UPPER DOOR INTERLOCK SWITCH (PRIMARY)
NO
COM
LOWER DOOR INTERLOCK SWITCH (SECONDARY)
NO
COM

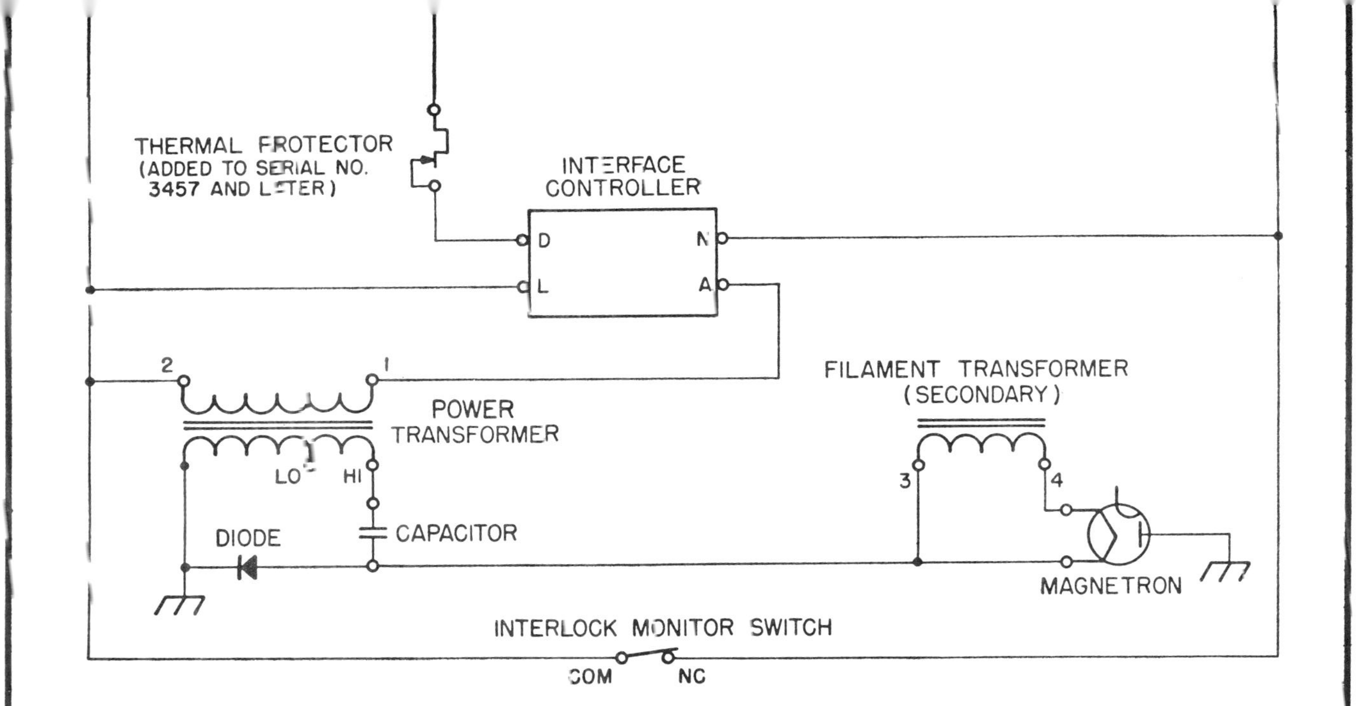

Fig. 7-3. Schematic diagram of L'tton Model 420.

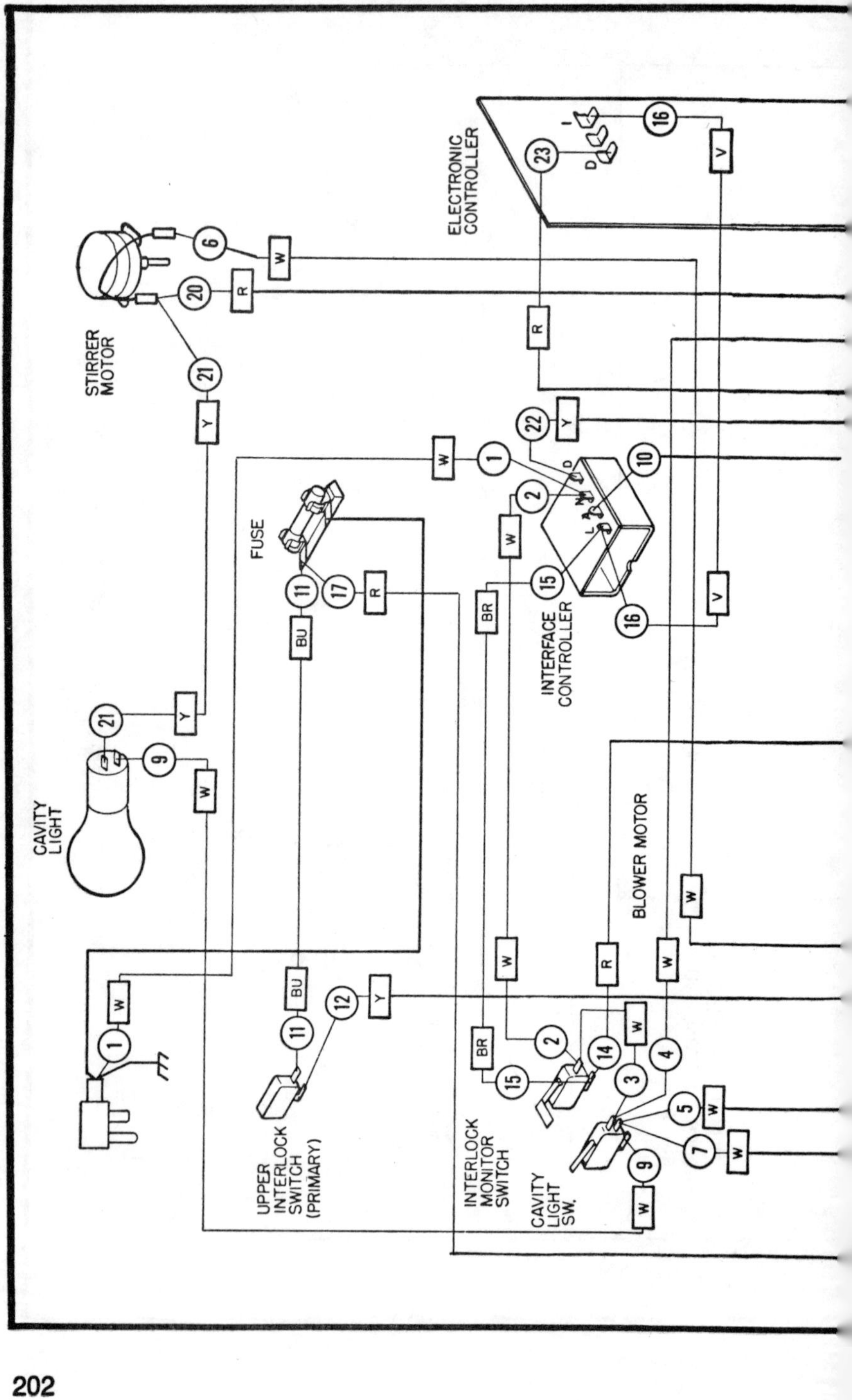
ELECTRONIC CONTROLLER
STIRRER MOTOR
FUSE
CAVITY LIGHT
INTERFACE CONTROLLER
BLOWER MOTOR
UPPER INTERLOCK SWITCH (PRIMARY)
INTERLOCK MONITOR SWITCH
CAVITY LIGHT SW.

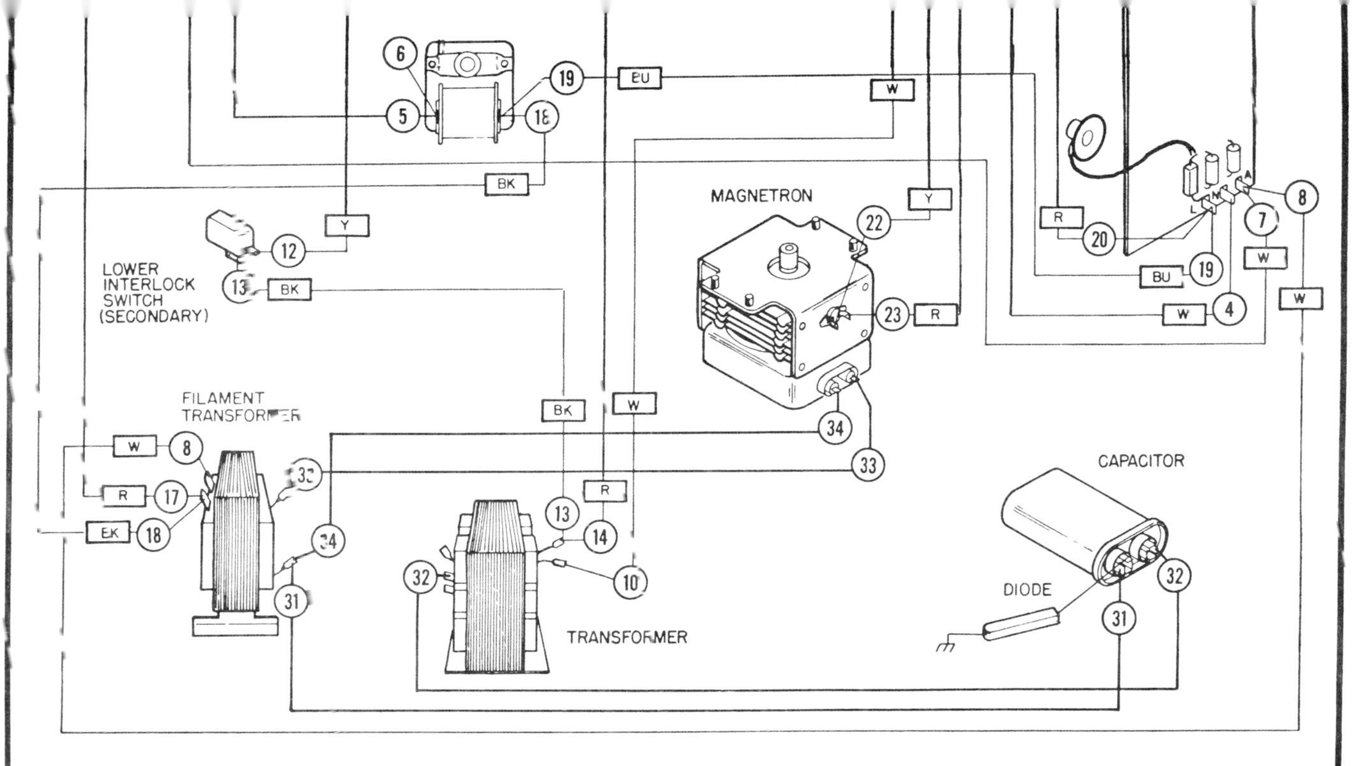

Fig. 7-4. Wiring diagram of Littor Model 420.

Full-Power Microwave Cooking

Touch the TIME button. The light above the TIME button will go on (.00 will appear in the display window).

Set the desired time setting by pushing the appropriate numbered buttons on the right side of the panel. For example, if you wanted to set 30 seconds, you would touch *3*, and then *0*. For 3 minutes, you would touch *3*, and then touch *0* twice. For 3.5 minutes you would touch 3, 3, and 0. The Memorymatic timer is a clock that is divided into 60 seconds. So one-half minute is equal to 30 seconds, and you would touch *30*, not 0.50 as with a calculator. The time you set will appear in the display window at the top of the panel.

You do not need to touch the VARI-COOK button when you are cooking with full power. However, you may touch the VARI-COOK button. The light above the VARI-COOK button will go on, and the word *HI* will appear in the display window. This is to indicate that the oven will automatically be operating under full microwave power unless a different Vari-Cook setting is selected.

Touch the START button, located in the lower-right corner of the timer panel. The oven will begin cooking, and the display window will show the time counting down by seconds. A tone sounds three times when the time is up, and the oven shuts off automatically.

Variable-Power Microwave Cooking

The *cookbook* tells which power setting is recommended for cooking food. To cook with variable power, first touch the TIME button. The light above the TIME button will go on.

Set the desired time by touching the appropriate numbers. The time you set will appear in the display window.

Touch the VARI-COOK button. The light above the VARI-COOK button will go on. Also, HI will appear in the display window until you enter a variable power setting. This is to remind you that the oven will always operate at full power unless you set a variable power.

Select the power setting you desire, and touch the number adjacent to it. For example, if the recipe called for cooking at the LOW setting, you would touch the number 2 (the number that is next to LOW on the panel). When you touch 2, the display window will read *20*. This means that in the LOW setting, the oven is operating at 20% of full power.

Touch the START button. The oven will begin cooking, and you will be able to see the time countdown. A tone will sound three times when the time is up, and the oven will shut off automatically.

Two-Phase Microwave Cooking

Two-phase microwave cooking lets you cook in one cooking phase at a certain power setting, and then in a second cooking phase at a different power setting. You can enter both phases at one time, start the oven, and let the Memorymatic timer do the rest. Simply follow these steps:

1. Touch the TIME button. The light above the TIME button will come on.
2. Touch the time setting for the first cooking phase.
3. Touch the VARI-COOK button. The light above the VARI-COOK button will come on.
4. Touch the variable power setting for the first cooking phase (if other than HI).
5. Touch the MEMORY TIME button. The light above the MEMORY TIME button will come on.
6. Touch the time setting for the second cooking phase.
7. Touch the VARI-COOK button.
8. Touch the power setting for the second cooking phase (if other than HI).
9. Touch the START button.

Minute Minder

To use as a conventional timer:

1. Touch the TIME button.
2. Touch in the desired time.
3. Touch the VARI-COOK button.
4. To use the Minute Minder, touch the *0* (the number that is next to Minute Minder). A *00* will appear in the display window, indicating that no microwave energy is present.
5. Touch the START button. The timer will begin counting down in the display window with no cooking or heating operation of the oven. A tone will sound three times when the time has elapsed.

Note: The oven is not operating during use of Minute Minder; however, the light and blower will continue to function.

A second way to use the Minute Minder is when you want to delay the start of the cooking cycle. You set the oven for a specified length of time with no cooking. Then the oven comes on automatically and begins cooking when the time has elapsed. Delayed cooking on a microwave oven does not have the same applications as on a conventional oven. You normally would not want to delay your fast microwave cooking for a long time. The microwave oven's delayed cooking is especially useful to allow your entire meal to complete cooking at the same time when you are using more than one cooking appliance. To use:

1. Touch the TIME button.
2. Touch in the time setting you want before you wish to start cooking.
3. Touch the VARI-COOK button.
4. To use the Minute Minder, touch the *0* (the number that is next to Minute Minder).
5. Touch the MEMORY TIME button. You are now ready to set the oven for the actual cooking period.
6. Touch in the time you want to food to cook.
7. Touch the VARI-COOK button.
8. Touch in the power setting you desire.
9. Touch the START button. The Minute Minder will count down the delay period. When this time is up, it will automatically change to the cook period and start cooking. You can see the time counting down in the display window during both phases. When the time is up, a tone will sound, and the oven will shut off automatically.

HOW MODEL 419 VARI-TEMP WORKS

Insertion of the temperature-probe plug into the probe jack, located in the oven-cavity wall, activates the Vari-Temp switch and timer-bypass-interlock switch. (See Figs. 7-5 and 7-6.) During Vari-Temp operation, the timer is bypassed to ensure that the cook cycle is controlled by the temperature of the load being heated, rather than the time set on the timer. The Vari-Temp switch places four major Vari-Temp components into operation when the door is closed.

- Vari-Temp transformer
- Temperature indicator
- Vari-Temp relay
- Temperature probe

Vari-Temp Transformer

A voltage of 120V AC is supplied to the primary winding, and 24V AC is produced by the secondary winding. The secondary winding has a centertap that allows splitting the 24V AC output into two separate 12V AC circuits, which provide voltage to the temperature-indicator assembly.

Temperature-Indicator Assembly

The components located on the assembly's PC board provide a voltage-rectifier, series-regulator circuit that converts the two 12V AC inputs supplied from the Vari-Temp transformer into two separate DC voltage outputs that are supplied to the temperature indicator.

Unregulated 12V DC is provided to terminal E5 through the normally closed (N.C.) contacts of the thermal relay, to terminal E1, and to the coil of the Vari-Temp relay. The contacts of the relay close to provide power to the HV section, and the oven begins cooking.

Regulated 6.8V DC is supplied to terminal E4 to provide voltage to the heater of the temperature-indicator contacts, which is in series with thermistor of the temperature probe. The resistance of the probe thermistor decreases as temperature increases. The resultant increase in current through the heater of the temperature-indicator contacts causes the contacts causes the contacts to close.

The heater of the thermal relay is in series with the temperature-indicator contacts. When the pointer reaches the selected temperature, the temperature-indicator contacts close to provide voltage to the thermal-relay heater, and the thermal-relay contacts open. Voltage is removed from the coil of the Vari-Temp relay. The Vari-Temp relay contacts open the circuit to the HV section and now provide a circuit to the HOLD indicator light.

TROUBLESHOOTING AND TESTING

The troubleshooting charts for models 419, 420, and 650 are given as Tables 7-1, 7-2, and 7-3. These charts provide a logical

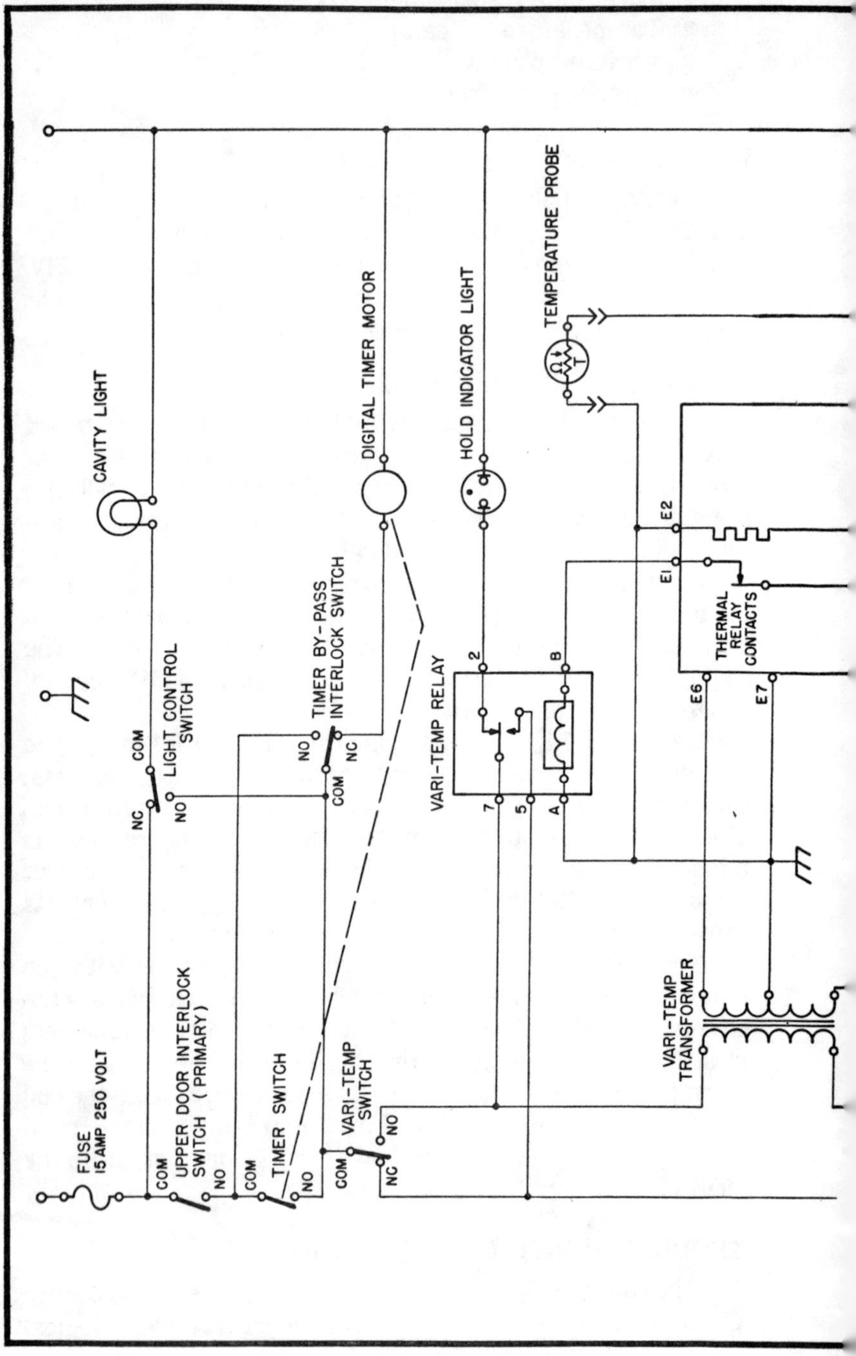
FUSE
15 AMP 250 VOLT
UPPER DOOR INTERLOCK SWITCH (PRIMARY)
COM
NO
TIMER SWITCH
COM
NO
VARI-TEMP SWITCH
COM
NC
NO
CAVITY LIGHT
LIGHT CONTROL SWITCH
NC
COM
NO
DIGITAL TIMER MOTOR
TIMER BY-PASS INTERLOCK SWITCH
NO
NC
COM
HOLD INDICATOR LIGHT
VARI-TEMP RELAY
2
7
5
A
B
TEMPERATURE PROBE
E2
E1
THERMAL RELAY CONTACTS
E6
E7
VARI-TEMP TRANSFORMER

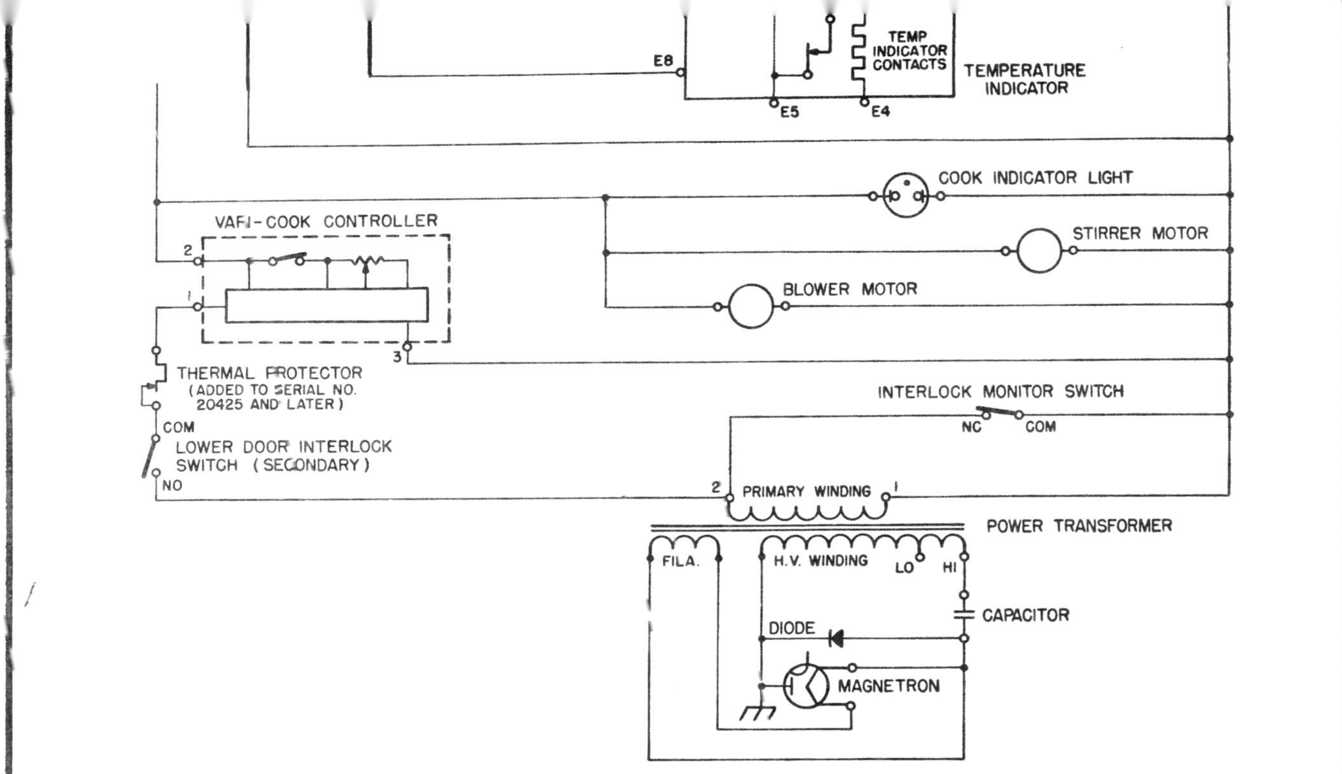

Fig. 7-5. Schematic diagram of Litton Model 419.

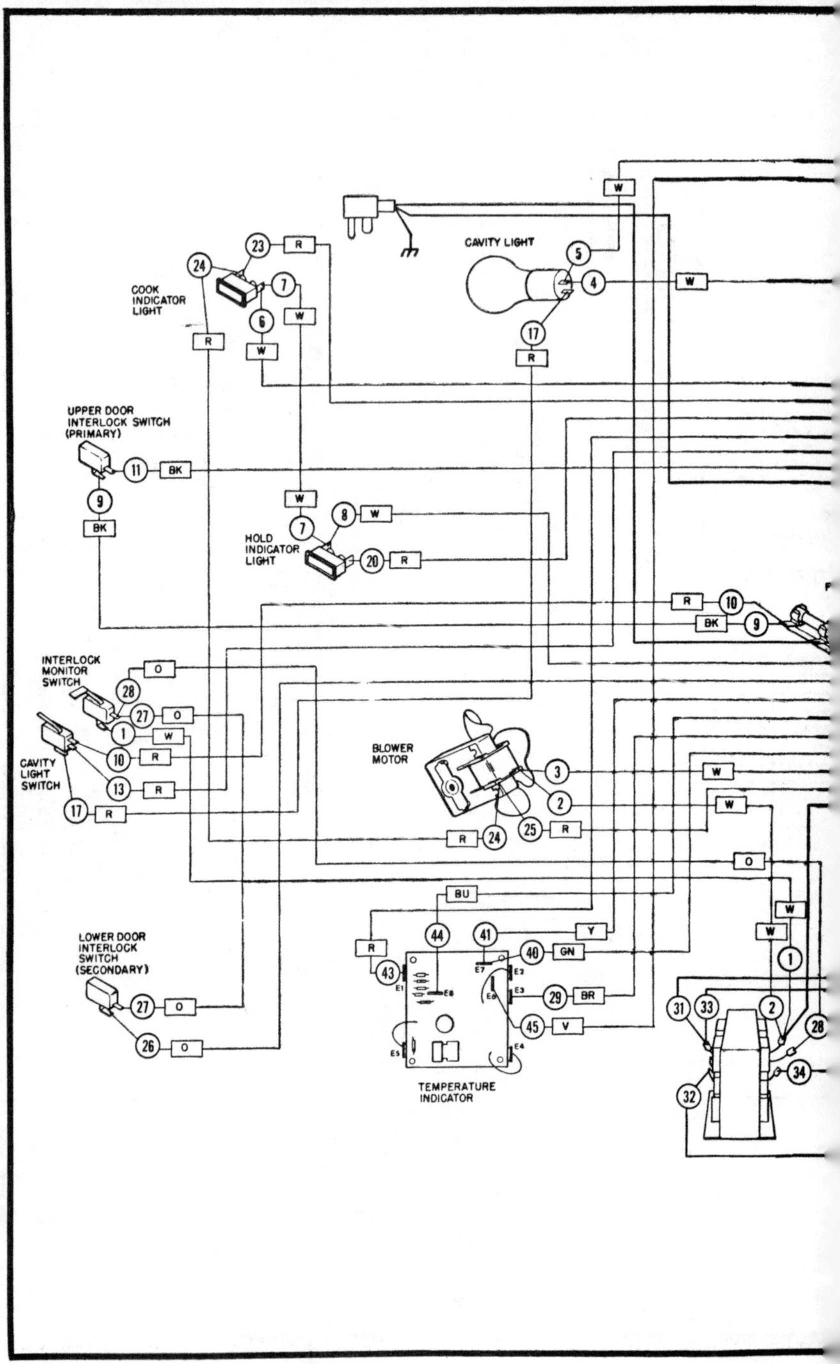

CAVITY LIGHT
COOK INDICATOR LIGHT
UPPER DOOR INTERLOCK SWITCH (PRIMARY)
HOLD INDICATOR LIGHT
INTERLOCK MONITOR SWITCH
CAVITY LIGHT SWITCH
BLOWER MOTOR
LOWER DOOR INTERLOCK SWITCH (SECONDARY)
TEMPERATURE INDICATOR

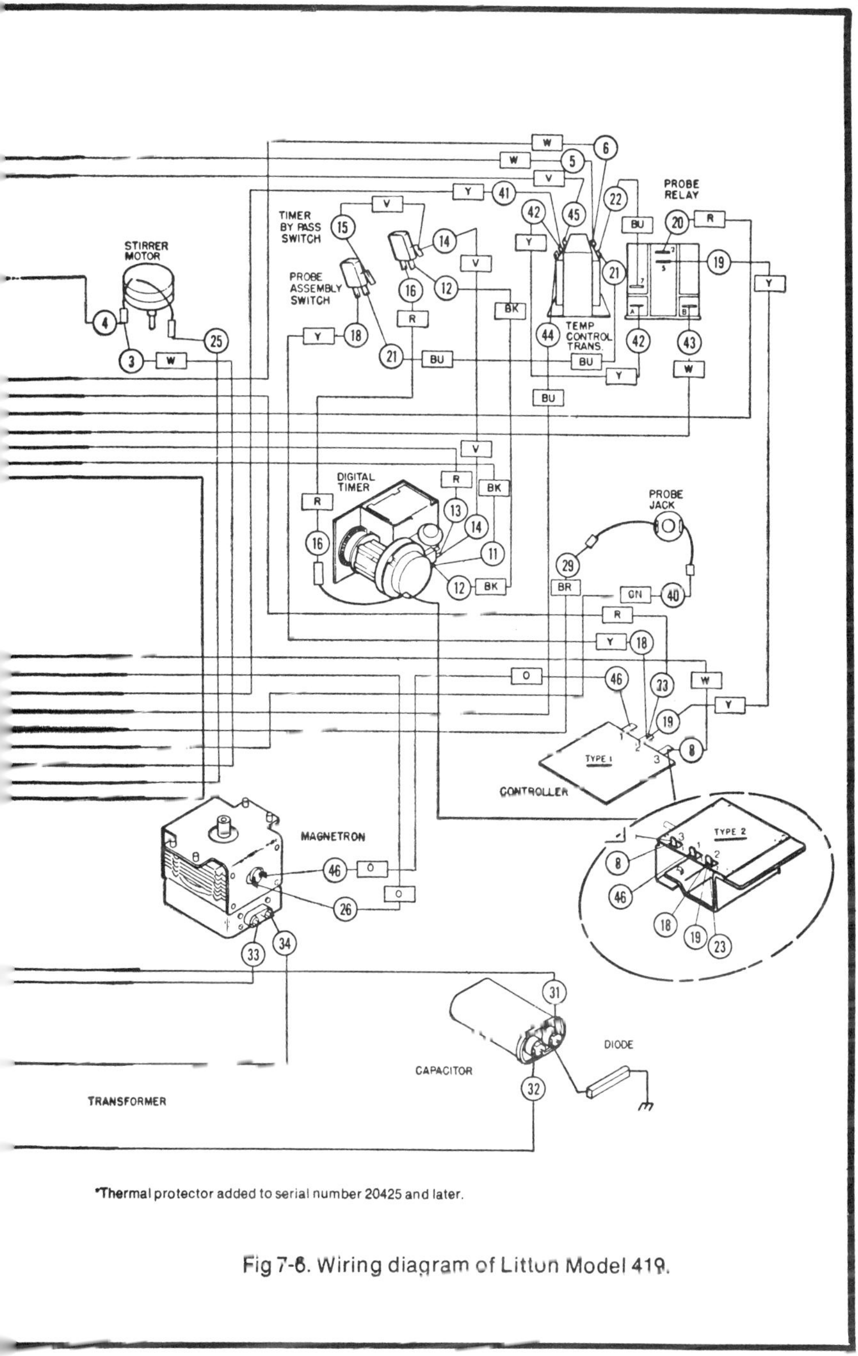

*Thermal protector added to serial number 20425 and later.

Fig 7-6. Wiring diagram of Littun Model 419.

Table 7-1. Troubleshooting the Litton Model 419

CAVITY-LIGHT-ON CONDITION	
Setup: plug oven in, turn timer off, open oven door. Normal Operation: Cavity light should be on.	
PROBLEM	**POSSIBLE CAUSES**
Cavity light does not come on; no other components operate.	1. Blown wall fuse or tripped circuit breaker. 2. Blown oven fuse. 3. Broken or loose wire connection.
Cavity light only does not come on.	1. Defective light-control switch. 2. Defective cavity switch. 3. Broken or loose connection.

COOK CONDITION	
Setup: 1) Open oven door. 2) Turn Vari-Cook dial to HIGH. 3) Place container of water in oven cavity. 4) Push in timer knob and rotate to 2 minutes. 5) Close oven door. Normal Operation: 1) Cavity light is on. 2) Timer motor operates. 3) Cook light is on. 4) Stirrer motor operates. 5) Blower motor operates. 6) Oven heats product placed in cooking cavity. 7) Timer shuts oven off at end of selected time setting.	
PROBLEM	**POSSIBLE CAUSES**
Cavity light comes on with door open, but no components operate when door is closed.	1. Defective upper door interlock switch. 2. Defective timer switch. 3. Broken or loose wire connection.
Cavity light comes on, timer times out, but no other components operate.	1. Defective Vari-Temp switch. 2. Broken or loose wire connection.
Timer does not advance, but other operation is normal.	1. Defective timer-bypass switch. 2. Defective timer assembly. 3. Binding timer assembly. 4. Broken or loose wire connection.
COOK indicator light does not come on, but other operation is normal.	1. Defective COOK indicator light. 2. Broken or loose wire connection.
Stirrer blade does not rotate at top of oven cavity, but other operation is normal.	1. Defective stirrer motor. 2. Stirrer blade binding. 3. Broken or loose wire connection.
Blower motor does not operate, but other operation is normal.	1. Defective blower motor. 2. Blower fan binding. 3. Broken or loose wire connections.
Oven appears to operate normally, but oven heats very slowly.	1. Vari-Cook control turned to WARM. 2. Low line voltage (should be at least 110V AC. 3. Defective solid-state controller. 4. Overheat condition causing thermal protector to open. 5. Defective magnetron. 6. Defective capacitor. 7. Defective power transformer. 8. Broken or loose wire connection.

Table 7-1 cont.

PROBLEM	POSSIBLE CAUSES
Oven appears to operate normally, but does not heat at all.	1. Defective Vari-Cook controller. 2. Defective thermal protector. 3. Defective lower-door interlock switch. 4. Defective power transformer. 5. Defective capacitor. 6. Defective diode. 7. Defective magnetron. 8. Broken or loose wire connection.

VARI-TEMP CONDITION

Before check Vari-Temp condition, check that the cook condition is normal.

Setup: 1) Open oven door. 2) Turn Vari-Cook dial to HIGH. 3) Place a container with 1 pint of water in the center of the oven cavity. 4) Insert the temperature-probe plug into the temperature-probe jack. 5) Place tip of temperature probe into container of water in oven cavity. 6) Close the door.

Normal Operation: 1) Cavity light on. 2) Timer does not operate. 3) COOK light on. 4) Stirrer motor operates. 5) Blower motor operates. 6) Oven heats product placed in oven cavity. 7) When product reaches set temperature, the oven stops cooking and the HOLD indicator light comes on.

PROBLEM	POSSIBLE CAUSES
Cavity light comes on, but no other components operate. Oven does not heat, and HOLD indicator light is not on.	1. Defective Vari Temp switch. 2. Broken or loose wire connection.
Oven will not go into Vari-Temp cook. Cavity light and hold indicator light comes on, but no other components operate and oven does not heat.	1. Defective Vari-Temp relay. 2. Defective Vari-Temp transformer. 3. Defective temperature indicator. 4. Broken or loose wire connection.
Oven doesn't stop cooking when Vari-Temp needle reaches the selected Vari-Temp setting.	1. Defective Vari-Temp relay. 2. Defective temperature indicator. 3. Broken or loose wire connection.
HOLD indicator light doesn't come on at the end of a Vari-Temp cycle.	1. Defective Vari-Temp relay. 2. Defective HOLD indicator light.

sequence for making tests. The tests themselves are given in the sections immediately following. General testing—that is, power-output and microwave-leakage tests—should be made according to the instructions given in an earlier chapter.

High Voltage Components Tests

The following procedures should be followed when little or no heat is produced by the oven, but all other operations, including COOK-light operation, appear normal. This operation should be tested with the oven in a cook condition.

Table 7-2. Troubleshooting the Litton Model 420

OVEN PROBLEMS	
PROBLEM	**POSSIBLE CAUSES**
Oven totally inoperative.	1. Blow oven fuse or tripped breaker. 2. Broken or loose wire connection within power cord or wire harness.
House fuse blows or circuit breaker trips when oven is put into a COOK cycle.	1. Overloaded circuit. 2. Shorted component or shorted harness wire.
Cavity lamp does not come on; other oven operation normal.	1. Loose or burned out cavity lamp. 2. Defective lamp socket. 3. Defective light-control switch. 4. Broken or loose wire connection.
Cavity lamp illuminates with door open, but light goes out when door is closed. With oven programmed to cook, oven will not go into a cook cycle.	1. Defective upper door-interlock switch. 2. Defective lower door-interlock switch. 3. Defective electronic-control panel. 4. Defective light-control switch. 5. Broken or loose wire connection.
Oven stops during a cook cycle and will not operate again until oven has cooled down. **Note**: Thermal protector added to serial No. 3457 and later.	Overheating condition causing thermal protector to open. Check blower motor operation and air flow.
Oven appears to operate normally but heats very slowly.	1. Low line voltage (should be at least 110V AC). 2. Defective electronic-control panel. 3. Defective magnetron. 4. Defective capacitor. 5. Defective power transformer. 6. Broken or loose wire connection.
Oven appears to operate normally but does not heat at all.	1. Defective electronic-control panel. 2. Defective thermal protector. 3. Defective interface controller. 4. Defective power transformer. 5. Defective filament transformer. 6. Defective diode. 7. Defective capacitor. 8. Broken or loose wire connection.

Diode Test. The diode is located on the bottom of the base, near the capacitor. (Fig. 7-7). Since the diode is installed in parallel with the magnetron tube, the diode and tube circuits must be isolated to determine where the problem exists.

Unplug the power cord, remove the outer case, and discharge the HV capacitor. Visually check the HV wiring for proper connections. Isolate the diode from the tube circuit by disconnecting it from the capacitor. With the ohmmeter set on the highest resistance scale, measure the resistance between its two terminals. Reverse the meter leads and again observe the resistance reading.

Table 7-3. Troubleshooting Chart for the Litton Model 650 Microwave Section

PROBLEM	POSSIBLE CAUSES
Microwave totally inoperative; range section operates normally except for cleaning operation.	1. Blown microwave fuse. 2. Defective left interlock switch. 3. Defective lock switch. 4. Defective right interlock switch. 5. Defective microwave timer. 6. Broken or loose wire connection.
Line fuse blows or circuit breaker trips when microwave is put into a cook cycle.	1. Overloaded circuit. 2. Defective interlock-monitor switch. 3. Shorted component or wire harness.
Microwave goes into a cook cycle, but blower motor doesn't operate.	1. Defective CLEAN switch. 2. Defective blower motor. 3. Blower fan binding. 4. Broken or loose wire connection.
Microwave appears to go into a cook cycle, but microwave indicator light does not illuminate and microwave oven does not heat.	1. Defective timer switch. 2. Broken or loose wire connection.
Microwave timer does not advance, but other operation is normal.	1. Defective timer assembly. 2. Binding timer assembly 3. Broken or loose timer assembly.
Microwave indicator light doesn't come on, but other operation is normal.	1. Defective indicator light 2. Broken or loose wire connection.
Microwave cook cycle operates normally, but with selector switch turned to microwave DEFROST, oven doesn't heat at all or heats at full cook instead of defrost.	1. Defective selector switch. 2. Defective defrost cycler switch. 3. Defective defrost timer motor. 4. Broken or loose wire connection.
Microwave goes into cook cycle but heats very slowly.	1. Full microwave power out-put is approximately 300 to 350W. Perform power-output test. 2. Low line voltage (should be at least 110V AC). 3. Defective selector switch, keeping microwave in defrost condition. 4. Problem in HV section (see specific tests for components except the diode) Note: If all component tests indicate normal, replace magentron and check oven for proper operation.
Microwave goes into cook or defrost cycle but does not heat at all.	Problem in HV section (see specific tests for HV components) Note: if all component tests indicate normal, replace magnetron and check oven for proper operation.

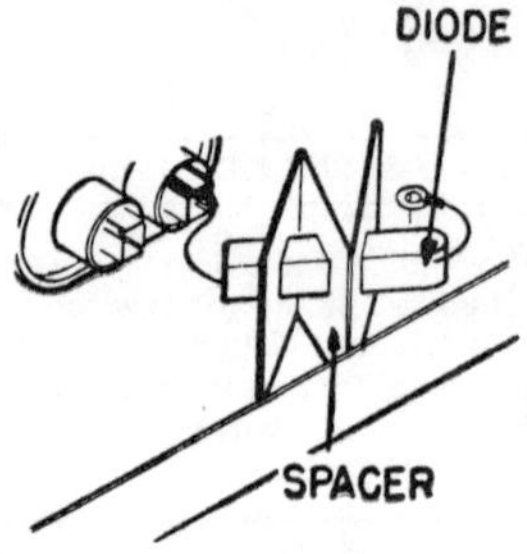

Fig. 7-7. Diode and spacer. Do not remove spacer from diode.

Note: Meters with less than a 6V battery are usually not adequate for checking the front-to-back resistance of the diode. The meter should be checked with a diode known to be good before judging a diode to be defective, since low battery voltage can give an indication of infinite resistance in both directions.

A normal diode should read infinite resistance or open in one direction, and approximately 50K or more when the meter leads are reversed. If the diode test is normal, connect the diode and proceed to *Magnetron Test* in this section. If continuity is indicated in both directions, or if an infinite resistance is read in both directions, the diode is probably defective and should be replaced.

Magnetron Test

1. Unplug the power cord, remove the outer case, and discharge the HV capacitor.
2. Disconnect the HV leads from the filament terminals (Fig. 7-8).
3. Measure the resistance across the two filament terminals (3 and 4) with the ohmmeter set on its lowest resistance scale (R × 1). If a normal resistance of less than 1Ω between magnetron terminals is indicated on the meter, proceed to step 4. If high resistance or infinite resistance is indicated between the magnetron terminals, replace the magnetron assembly and check the oven for proper operation.
4. With the ohmmeter set on the highest resistance scale, measure the resistance between each magnetron-

filament terminal and magnetron chassis (ground). If the ohmmeter indicates infinite resistance, connect the HV lead to the magnetron and proceed to *Capacitor Test*. If continuity is read with the meter, replace the magnetron assembly and check the oven for proper operation.

Capacitor Test

If the capacitor is open, no HV is available to the magnetron. A shorted capacitor normally causes high line current, which should trip the house circuit breaker or blow a line fuse. An ohmmeter can be used to check for a shorted or open capacitor.

Note: The ohmmeter should have a 6V minimum battery voltage and should be set on the high-resistance scale when making a capacitor check.

Unplug the oven, discharge the capacitor, and remove the leads from the capacitor terminals. With an ohmmeter set on the highest scale, measure the resistance between the two capacitor terminals. The meter needle should *momentarily* deflect upward to indicate continuity, then should give an infinite indication. If the ohmmeter indicates continuity

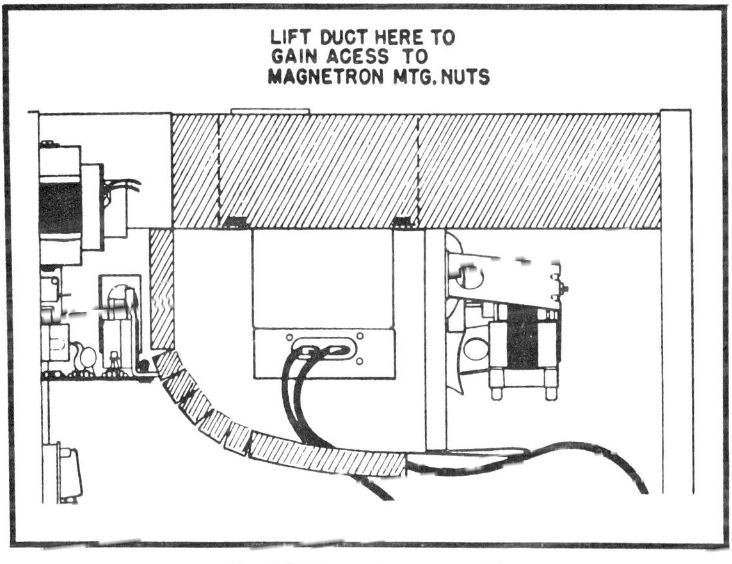

Fig. 7-8. Magnetron assembly.

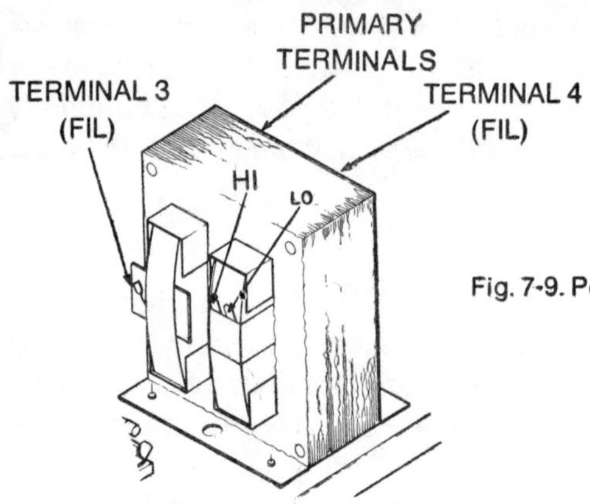

Fig. 7-9. Power transformer.

between the capacitor terminals at all times, or if no meter deflection occurs at all, the capacitor should be replaced.

Power-Transformer Test

Filament-Voltage Test. An operational test can be made in order to determine the amount of filament voltage produced by the power transformer.

Warning: HV up to 6100V AC can be present at the HV terminal of the power transformer during a cook cycle. Observe caution at all times. Proceed as follows:

1. With the power cord unplugged and the outer case removed, discharge the HV capacitor with an insulated screwdriver.
2. Completely remove the HV lead that connects the capacitor to one of the HV terminals marked HI or LO (see Fig. 7-9).
3. With alligator clips, connect a meter capable of measuring up to 10V AC across the filament terminals (3 and 4).
4. Apply power to the oven. Put the oven into a cook cycle to obtain a filament-voltage reading. A normal indication should be approximately 3V AC.
5. Unplug the oven and disconnect the meter leads. If a normal 3V AC reading was indicated on the meter, go on to *Door-Interlock Switch Test*. If no voltage was

indicated on the meter, set the meter to the proper scale to check the 120V AC primary input to the power transformer to see if it is present during a cook cycle. If the input voltage is normal but no filament voltage is present, replace the power transformer.

Transformer-Continuity Test

One side of the power transformer's secondary winding is connected to the oven's chassis ground. The other end of the HV secondary has two terminals marked HI and LO to provide a method of increasing or decreasing the amount of power produced by the magnetron (see Fig. 7-9). Normally, an continuity check of the HV secondary winding will be sufficient to determine the condition of the secondary winding.

With the oven unplugged, discharge the HV capacitor and disconnect the HV lead from the power transformer's HV secondary.

With an ohmmeter set on the low-resistance scale (R × 1), measure the resistance between each HV terminal (marked HI and LO) of the power transformer and chassis ground. The meter should indicate approximately 65Ω between the oven's chassis ground and the transformer HV terminal marked LO, and it should indicate approximately 75 ohms between oven-chassis ground and the transformer high-voltage terminal marked HI.

If extremely low resistance is indicated, or if extremely high or infinite resistance is indicated with the continuity test, replace the power transformer.

Door-Interlock Switch Test

The upper and lower door-interlock switches are mounted on a latch-slide assembly that moves up when the door is closed (Fig. 7-10). As the slide moves up, the switches are activated by the latch hooks mounted on the door assembly. The door-interlock switches are mounted on adjustable strikes, which should be aligned to provide maximum switch activation by the door hooks when the door is closed The switches can be tested with an ohmmeter. Proceed as follows:

1. Unplug the oven power cord, remove the outer case, and discharge the HV capacitor.

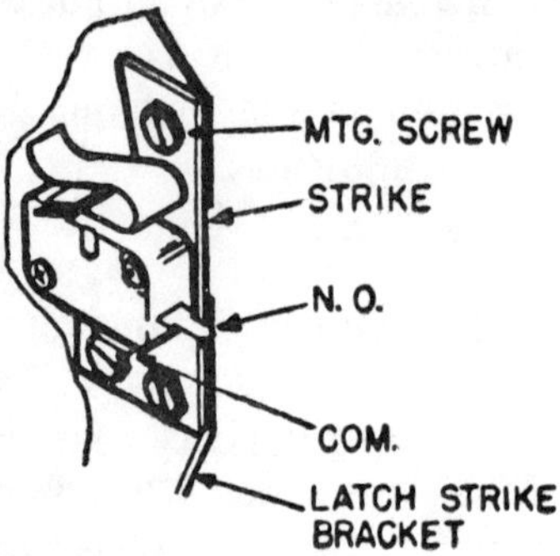

Fig. 7-10. Door-interlock switch.

2. Disconnect the harness leads from the switch terminals.
3. With the oven door closed, use an ohmmeter to check continuity between the switch terminals marked N.O. and COM. Since the switches are activated with the the door closed, the ohmmeter should indicate continuity.
4. Open the oven door and check continuity between terminals N.O. and COM. The ohmmeter should indicate an open circuit with the door opened.
5. If improper indications are given, make the necessary switch adjustments or replace the defective switch. Connect the harness leads to the switch terminals.

Interlock-Monitor Switch Test

The interlock-monitor switch, mounted on a bracket near the cavity rollers, is designed to ensure proper operation of the safety interlock systems. The interlock-monitor switch can be tested with an ohmmeter. Proceed as follows:

1. Unplug the oven power cord, remove the outer case, and discharge the HV capacitor.
2. Disconnect the harness leads from the interlock-monitor switch.
3. With the *door closed*, a continuity check between the switch terminal marked COM and the switch terminal marked N.C. should indicate an open circuit.

4. Open the oven door. A continuity check between the COM terminal and N.C. terminal should indicate continuity with door opened.
5. If improper indications are given, replace the defective switch.
6. Connect the leads to the interlock-monitor-switch terminals.

Light-Control Switch Test

The light-control switch, mounted on a bracket near the cavity rollers, controls the cavity-light operation. The switch operation can be checked with an ohmmeter. Proceed as follows:

1. Unplug the oven power cord, remove the outer case, and discharge the HV capacitor.
2. Disconnect the harness connector from the light-control switch.
3. With the door closed, a continuity check between the switch terminals COM and N.O. should indicate continuity. A continuity check between terminals marked COM and N.C. should indicate an open circuit. See Fig. 7-12.
4. Open the oven door. A continuity check between switch terminals marked COM and N.O. should now indicate an open circuit, and a continuity check between terminals marked COM and N.C. should indicate continuity with the oven door opened.

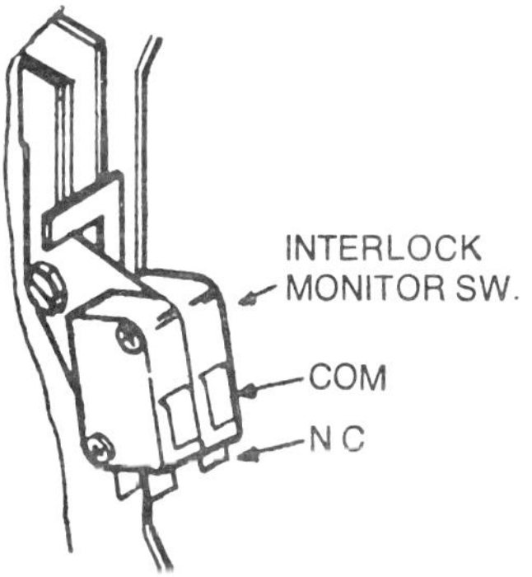

Fig. 7-11. Interlock-monitor switch.

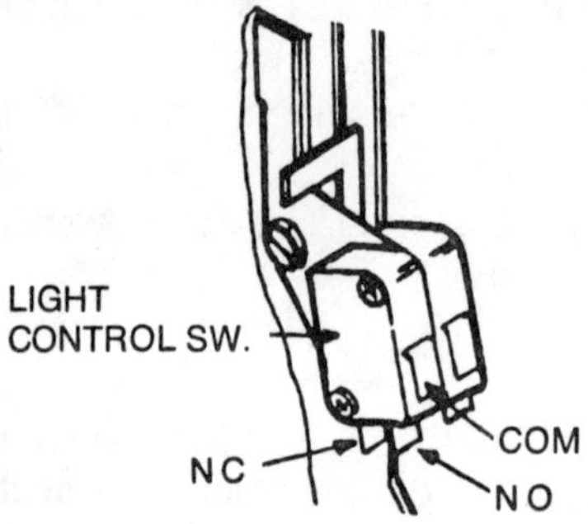

Fig. 7-12. Light-control switch.

5. If improper indications are given, replace the defective switch and reconnect harness leads.

Model 419 Timer Assembly Test

The Model 419 has a digital timer. If the timer motor does not operate with 120V AC applied to the motor leads, the timer should be replaced. The timer switch can be checked with an ohmmeter. Proceed as follows:

1. Unplug the oven power cord, remove the outer case, and discharge the HV capacitor.
2. Disconnect the wire leads from the timer switch terminals.
3. Check continuity between the timer switch terminal marked COM and N.O. (Fig. 7-13). With the timer in the OFF position, the ohmmeter should indicate an open circuit. With the timer turned to some time setting, the ohmmeter should indicate continuity.
4. Replace the timer if defective, and install the harness leads.

Model 419 Solid-State Controller Test

An operational test can be made while bypassing the solid-state controller to determine whether abnormal oven operation is caused by a defective controller.

Warning: HV up to 6100V AC can be present at the HV terminal of the power transformer during a cook cycle. Observe caution at all times.

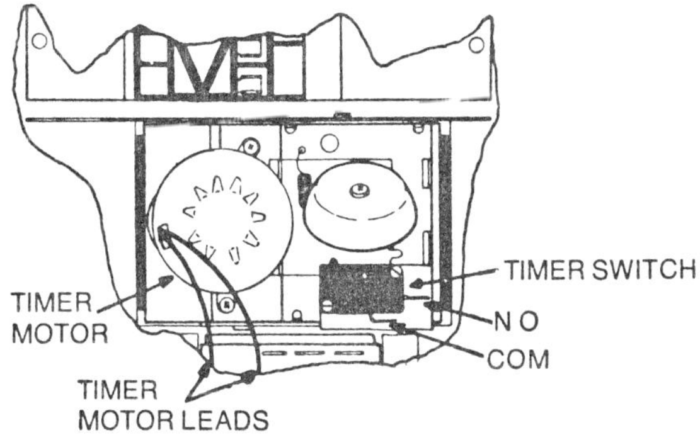

Fig. 7-13. Timer assembly.

Unplug the power cord, remove the outer case, and discharge the HV capacitor. Two types of controllers are used in production. Check Fig. 7-14 for the type of controller installed in oven before performing test.

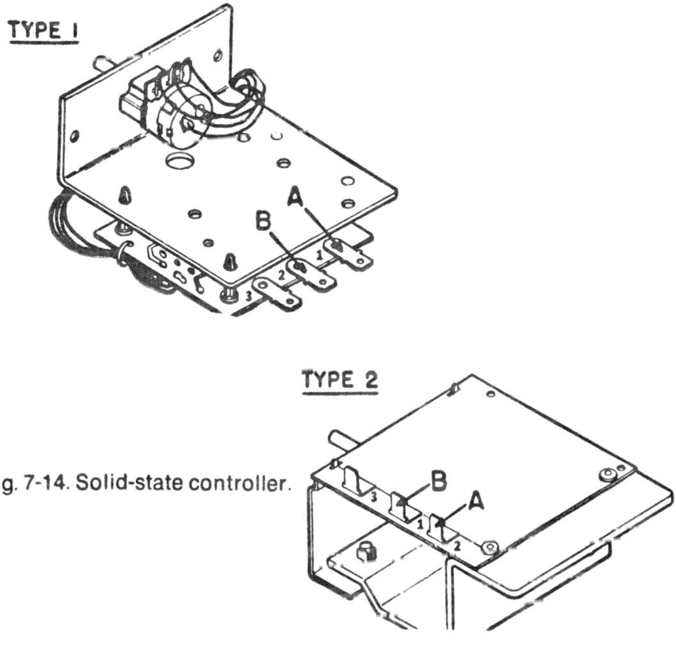

Fig. 7-14. Solid-state controller.

Disconnect the wire leads from terminals *A* and *B*. Connect the wire leads removed from terminals *A* and *B* together with an in-line connector or jumper wire. Insulate the connections, if necessary, to prevent an electrical short.

Perform an operational test to determine if the oven is operating normally at full power output.

- If the oven operates normally with full-power output under these conditions, replace the solid-state controller and test oven for proper operation.
- If the abnormal condition still exists with the controller bypassed, perform tests on other appropriate oven components to determine the cause of the abnormal condition.

Model 419 Timer-Bypass & Vari-Temp-Switch Tests

The timer bypass and Vari-Temp switches, mounted on a bracket near the upper door-interlock switch, control Vari-Temp operation. The switch operation can be checked with an ohmmeter. Proceed as follows:

1. Unplug the oven power cord, remove the outer case, and discharge the HV capacitor.
2. Disconnect the harness leads from the timer-bypass and Vari-Temp switches.
3. With the temperature probe plugged in, a continuity check between the switch terminals COM and N.O. should indicate continuity. A continuity check between terminals marked COM and N.C. should indicate an open circuit.
4. Remove the temperature probe. A continuity check between switch terminals marked COM and N.O. should now indicate an open circuit, while a continuity check between terminals marked COM and N.C. should indicate continuity with the temperature probe removed.
5. If improper indications are given, replace the defective switch and connect the harness leads.

Model 419 Temperature-Probe Test

The temperature probe can be tested with an ohmmeter. With an ohmmeter set on the lowest resistance scale, connect the leads to the temperature probe as shown in Fig. 7-15.

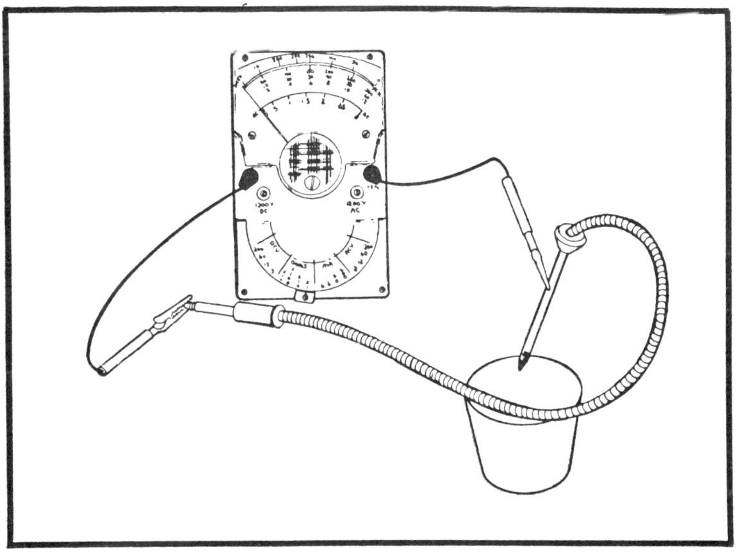

Fig. 7-15. Temperature-probe test.

1. If resistance is indicated on the meter, go on to step 3. If no resistance is indicated on the meter, replace the temperature probe.
2. With the ohmmeter connected to the temperature probe, immerse the tip of the probe into a container of hot water.
3. If the ohmmeter resistance decreases with the probe immersed in the hot water, go on to the *Temperature-Indicator Test*, described later in this chapter. If the resistance does not decrease when the temperature probe is immersed in the hot water, replace the temperature probe.

Model 419 Thermal Protector Test

The thermal protector was added to units with serial number 20425 and later. Unless an overheat condition exists, a continuity check across the thermal-protector contacts should indicate a closed circuit. The thermal opens at approximately 300°F and reset automatically at approximately 260°F.

Model 419 Vari-Temp Relay Test

When the oven is put into a Vari-Temp cycle, the Vari-Temp relay contacts close to provide power to the power

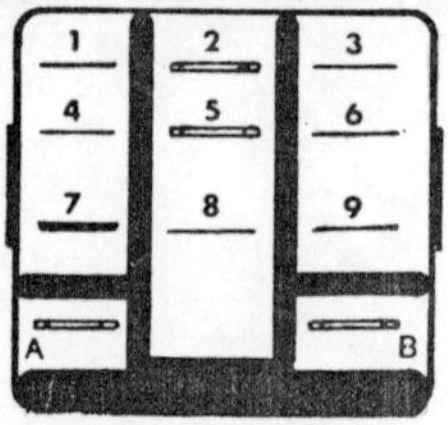

Fig. 7-16. Vari-Temp relay coil.

transformer. At the end of the Vari-Temp cycle, the contacts open to interrupt voltage to the power transformer, while, at the same time, completing a circuit to the HOLD indicator light.

The Vari-Temp relay coil can be checked with an ohmmeter. Unplug the power cord, remove the outer case, and discharge the capacitor. Disconnect the wire leads from terminals *A* and *B* (Fig. 7-16). With an ohmmeter, make a continuity test between terminals *A* and *B*. The ohmmeter should indicate approximately 120Ω. If improper indications are given, replace the defective Vari-Temp transformer/relay assembly; then connect the harness leads.

Model 419 Vari-Temp Transformer Test

An operational test can be made in order to determine the amount of voltage produced by the Vari-Temp transformer.

Warning: HV up to 6100V AC can be present at the HV terminal of the power transformer during a cook cycle. Observe caution at all times.

Unplug the oven power cord, remove the outer case, and discharge the capacitor with an insulated screwdriver. Disconnect the wire leads from terminals *C*, *D*, and *E*, shown in Fig. 7-17.

With alligator clips, connect a meter capable of measuring up to 120V AC across the terminals of the Vari-Temp transformer shown in Fig. 7-17. Apply power to the oven and put the oven into a Vari-Temp condition to get an approximate voltage reading as follows:

Vari-Temp Condition

Contacts	Voltage
A-B	120V
C-D	12V
D-E	12V
C-E	24V

If the proper AC voltage is indicated on the meter, go on to the *Vari-Temp Indicator Test.* If no voltage is indicated on the meter replace the Vari-Temp transformer and relay.

Model 419 Temperature-Indicator Test

An operational test can be made to determine if the Vari-Temp indicator circuit is operating. Proceed as follows:

1. Unplug the power cord, remove the outer case, and discharge the capacitor.
2. Plug the temperature probe into the temperature-probe jack.
3. Place the tip of the temperature probe into a small container of water.
4. Turn the Vari-Temp selector to 150°.
5. Open the oven door.
6. Use alligator clips to connect a DC voltmeter to the test points listed in the following table. Attach the negative (black) voltmeter lead to test point E7 and the positive lead to the indicated test point.

   Warning: Unplug the oven and discharge the capacitor when changing test points, to avoid electrical shock.

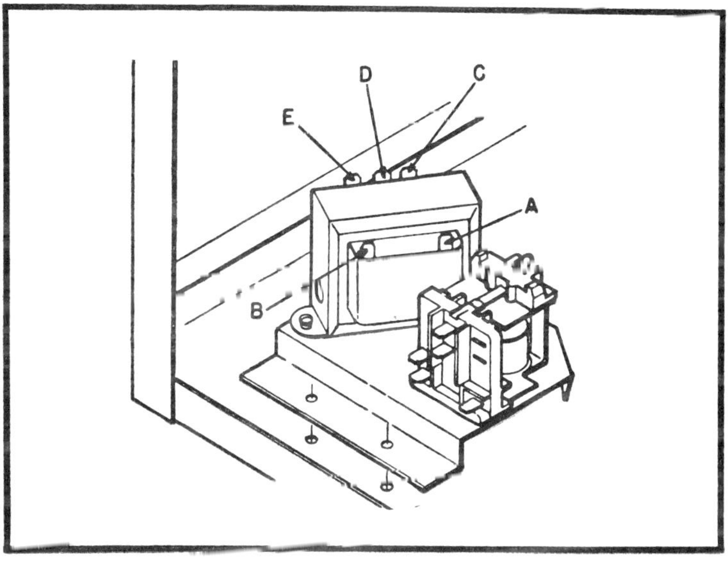

Fig. 7-17. Vari-Temp transformer.

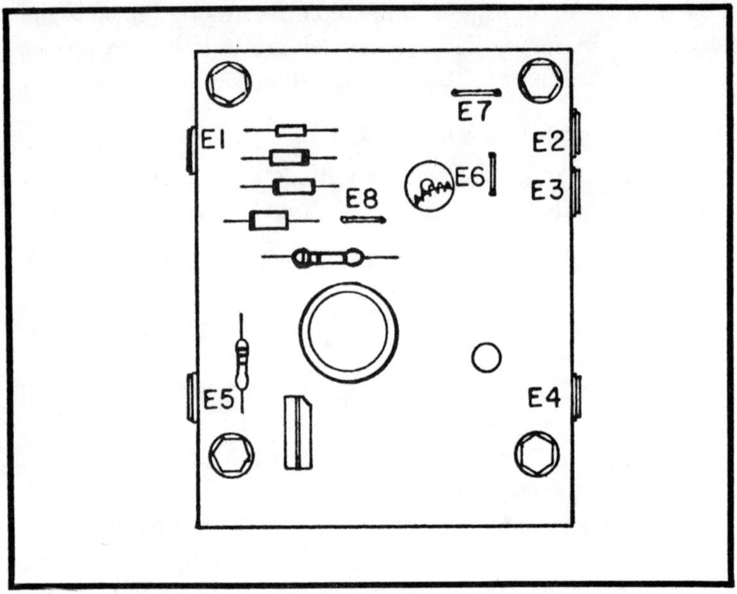

Fig. 7-18. Test points for temperature-indicator test.

Test Points	Approximate Voltage
E5	+12V DC
E4	+7V DC
E1 (Temperature indicator needle below selected temperature setting.)	+12V DC
E1 (Temperature indicator needle reaches selected temperature setting.)	0V DC

If the proper voltage was indicated on the meter as shown in Fig. 7-18, proceed to the *Vari-Temp Indicator Test.*

If improper voltage was indicated on the meter, set the meter to the proper scale. Ensure that the secondary output from the Vari-Temp transformer is present during Vari-Temp cook condition.

Model 420 Interface-Controller Test

Caution: Do not touch oven components or wiring during oven operation. Attach the meter leads with alligator clips when making operational tests. Before touching any oven components or wiring, always unplug the oven from its power source; discharge the capacitor by shorting across the capacitor terminals with an insulated screwdriver.

1. Connect a meter capable of measuring 120V AC to the controller as shown in Fig. 7-19. With the oven door closed and timer off, the meter should indicate line voltage. With the oven in a cycle, the meter should indicate 0V. With the oven door open, the meter should indicate 0V.
   a. If normal indications are given as described, proceed to *High-Voltage Components Tests*.
   b. If abnormal meter indications are given, proceed to step 2.
2. Connect a meter capable of measuring 12V DC to the controller as shown in Fig. 7-20. With the oven in a cook cycle, the meter should indicate 12V DC.
   a. If a normal 12V DC indication is given on the meter, replace the interface controller if step 1 gave abnormal indications.
   b. If abnormal indications are given on the meter for step 2, go on to step 3.
3. Connect a meter capable of measuring 12V DC to the electronic control panel and interface controller as

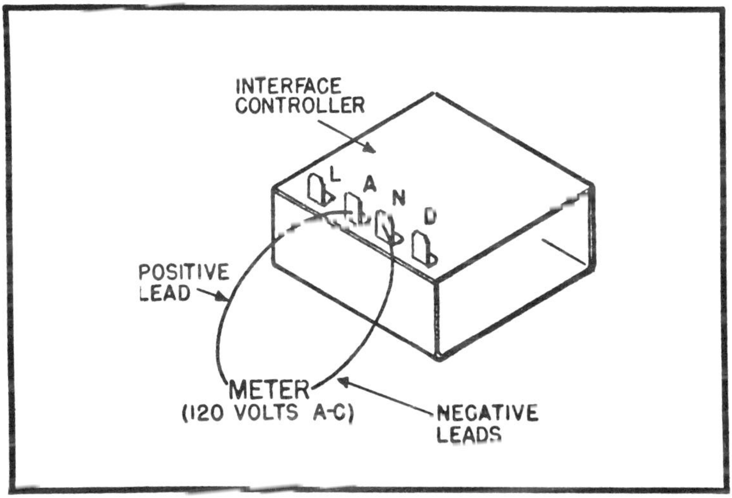

Fig. 7-19. Interface-controller test A.

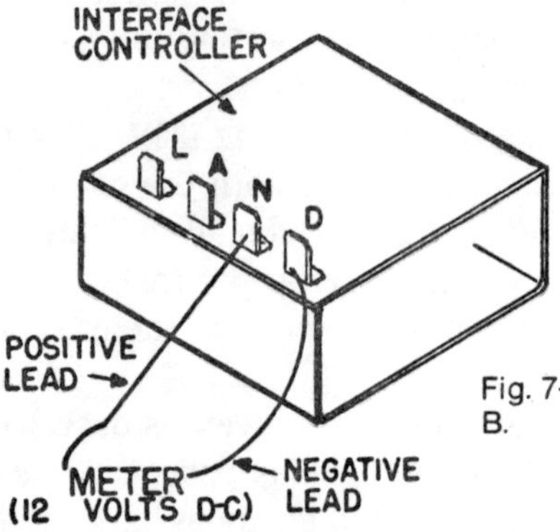

Fig. 7-20. Interface-controller test B.

shown in Fig. 7-21. With the oven in a cook cycle, the meter should indicate 12V DC.

a. If normal indications are given on the meter, check the thermal protector and wiring.
b. If abnormal indications are given on the meter, proceed to *Electronic-Control-Panel Test*.

Model 420 Electronic-Control-Panel Test

Caution: Do not touch oven components or wiring during oven operation. Attach meter leads with alligator clips when making operational tests.

1. Connect a meter capable of measuring 120V AC to the electronic control panel and interface controller as shown in Fig. 7-22. With the door closed, the meter should indicate 120V AC. With the door open, the meter should indicate 0V.
   a. If normal indications are given on the meter and the oven does not heat, proceed to step 2.
   b. If abnormal indications are given on the meter, check the upper and lower door-interlock switches.
2. Move the positive meter lead to point *A* on the electronic control panel (Fig. 7-22). With the oven plugged in and the door closed, the meter should

indicate 120V. With the oven in a cook cycle, the meter should indicate 0V.

a. If normal indications are given on the meter, proceed to step 3.
b. If abnormal indications are given on the meter, replace the electronic control panel.

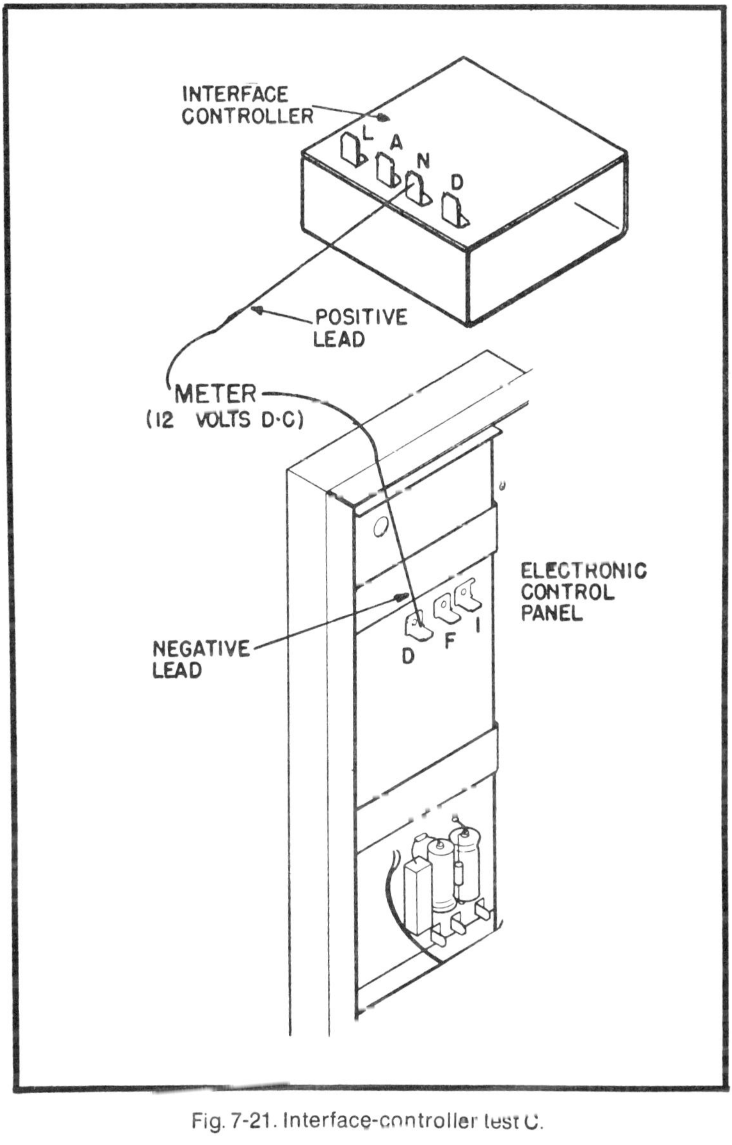

Fig. 7-21. Interface-controller test C.

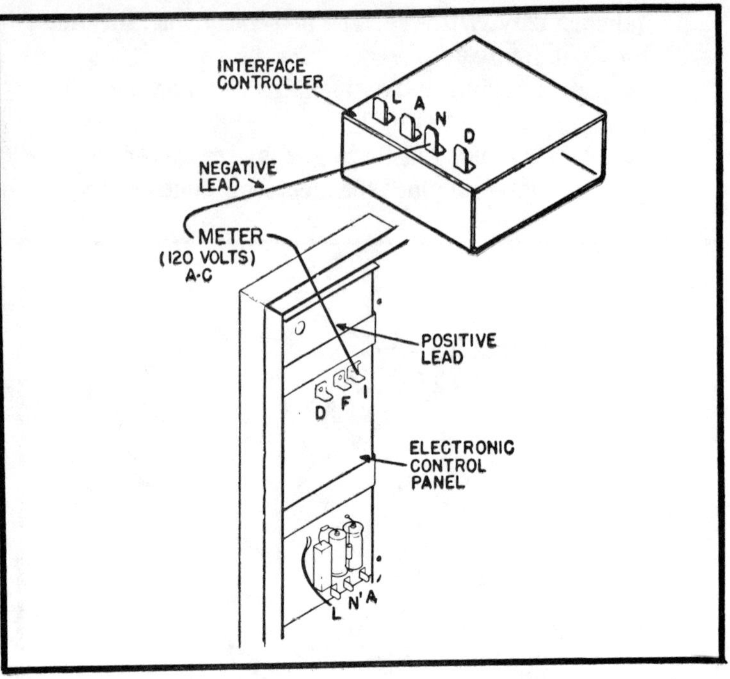

Fig. 7-22. Electronic-control-panel test A.

3. Move the positive lead to point *D* on the electronic control panel (Fig. 7-22). With the oven in a cook cycle, the meter should indicate 12V DC.
   a. If normal indications are given on the meter, refer to *Interface-Controller Test.*
   b. If abnormal indications are given on the meter, replace the electronic control panel.

Model 650 Door-Interlock Switches

The left and right interlock switches are activated by the door-handle hooks when the door-handle is pushed down (Fig. 7-23). When the door-handle is pushed down, an audible click should be heard to indicate the switches are activated. The switches are tested as follows:

1. Place the cook top into service position as follows: Open oven door and remove three decorative-panel screws and two cook-top screws. Pull the cook top forward 2 inches, and prop it up 10 inches.

2. Disconnect wire leads from the switch being tested.
3. With the oven door closed and the door handle pushed down, use an ohmmeter to check continuity between the switch terminals. Since the switches are activated with the door handle down, the ohmmeter should indicate continuity.
4. Lift the door handle up and check continuity between the switch terminals. The ohmmeter should indicate an open circuit with the door handle up.
5. If improper indications are given, make the necessary switch adjustments or replace the defective switch; connect the harness leads to the switch terminals.

Model 650 Interlock-Monitor Switch Test

The interlock-monitor switch, mounted on the right switch assembly (Fig. 7-23), is activated by the door-handle guide. To check the interlock-monitor switch, remove the right switch assembly. With an ohmmeter, check for continuity between the switch terminals. When the switch lever is pushed against the switch body, the meter should indicate an open circuit. When the switch lever is away from the switch body, the ohmmeter should indicate continuity. If improper indications are given, replace the defective switch.

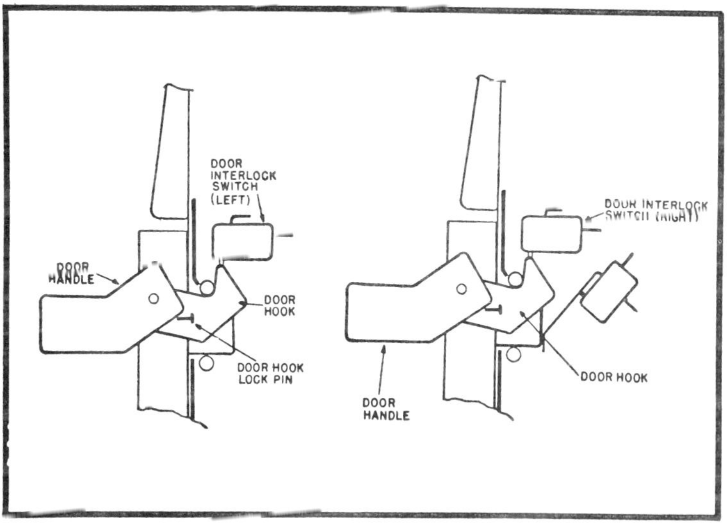

Fig. 7-23. Door-interlock switches—Model 650.

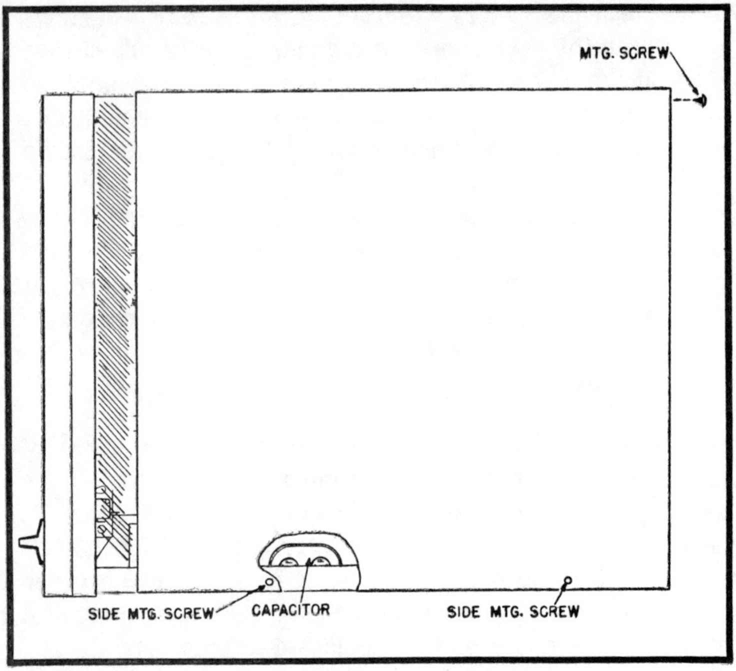

Fig. 7-24. Illustration for outer-case removal.

COMPONENT REPLACEMENT & ADJUSTMENT

This section describes the replacement of certain components of the Litton microwave ovens. The operations relating to the door assembly are especially important; they must be made properly to ensure that the oven operates with the safety designed and built into it.

Outer Case Removal

Turn the timer to OFF and disconnect the power cord. Discharge the capacitor by shorting the two capacitor terminals together with an insulated screwdriver. Raise the rear of the outer case approximately an inch to slide the outer case back. (Refer to Fig. 7-24.) When you are installing the outer case to the oven, be careful to properly align the gasket over the cavity exhaust duct.

Door-Interlock Switch Adjustment

1. Unplug the oven power cord, remove the outer case, and discharge the capacitor.

2. Loosen the three screws that mount the strike assembly to the latch-slide bracket.
3. With the oven door closed, align the strike assembly to the door hook to provide maximum activation of the door interlock switch.
4. Tighten the strike-mounting screws.
5. Check the oven for proper door closure and switch operation.
6. Install the outer wrapper and check the oven for proper operation. Check the oven-door operation for RF leakage around the door.

Stirrer Motor Removal

1. Unplug the oven power cord, remove the outer case, and discharge the capacitor.
2. Remove the stirrer cover.
3. Remove the tinnerman clip from the stirrer shaft. Pull down on the bottom spacer that is mounted directly under the stirrer blades (Fig. 7-25). *Note:* The tinnerman clip and spacer can be twisted downward for easier removal.

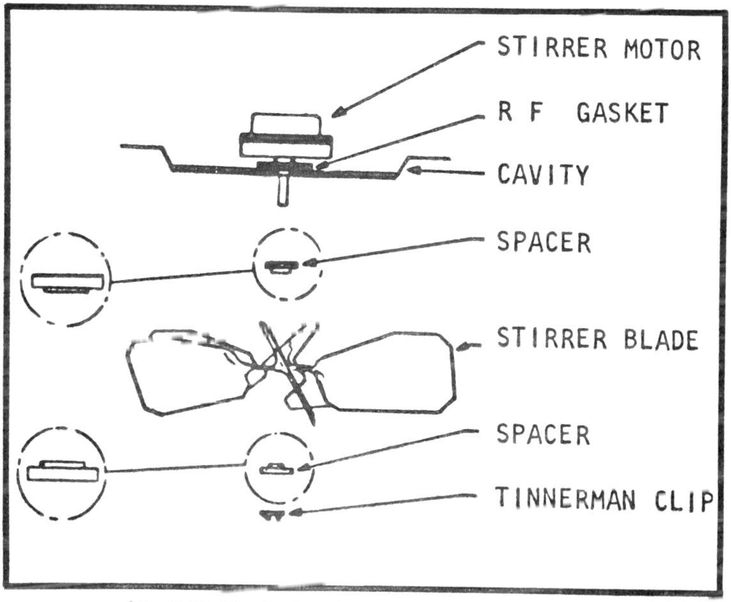

Fig. 7-25. Illustration for stirrer-motor removal.

4. Remove the stirrer-blade assembly, being careful not to bend the blade assembly as it is pulled from the shaft.
5. Disconnect the stirrer-motor leads from the wire harness.
6. Remove the two mounting screws and nuts that hold the stirrer motor to the top of the cavity feedbox.
7. Remove the stirrer motor and shaft.
8. When installing the stirrer-blade assembly to the stirrer shaft, observe the following procedures:
   a. Install the top spacer on the shaft with the smaller side down for blade M35D2, up for blade M35D11.
   b. Install the stirrer blade as shown in Fig. 7-25.
   c. Install the bottom spacer with the smaller side up.
   d. Install the tinnerman clip on the stirrer motor shaft.
   e. Turn the stirrer blade by hand to see that the blade turns easily between the spacers.

Model 650 Door-Interlock Switch Adjustment

The left and right door-interlock switches are activated by the door-handle hooks when the door is closed tightly and the door handle is pushed down (Fig. 7-23). When the door handle is pushed down, an audible click should be heard as the switches are activated.

To adjust, place the cook top in a service position as described earlier. Adjust the door interlock-switch, using one or a combination of the following procedures:

a. Loosen the switch-bracket screws, move the bracket up or down as required, then tighten the mounting screws.
b. Loosen one or both of the door-assembly screws, slide the door up or down, then tighten the mounting screws.

Check for RF leakage around the door.

MICROWAVE-COMPONENT-TRAY REMOVAL

Refer to Fig. 7-26. Proceed as follows:

1. Disconnect the power to the range (circuit breaker or fuse).

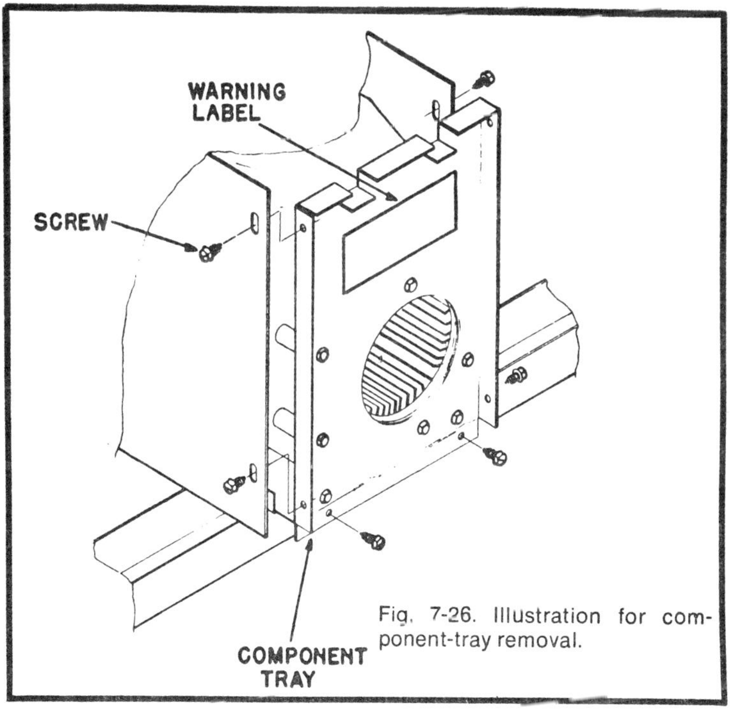

Fig. 7-26. Illustration for component-tray removal.

2. Remove the two kick-panel mounting screws and remove kick panel.
3. Remove the two bottom front-access panel screws and two top front-access panel screws.
4. Remove the six component-tray mounting screws. *Do not pull the component tray out more than 7 inches or wires may be damaged.*
5. Pull the component tray partially from range. Discharge the capacitor by shorting across the capacitor terminals with an insulated screwdriver.
6. Disconnect wire leads from the blower motor and air-sensor switch. Remove the magnetron HV lead from the capacitor.
7. Remove the harness wires from its retainer.
8. Pull the component tray partially out of the range; reach into the opening to disconnect the transformer filament leads from the magnetron
9. Disconnect the primary leads from the transformer. Remove the component tray.

Chapter 8
Servicing Magic Chef Microwave Ovens

This chapter describes the adjustment and servicing of Magic Chef[1] countertop and built-in microwave ovens in the following series: MW-36, MW-56, MW-66, MW-307, and MW-317. Some procedures for some models may deviate slightly from the instructions given here, but the extensive information given here—including complete schematic diagrams for the above models—should enable a qualified electronics or appliance technician to make the necessary adaptations. Exceptions that may cause special problems are noted.

HOW MW-36, MW-56 & MW-66 WORK

The N.C. contacts of interlock-monitor switch S11 are connected from the transformer side of secondary interlock switch S2 N.O. terminal to neutral. (Refer to Fig. 8-1, 8-2, or 8-3.) The door-interlock switches actuate with essentially no door movement but depend on movement of the door latch. When the door is latched, secondary interlock switch S2 closes with primary interlock switch S1. When the door is unlatched, primary interlock switch S1 opens with secondary interlock switch S2. Monitor switch S11 depends on door movement for actuation. As the oven door is closed, a blade on the door opens monitor switch S11 contacts before the interlock switches are

1 All of the illustrations in this chapter are courtesy of Magic Chef, Inc.

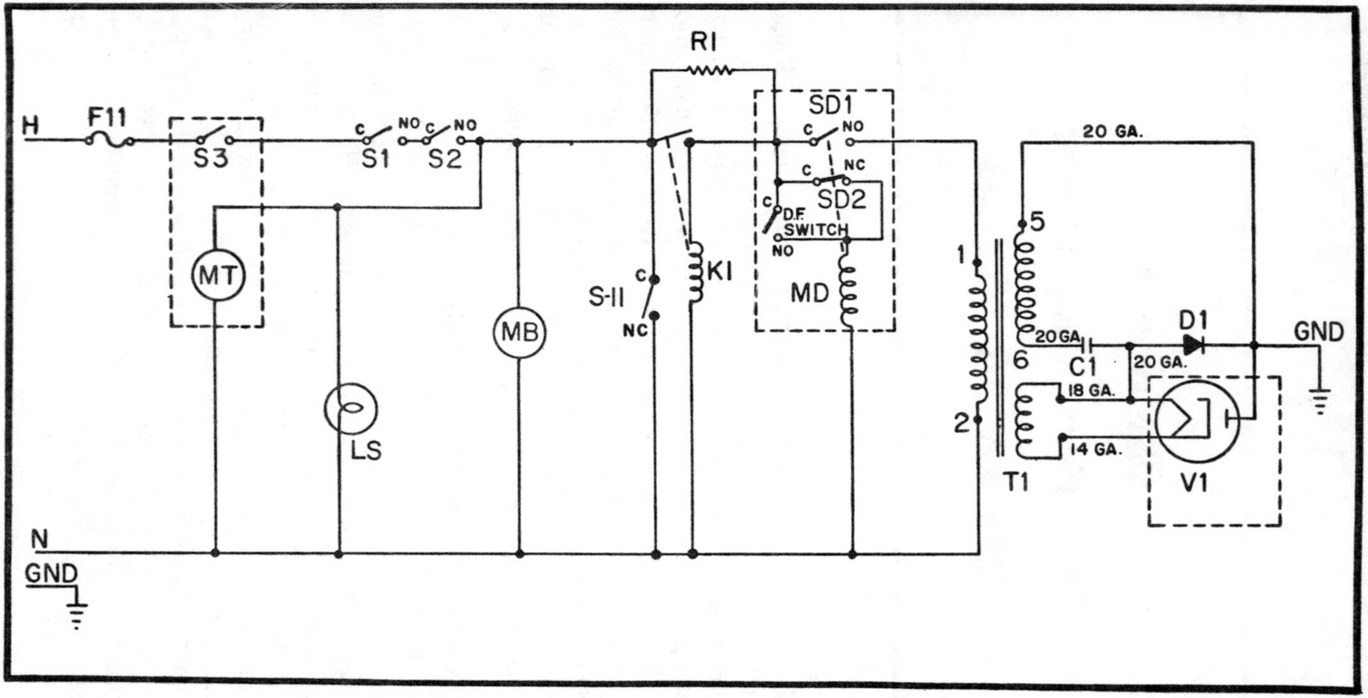

Fig. 8-1. Schematic diagram for Magic Chef Model MW-36.

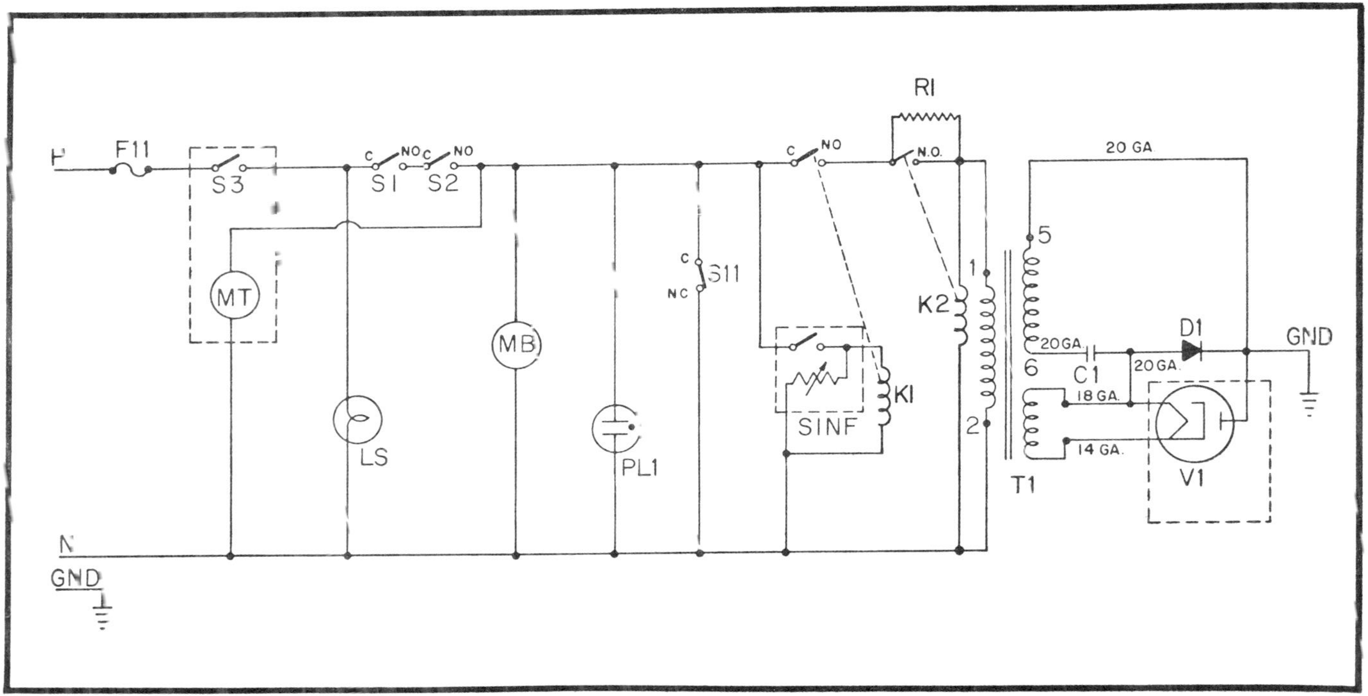

Fig. 8-2. Schematic diagram for Magic Chef Model MW-56.

closed. As the door is opened, the interlock switches open before the monitor-switch contacts close. Voltage is not present across the monitor-switch contacts when its contacts are closed. If, however, the door-interlock circuit has failed, voltage remains present across monitor switch S11 contacts. In the door-closed position the N.C. contacts of this switch are open. As the door is opened, the malfunctioning interlock system causes voltage to remain across the monitor switch contacts. As soon as the monitor switch contacts close, a short circuit then exists through the interlock circuit and monitor switch S11. Fuse F11 will then open to disable the oven.

General Operation

The following discussion is specifically for the MW-36, but it can also be applied to the MW-56 and MW-66 models. With the door open and the timer on or off, no components are energized. With the door closed and the timer *off*, no components are energized. With the door closed and the timer *on*, voltage is applied through interlock-monitor fuse F11, timer switch S3 contacts, and door-interlock switches S1 and S2 N.O. contacts to timer motor MT, oven lamp LS, blower motor MB, the *C* contact of monitor switch S11, through 10Ω surge resistor R1 to the *C* contacts of defrost timer switches SD1 and SD2, and defrost OFF-ON switch *C* contact. Voltage is also applied to the coil of cook relay K1, closing the contacts and bypassing the 10Ω resistor. With the defrost OFF-ON switch in the OFF or open position, voltage is applied through the SD2 contacts to the cycling-switch motor. The timer motor then operates to rotate a cam, which causes SD1 contacts to close and SD2 contacts to open, resulting in a cease of cam rotation. When switch SD1 is closed and switch SD2 is open, (manual defrost switch off), no rotation of the timer cam is necessary for normal operation. Placing the manual defrost switch in the ON or closed position causes defrost-timer motor MB to remain energized, causing defrost-timer contacts SD1 to open and close continuously for the defrost mode of operation. The closing of defrost-timer contacts SD1 applies voltage to the primary winding of HV transformer T1. The secondary windings of T1 energize the HV circuit to operate the oven.

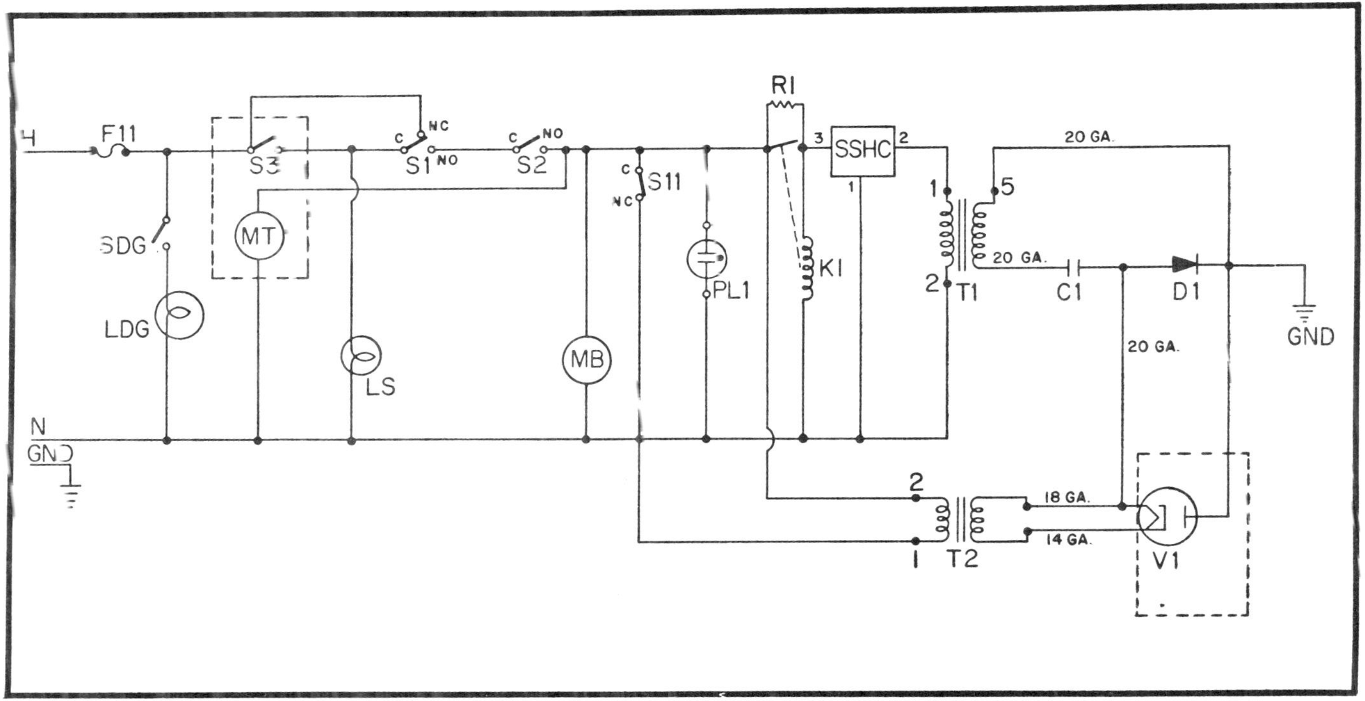

Fig. 8-3. Schematic diagram for Magic Chef Model MW-66.

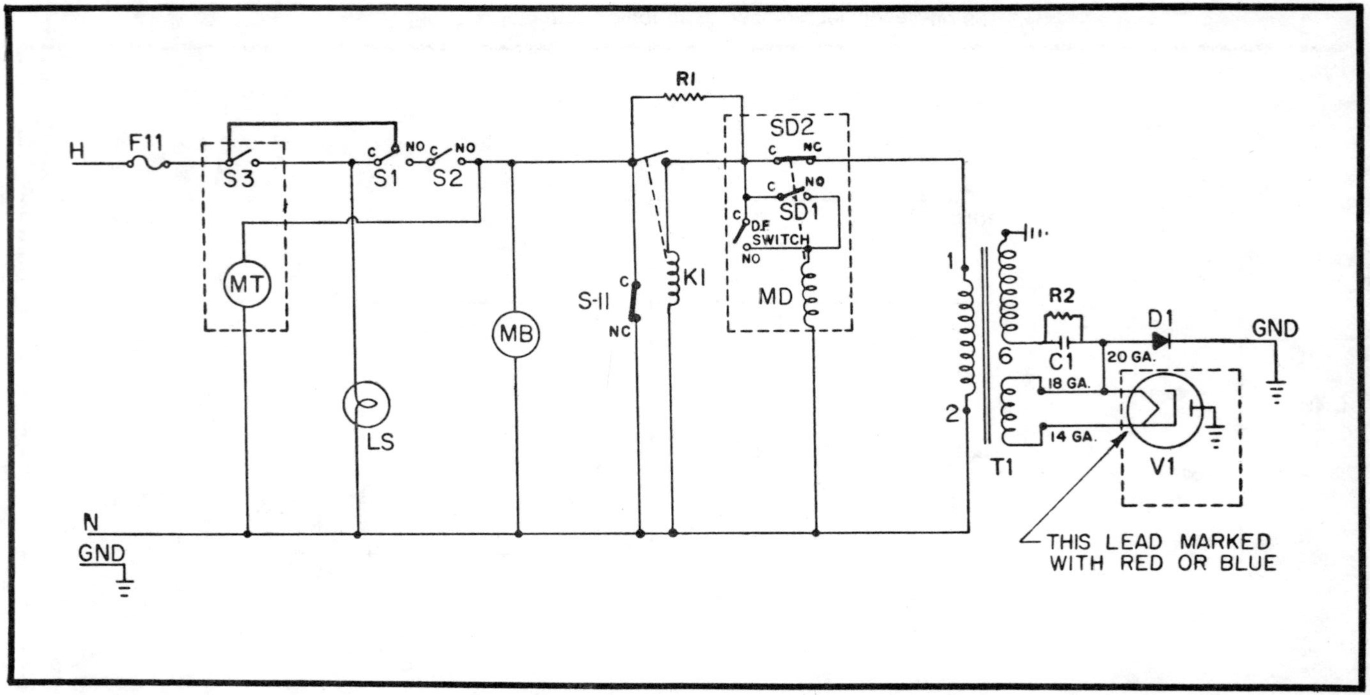

Fig. 8-4. Schematic diagram for Magic Chef MW-333-2 series.

HOW MW-307 & MW-317 WORK

The N.C. contacts of the interlock monitor switch S11 are connected from the transformer side of secondary interlock switch S1 N.O. terminal to neutral. The door-interlock switches actuate with essentially no door movement, but depend on movement of the door latch. When the door is latched, secondary interlock switch S2 closes with primary interlock switch S1. When the door is unlatched, primary interlock switch S1 opens with secondary interlock switch S2. Monitor switch S11 depends on door movement for actuation. As the oven door is closed, a blade on the door opens monitor switch S11 contacts before the interlock switches are closed. As the door is opened, the interlock switches open before the monitor-switch contacts close. Voltage is not present across the monitor switch contacts when its contacts are closed. If, however, the door interlock circuit has failed, voltage remains present across monitor switch S11 contact. In the door-closed position, the N.C. contacts of this switch are open. As the door is opened, the malfunctioning interlock system causes voltage to remain across the monitor switch contacts. As soon as monitor switch S11 contacts close, a short circuit exists through the interlock circuit and monitor switch S11. Fuse F11 will then open to disable the oven.

These two basic microwave oven designs are produced in several versions, such as the -2, -3, -5, -6, and -6P. Specifics for these versions are as follows:

Operation of -2 & -3 MW-300 Series

Door Open, Timer Off or On. Oven lamp LS is energized by voltage applied through F11 to the *C* terminal of timer switch S3 and through the N.C. contacts of interlock switch S1. Refer to Fig. 8-4 or 8-5.

Door Closed, Timer Off. No components are energized.

Door Closed, Timer On. Voltage is applied through interlock monitor fuse F11, timer switch S3 contacts, and door-interlock switches S1 and S2 N.O. contacts to timer motor MT, oven lamp LS, blower motor MB, the *C* contact of monitor switch S11, through 10Ω surge resistor R1 to the *C* contact. Voltage is also applied to the coil of cook relay K1, closing the contacts and bypassing the 10Ω resistor. With the defrost OFF-ON switch in the OFF or open position, voltage is applied

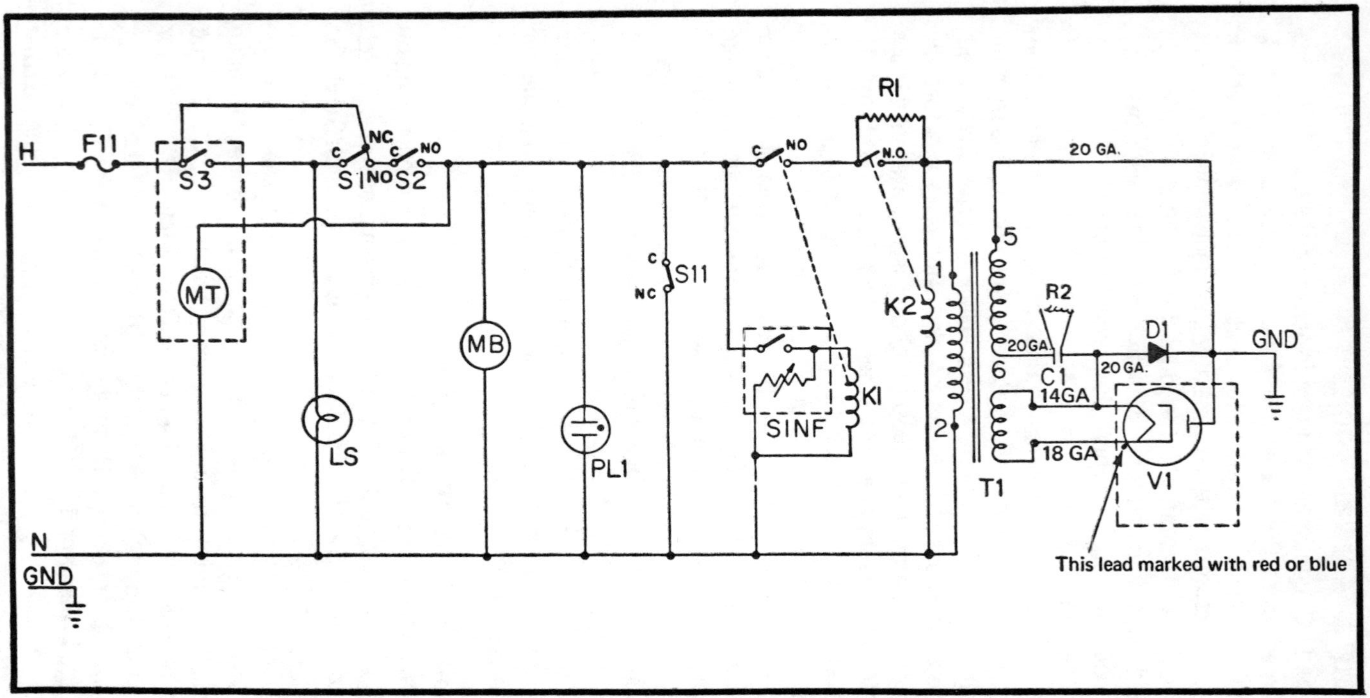

Fig. 8-5. Schematic diagram for Magic Chef MW-300-3 series.

through SD2 contacts to the cycling-switch motor. The timer motor then operates to rotate a cam, which causes SD1 contacts to close and SD2 contacts to open, which causes cam rotation to cease. When switch SD1 is closed and switch SD2 is open (manual defrost switch off), no rotation of the timer cam is necessary for normal operation. Placing the manual defrost switch in the ON or closed position causes defrost timer motor MB to remain energized, causing defrost timer contacts SD1 to open and close continuously for the defrost mode of operation. The closing of defrost timer SD1 contacts applies voltage to the primary winding of HV transformer T1. The secondary winding of T1 energizes the HV circuit to operate the oven.

Operation of -5 Series

Oven Connected to Voltage Source. Defrost guide lamp LDG is energized by voltage applied through the interlock monitor (Fig. 8-6). Fuse F11 and momentary switch SDG are held in the closed position.

Door Open Timer Off or On. Oven lamp LS is energized by voltage applied through F11 to the *C* terminal of timer switch S3 and through the N.C. contacts of interlock switch S1.

Door Closed, Timer Off. No components are energized.

Door Closed, Timer On. The oven lamp is energized by voltage applied through fuse F11 and closed timer switch S3. Voltage is then applied through the N.O. contacts of door-interlock switches S1 and S2 to energize blower motor MB, pilot lamp PL, magnetron filament transformer T2, and timer motor MT. Voltage is also applied to the *C* contact of the monitor switch and through 10Ω surge resistor R1 to the coil of cook relay K1, energizing the relay and closing the relay contacts, bypassing resistor R1. Voltage is then applied to terminal 3 of solid-state pulse control SSHC. The voltage output of the solid state control, terminal 2, may be continuous or may be of a pulsed nature, depending on the setting of its variable control. The output voltage is applied to HV transformer T1's primary winding. The secondary winding of T1 supplies the HV circuitry.

Operation of -6 & -6P Series

The N.C. contacts of the interlock monitor switch S4 are connected from the transformer side of primary interlock switch S2 N.O. terminal to neutral. Refer to Figs. 8-7 and 8-8.

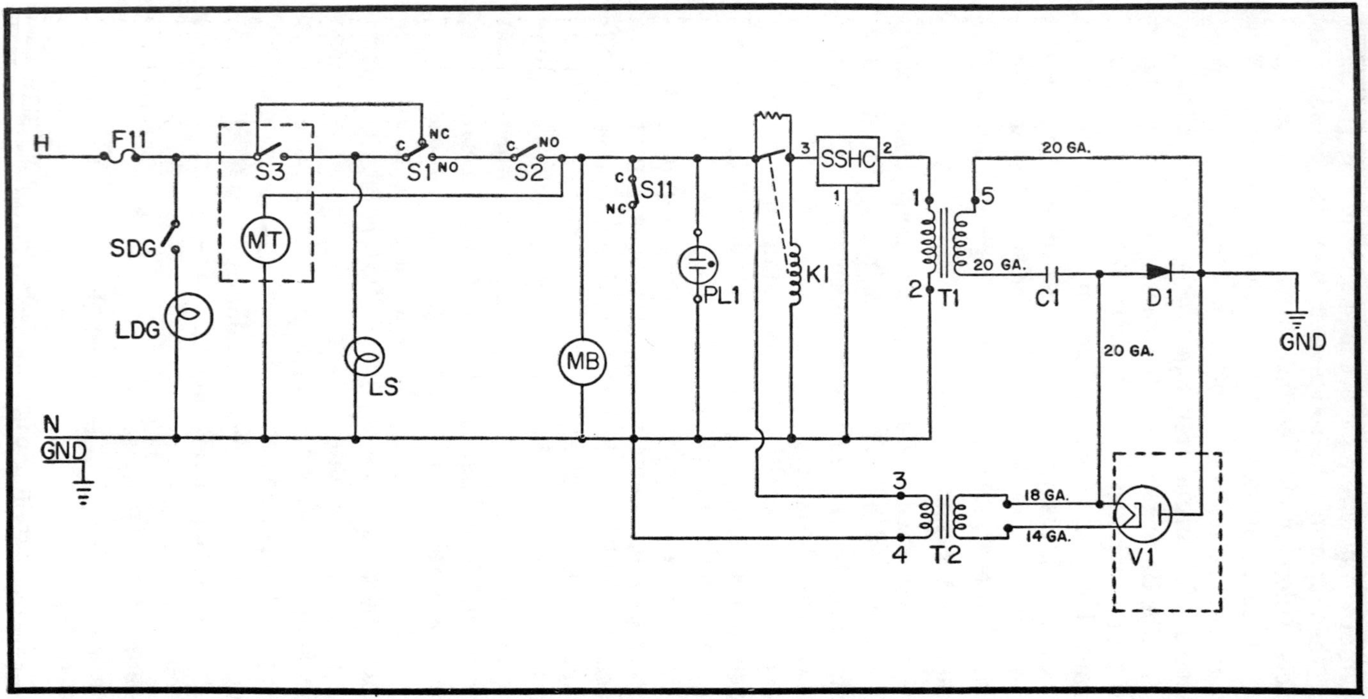

Fig. 8-6. Schematic diagram for Magic Chef MW-300-5 series.

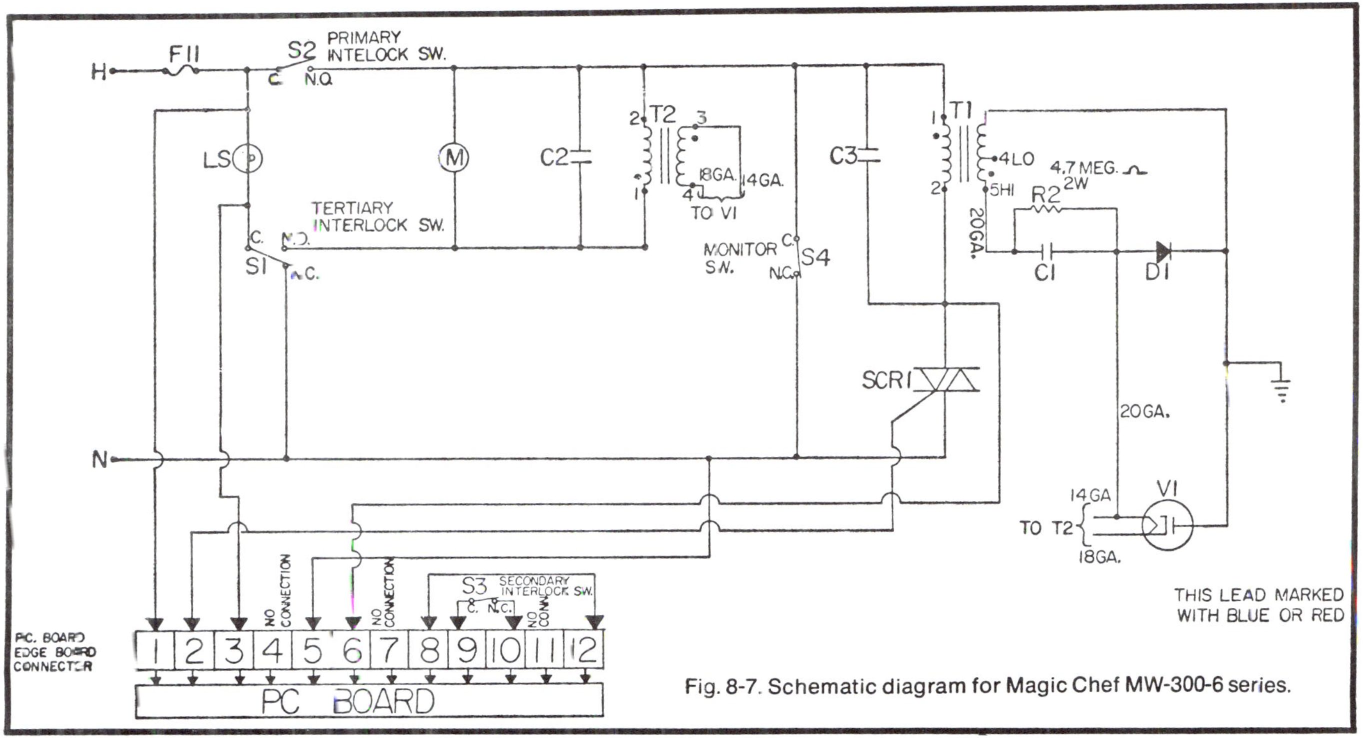

Fig. 8-7. Schematic diagram for Magic Chef MW-300-6 series.

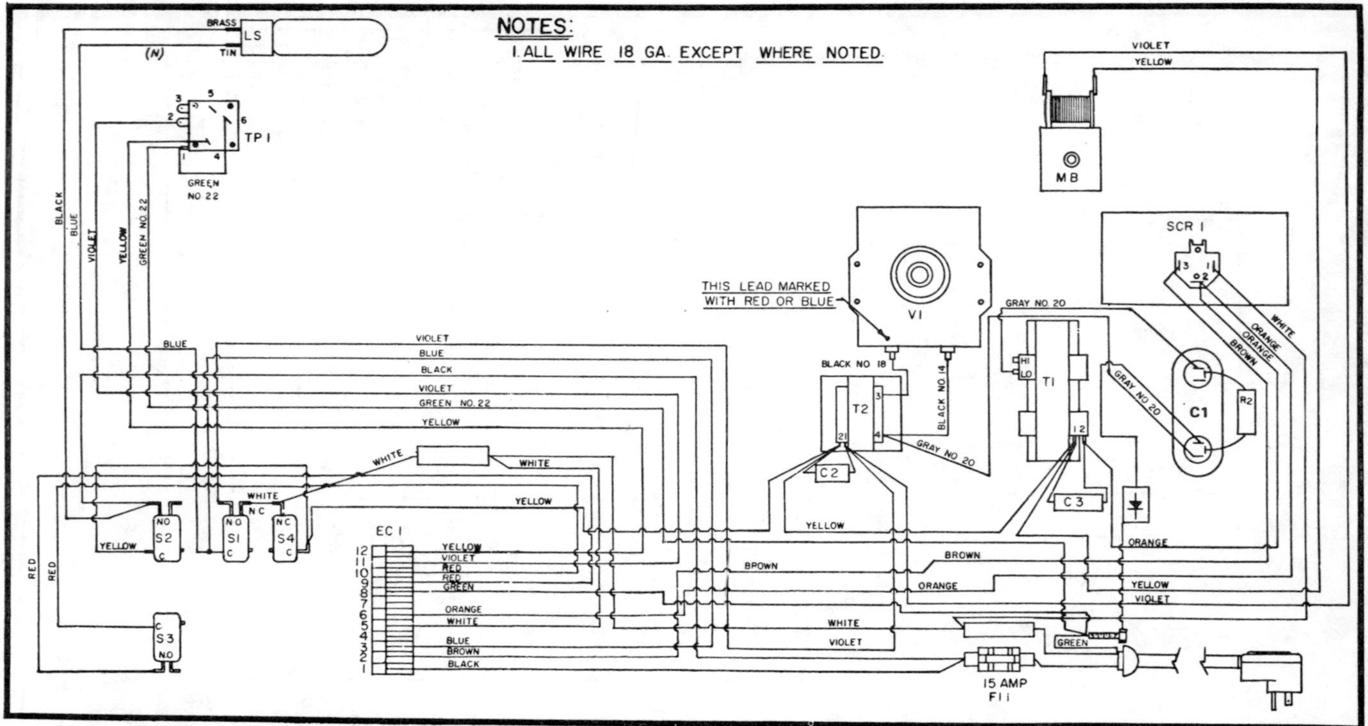

Fig. 8-8. Wiring diagram for Magic Chef MW-300-6P series shows physical appearance of components of this and other Magic Chef models.

The door-interlock switches actuate with essentially no door movement, but depend on movement of the door latch. When the door is latched, secondary interlock switch S3 closes at the same time as primary interlock switch S2. When the door is unlatched, primary interlock switch S2 and secondary interlock switch S3 open before tertiary interlock switch S1. Monitor switch S4 depends on door movement for actuation. As the oven door is closed, a blade on the door opens monitor switch S4 contacts before the primary interlock switch is closed. As the door is opened, the primary interlock switch opens before the monitor switch contacts close, so no voltage is across the N.C. contacts. If, however, the door interlock circuit has failed, current would pass through monitor switch S4 contacts when the door is open and the N.C. contacts close. As soon as monitor switch S4 contacts close, a short circuit exists and fuse F11 will open to disable the oven.

TROUBLESHOOTING & TESTING

Make certain the oven in unplugged before performing any continuity checks. Using an insulated screwdriver, momentarily short together the terminals of the capacitor to discharge it. Always disconnect the wiring to any component before testing it.

For general troubleshooting instructions, refer to Table 8-1. For troubleshooting the touch-programming panel and solid-state board of models so equipped, follow the logical program listed below.

Caution: Do not touch the solid-state board or the contact fingers of the board. Static electricity from the body can damage the board beyond repair.

Proceed as follows:

1. Oven does not operate properly.
2. If LED readout is lighted, go to step 11.
3. Verify 102–132V AC and a good ground connection at the receptacle.
4. Unplug the oven.
5. If the oven fuse blown, go to step 13.
6. Unplug and replug the oven.
7. If display reads *12:00*, go to step 16.
8. Check seating of edgeboard connector.

Table 8-1. Troubleshooting Guide for Magic Chef Microwave Ovens

SYMPTOM	POSSIBLE CAUSE	REMEDY
Oven lamp does not light, although blower runs.	Defective oven lamp.	Replace.
Blower doesn't run, although oven lamp lights.	Defective blower motor.	Test with ohmmeter; should read about 12Ω.
Blower and oven lamp stay on with door closed and timer not on; red indicator lamp is on.	Timer switch S3A stuck closed.	Check continuity across switch. Operate the timer back and forth to zero. When zero is reached, the meter should read infinity If meter reads zero, switch is stuck shut. Replace entire timer.
With door closed and timer set, indicator lamp doesn't light and oven doesn't cook or time out.	Door interlocks S1 and S2 not closing; Timer switch S3B not closing.	Test as outlined under **Interlock Tests** Check continuity of S3A (the top switch on the timer assembly).
Oven cooks and indicator lamp lights, but timer doesn't time out and shut off oven.	Defective timer motor MT; Timer knob binding.	Check continuity of MT; Should be about 1250Ω Adjust.
Indicator lamp lights, but oven does not cook.	1. Defective diode CR1 If diode is shorted, magnetron may be defective and may cause replacement diode to short.	Check resistance across diode. Reverse meter leads and check resistance again. Should read infinit; one way, and 50K – 200K the other way.
	2. Defective magnetron V1.	Check plate current as outlined in Plate Current Test.
	3. Defective HV transformer T1.	Check continuity as follows: Primary terminals 1 and 2; secondary terminals 3 and 4; secondary terminal 6 to ground.
	4. Defective capacitor C1.	Check with ohmmeter on high (R × 10) resistance scale. Meter should arc toward zero, then drift back toward infinity. Reverse leads and same results should be noted.
Unit fails to operate, or blows fuse on settings other than HIGH on MW-300-5 series (solid-state control).	Defective solid-state heat control SSHC.	Put jumper wire between terminals 3 and 2 on the solid-state heat control, and if the unit operates, it indicates the solid-state pulse control is defective. After replacing solid-state control and fuse, test the unit with control set on HIGH, then on lower heats also.

9. Replace PC board assembly and check it for proper operation.
10. Go to step 51.
11. If automatic thermometer is plugged in, go to step 18.
12. Go to step 6.
13. Replace the fuse and four interlocks and check oven for proper operation.
14. If oven fuse blows, go to step 20. (Do not jumper fuse.)
15. Go to step 6.
16. If the display increments from 12:00 to 12:01, go to step 25.
17. Go to step 9.
18. Check the resistance of the automatic thermometer (at room temperature); resistance reading should be 45 to 65 K.
19. Go to step 6.
20. Check interlock switches and monitor switch for proper operation.
21. Check polarity of AC receptacle and the presence of a good ground.
22. Check HV circuit, including diode, capacitor, and magnetron.
23. Check low-voltage circuit, including blower motor.
24. Go to step 6.
25. If all pads on the touch panel operate properly, go to step 29.
26. Ensure that each metal finger makes contact with the rear of the touch panel. See previous *Caution* note.
27. Replace the touch panel and PC board.
28. Go to step 51.
29. Program a cook sequence (with a water load). Check for 120V across the transformer's primary leads.
30. If the magnetron duty cycles, go to step 36.
31. Unplug the oven.
32. Check the seating of the PC edgeboard connector.
33. Check the quick-connects on the rear (power) triac.
34. Replace the rear (power) triac heatsink assembly and test for proper operation.
35. Go to step 25.
36. If the cavity light and blower motor operate during cooking, go to step 40.

37. Check the seating of the PC edgeboard connector.
38. Replace the control.
39. Go to step 51.
40. If the end-of-cycle tone beeps at end of the cook sequence, go to step 43.
41. Replace the control.
42. Go to step 51.
43. Insert the automatic thermometer into its jack.
44. If the colon (:) goes out, go to step 49.
45. Check the automatic thermometer jack (including the jack switch) for proper operation.
46. Check the seating of the PC edgeboard connector.
47. Replace the control.
48. Go to step 51.
49. Immerse at least one-half of the automatic thermometer shaft into the water load and program a cooking sequence, using the temperature mode of operation.
50. Go to step 18.
51. If cooking stops when the door is opened, go to step 53.
52. Check the interlock switches and monitor switch for proper operation.
53. Check for oven leakage.

Door Adjustments

The oven door and its hinge are factory adjusted and should require no further attention unless the entire door assembly is removed or replaced. To adjust the oven door assembly and hinge, proceed as follows:

1. Open the door to its fully open position.
2. Loosen the eight screws that secure the door hinge to the oven front frame.
3. Move the oven-door assembly so that its edge is parallel to, and just even with, the left edge of the front frame.
4. Tighten one top and one bottom screw snugly.
5. Close the oven door and check the top edge of the door assembly so it is paralled to, and even with the top edge of the front frame.

6. Adjust as necessary and tighten all screws in the hinge.
7. Close the oven door and note how the door latches. Latching should be smooth and almost effortless. If the door binds on the hinge side (as evidenced by the door being difficult to close and springing open as soon as the latch is tripped), then the door hinge must be adjusted outward slightly.
8. Once the door is properly adjusted, the oven should be tested for microwave leakage as described in Chapter 3.

Oven Cabinet Removal

Be certain that the oven is unplugged before attempting to disassemble. Remove the two screws from the bottom flange of the cover, and the seven screws from the back flange. Slide the cover back to free the notched front edge from the front frame. Spread the sides slightly from the bottom and lift the cover off.

Interlock Tests

Make certain that the oven is unplugged from AC line for the following tests and that the capacitor is discharged.

Disconnect the leads from the *C* terminal of the interlock switch. Set an ohmmeter to the low (R × 1) resistance scale. Connect the meter to the *C* and N.O. terminals of the switch to be tested. Open the oven door and note that the meter now registers infinity. Repeat the test with door closed and meter should read zero. If meter does not register infinity with door open, disconnect all leads to the switch in question to retest as above. If the switch still registers less than infinity with the door open, switch must be replaced.

Magnetron Plate Current Test

To check the magnetron plate current, it is necessary that you have a 10Ω, 10W wirewound resistor with lead wires attached as shown in Fig. 8-9. These resistors are available from your Magic Chef parts distributor at a nominal cost. To order, ask for a *plate-current test kit*. Do not attempt to check the magnetron plate current without this resistor.

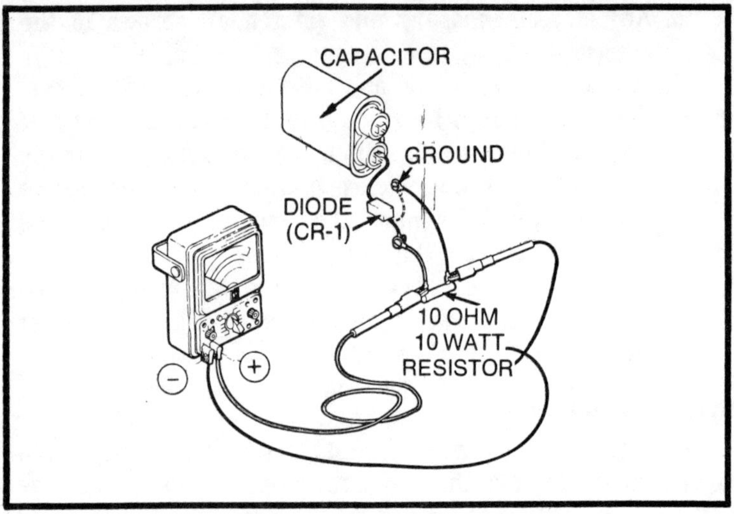

Fig. 8-9. Hookup for plate-current test.

Additionally, do not attempt to check magnetron plate current without following the following instructions:

1. Unplug the power cord.
2. Using an insulated screwdriver, short both terminals of the capacitor to each other momentarily.
3. Remove the ground lead from diode CR1 as shown in Fig. 8-10.
4. Attach the 10Ω, 10W resistor and diode to ground as shown in Fig. 8-9.
5. Attach an ammeter set to the 500mA DC range. Be certain that the polarity is correct.
6. Plug the power cord in.
7. Place a cup (8 ounces) of water into the oven cavity.
8. Close the oven door.
9. Activate the timer (touch panel on -6 series).
10. The meter should read between 280 and 350 mA. Be sure to use the correct scale on the meter (read the 0–to–50 *scale* and multiply by 10). The precise level is not important for proper operation. A magnetron tube should not be replaced merely because of a reading of 250 to 280 mA is obtained. The plate-current reading is only an indication of proper or improper operation. Watch for signs of fluctuating plate current

or a defective magnetron. If a reading much higher than 400 mA is obtained, the magnetron is possibly gassy or shorted and should be replaced. If a reading of about 150 to 200 mA is obtained, the magnetron is not oscillating and should be replaced. If the reading varies between 100 and 400 mA, with quite a bit of fluctuation, then the magnetron is *moding* and should be replaced.

11. When the test is complete, open the oven door.
12. Unplug the oven.
13. Discharge the capacitor as described in step 2.
14. Remove the 10Ω, 10W resistor.
15. Connect the ground lead of the diode.
16. Be sure the RF gasket is in place before installing magnetron (see Fig. 8-10).

Interlock Monitor Fuse F11 Failure

1. Replace both interlock switches (primary S2 and secondary S1), the interlock monitor switch (S11 or S4), and the interlock monitor fuse (F11); check that the timer switch (S3) is operating correctly and that

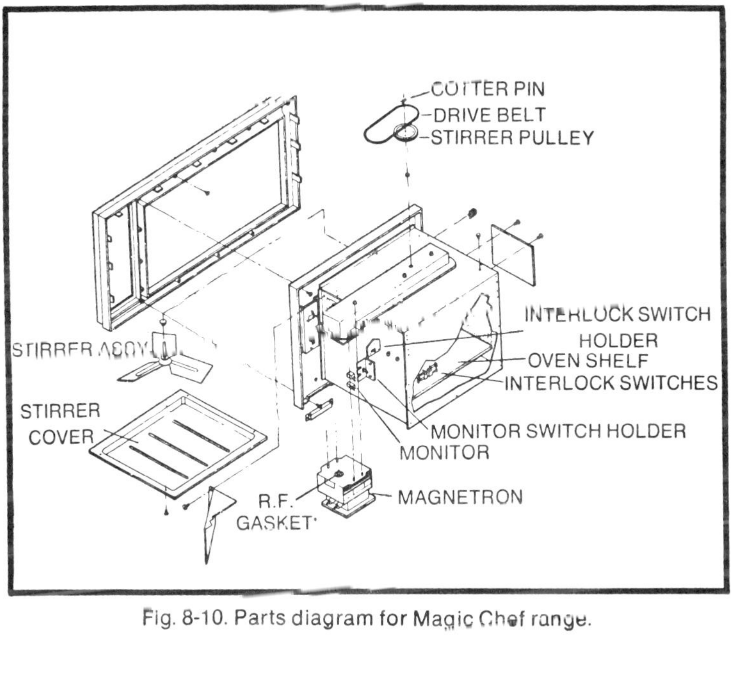

Fig. 8-10. Parts diagram for Magic Chef range.

the timer times out, causing the timer switch (S3) to open, If either the timer switch (S3) or the timer does not operate correctly, then replace the entire timer assembly also.

2. The component locations are shown in Fig. 8-10.
3. Run an RF leakage test and interlock-function test.
4. Return the switches, fuse, and timer if necessary to Magic Chef, Inc.

PARTS DESCRIPTION, LOCATION & TESTING

Primary and secondary interlock S1 and S2 are located on back of the oven front frame and are operated by the door latch. These switches assure that the oven cannot be operated with the door open. To test, refer to *Interlock Test.*

In replacing faulty switches, be sure that switch-holding tabs of nylon holder are not bent, broken, or otherwise deficient in their power to hold the switches. If they are deficient, replace the holder as well as switch. See Fig. 8-10. Exercise caution in removing and replacing terminals on switches to avoid breakage of switch housing.

Interlock Monitor S11 and Sr. This switch (Fig. 8-10) is operated by the blade in the door located just below the door latch. It ensures that the oven cannot be operated with the door open. It causes a short circuit should the other interlock system malfunction, thus blowing fuse F11.

Timer MT and S3. The timer is located on the front control panel (Fig. 8-11). It is a motor-driven switch assembly (S3). Setting the timer knob causes switch S3 to close, and the switch open when the knob returns to zero. To test, check the continuity across the switch with timer knob set to a given time. A resistance reading across timer motor MT should be about 1250Ω.

Heat Controls These controls enable the microwave oven to do far better cooking and defrosting operations. The -2 series has a defrost heat control providing equalizing (or heat transfer) by cycling energy on and off at a fixed rate. It has two settings: HIGH and DEFROST. To test, check the switches for continuity. When the defrost motor is energized, it has to complete its cycle even if the defrost switch is set back to OFF.

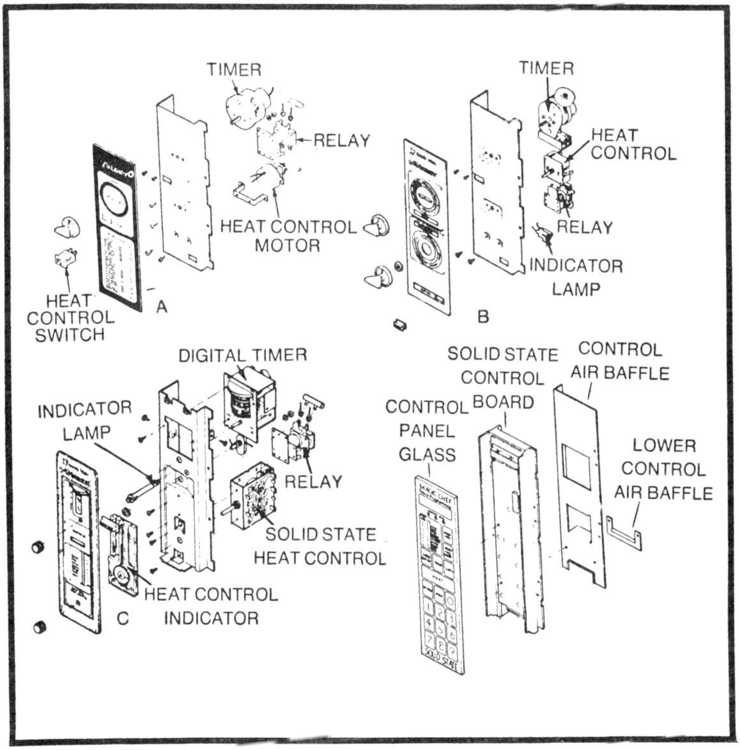

Fig. 8-11. Control panels of Magic Chef microwave ovens.

The -3 series has a variable heat control with numbers from 1 to 5, plus HIGH. These settings vary the percentage of *on* to *off* time, thereby changing the rate at which foods cook. To test it, check the continuity across the switch (H1 and H2) with the switch in the ON position. Then plug the unit in and set control in cycling position to see if the relay cycles off and on.

The -5 series has a solid-state heat control, making the cooking cycles faster and giving better control. Solid-state control also provides an infinite number of settings from 1 to 10 with shorter and more frequent pulses of energy to give more precise cooking control. To test it, proceed as follows: If the unit fails to operate, put a jumper wire between terminals 3 and 2 on the solid-state control; if the unit operates, the solid-state pulse control is defective.

The -6 series has a solid state touch-programming panel, giving the speed and control of solid-state cooking and much more. The microprocessor controls timing (cooking time and a TOD clock), the magnetron-filament transformer, the oven-cavity light, the blower motor, and turn-on time. It also controls an automatic temperature-probe circuit on the -6P model. To test it, refer to the troubleshooting section of this chapter.

Relay K1. The relay is mounted to the back of the control panel (Fig 8-11). To test it, visually check operation of K1 to see if it is closing during a cook cycle. Check the continuity through K1 contacts while manually holding relay closed.

Blower Motor MB. This component is located in the right-rear upper corner of the chassis (Fig. 8-12). The blower provides the necessary amount of air to cool the magnetron and oven cabinet; it also drives the stirrer by means of a special drive belt (Fig. 8-10). To test it, check the resistance of motor winding. A normal reading should be about 10Ω.

HV Transformer T1. This is located in the right rear of the oven chassis. The transformer supplies both filament voltage (about 3.9V AC) and HV (about 2000V AC) to the magnetron and HV supply on the -2 and -3 series. The -5 and -6 series have a separate filament transformer. To test it, see the troubleshooting section.

Capacitor C1. This is located at the right rear, behind T1. It provides voltage regulation in the HV supply. To test it, discharge the capacitor first. Then connect an ohmmeter across the terminals of C1, using a high (R × 10K) scale. The meter should move toward zero, then slowly back toward infinity. Reverse leads and note same results.

Diode CR1. This is attached between one contact of the capacitor and ground. The diode rectifies the 2000V AC input from the transformer. To test it, first discharge the capacitor. Then connect an ohmmeter, set to the high (R × 10K) resistance scale, across the diode terminals. Note the reading. Then reverse the meter leads and again note the reading. The meter should read infinity one way and about 50K to 200K the other way.

Magnetron V1. This part is located in the middle of the right side, mounted into the waveguide. The magnetron produces the RF energy necessary to cook. To test it, take a plate-current reading as described earlier.

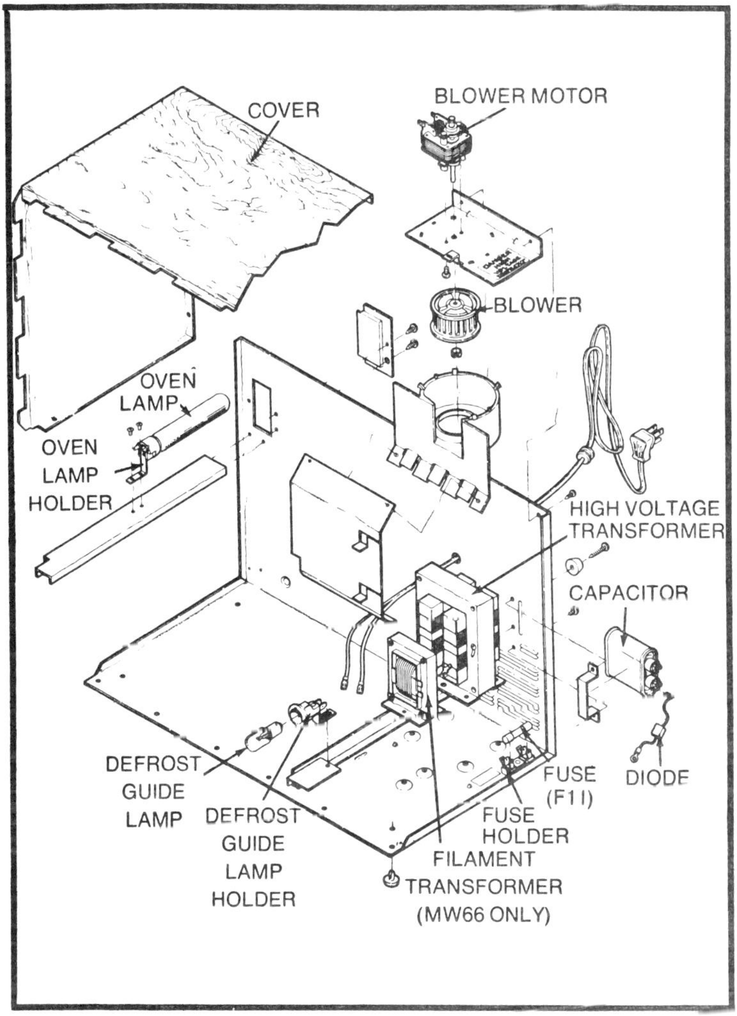

Fig. 8-12. Illustration of oven disassembly.

DISASSEMBLY PROCEDURES

Be certain that the oven is unplugged before attempting to disassemble. To remove the magnetron assembly (Fig. 8-10), remove the four nuts securing the magnetron to the waveguide, using a wrench or nutdriver while supporting the tube from the bottom. Remove the tube by pulling straight down until the dome is clear of the waveguide. Be sure the RF gasket is in place before installing the magnetron.

Oven Door Replacement

Open the door to its fully opened position. (Refer to Fig. 8-13). Remove the entire oven-door assembly. To install it, reverse the procedure. Make certain that the hinge is adjusted as outlined under *Door Adjustments* and that the oven is tested for microwave leakage as described in Chapter 3.

Door Seal Removal

1. Remove the oven-door assembly.
2. Remove the hinge from the oven-door assembly.
3. Remove the 18 countersunk screws from the outside rim of the oven door assembly.
4. Remove the outer door frame from the oven door assembly.
5. Remove the rubber door seal from around the door assembly.
6. Place the oven door on a flat surface with the seal plate facing down.
7. Carefully set aside the window assembly.
8. Using a nutdriver, remove the 22 hex nuts that secure the seal-plate assembly to the oven door.
9. Remove the door liner.
10. Remove the four inner-door supports.
11. Remove the seal-plate assembly.
12. Remove the four strips of aluminum tape from the seal-plate assembly.
13. Remove the perforated metal window and the plastic window. Be careful not to scratch the window.

New Door-Seal Plate Installation

1. Place the new seal plate face down on a flat surface so that the weld studs face upward.
2. Place the clear plastic window onto the seal plate, inside the 22 weld studs.
3. Place the perforated metal window over the plastic window so that the plastic window nests inside the recessed area of the perforated window.
4. Secure the perforated window with a couple of nuts until the tape (mentioned below) is applied.
5. Cut two strips of adhesive metal tape 7 inches long and two strips 13 inches long.

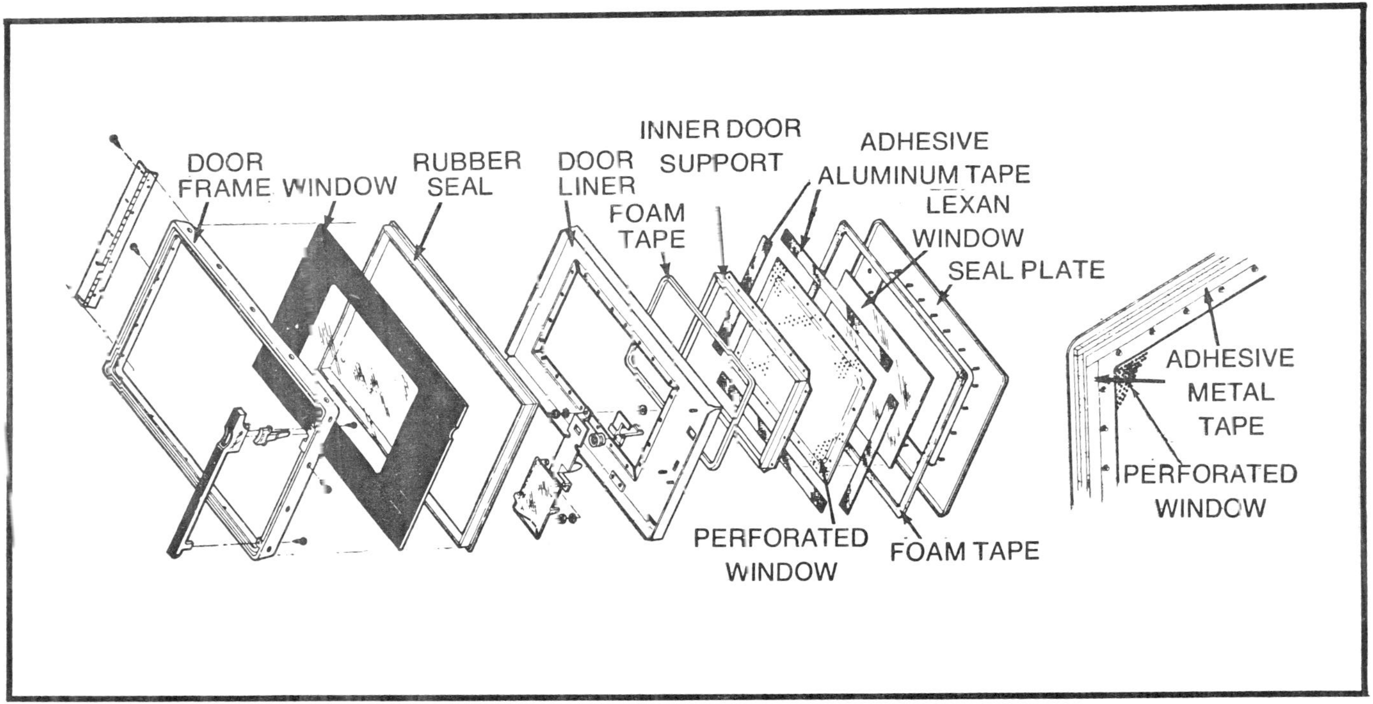

Fig. 8-13. Oven-door construction.

6. Place these strips of tape over the edges of the perforated window so that they bond it to the inner surface of the seal plate. (See Fig. 8-13). The tape should cover the edge of the perforated window up to, but not over, the 22 weld studs.
7. Install the four inner-door supports.
8. Replace the door liner.
9. Install the 22 nuts securing the assembly and tighten securely.
10. Replace the rubber door seal around the outer edge of the door liner.
11. Replace the window assembly.
12. Carefully fit the door seal around the outer edge of the door liner.
13. Install the 18 countersunk screws into the door frame.
14. Make certain that the hinge is adjusted as outlined under *Door Adjustments* and that the oven is tested for microwave leakage.
15. Install the hinge into the oven-door assembly.
16. Install the oven-door assembly into the oven.

Stirrer Cover Replacement

1. Open the oven door fully.
2. With a small screwdriver, carefully pry the six plastic fasteners loose from the baffle cover (Fig. 8-10).
3. Remove the stirrer cover through the oven-cavity front opening.
4. To replace, reverse the process, snapping the six plastic fasteners into place.

Oven Shelf Replacement

Shelf breakage, as a general rule, is due to improper use of the oven, such as running the oven without being properly cleaned. Grease deposits and food particles can carbonize if not cleaned out properly, causing the oven to arc. This could result in extreme heat on certain areas of the oven shelf, causing the glass to melt or crack. The same problem can be caused by improper use of aluminum foil in oven or the use of metal pans or containers.

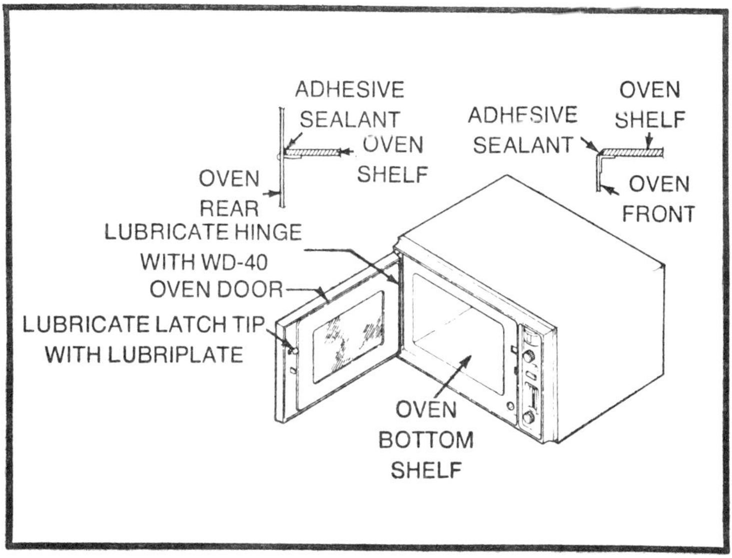

Fig. 8-14. Replacing the oven shelf.

1. Using a hook-type knife, such as a linoleum knife, cleanly cut sealant all around oven shelf. (Fig. 8-14). Remove the old shelf completely and remove all debris and chips that have fallen into the oven bottom. A piece of self-sticking tape is good to remove small chips.
2. Hook the deep-grooved edge of the plastic trim strip, supplied with the new shelf, over the turned-back flange on the edge of the oven front with the beveled edge up and toward the front.
3. Put a small bead of RTV-102 white silicone-rubber adhesive behind the beveled edge, forcing the RTV up along the edge and making a good seal.
4. Place the new shelf onto the support studs in back of the oven cavity.
5. Seal the shelf in place by applying a bead of RTV-102 white silicone-rubber adhesive around the back and side edges of the oven shelf.
6. Wipe away any excess adhesive with a damp cloth so that a smooth fillet is formed between the shelf and the walls of the oven cavity on sides and back. On front edge, wipe away any excess adhesive.

Stirrer Replacement

1. Remove the stirrer cover as previously explained.
2. Remove the drive belt from pulley and motor shaft.
3. Remove the cotter pin from the stirrer shaft, holding the stirrer in the oven cavity.
4. Remove the pulley and shaft, noting the positions of the washer, spacers, and bushings.
5. When replacing the stirrer, make certain that the blades are not out of line, which would cause them to hit the top of the oven or cover. Be sure the stirrer and pulley turn freely.
6. Be sure the cotter pin on top of pulley is pushed all the way in and the ends spread to lay flat on the pulley so it will not scrape on the cover.

Chapter 9
Servicing Panasonic Microwave Ovens

This chapter provides servicing information for a wide range of Panasonic[1] countertop microwave ovens. It covers recent as well as early productions. Complete wiring diagrams are provided for seven models. The detailed servicing instructions provided for four models will assist you in repairing all other models as well.

Energy-emission and cooking-power measurements follow the general procedures given in Chapter 3. These measurements are important and should be made carefully and accurately, following the safety precautions given earlier.

HOW THEY WORK

The structure and functioning of parts are illustrated by Fig. 9-1. The schematics of the oven circuits are given as Figs. 9-2 through 9-8. General operation of these ovens is described next. Succeeding paragraphs provide detailed descriptions of the circuits and point out exceptions for specific models. The emphasis of the detailed circuit descriptions is on these models: NE-5300, NE-6300, NE-6400, and NE-6700.

General Operation

Cook Timer Set. The fan motor and stirrer motor rotate. The fan motor cools the magnetron by blowing air over the

1. All of the illustrations in this chapter are courtesy Panasonic Company.

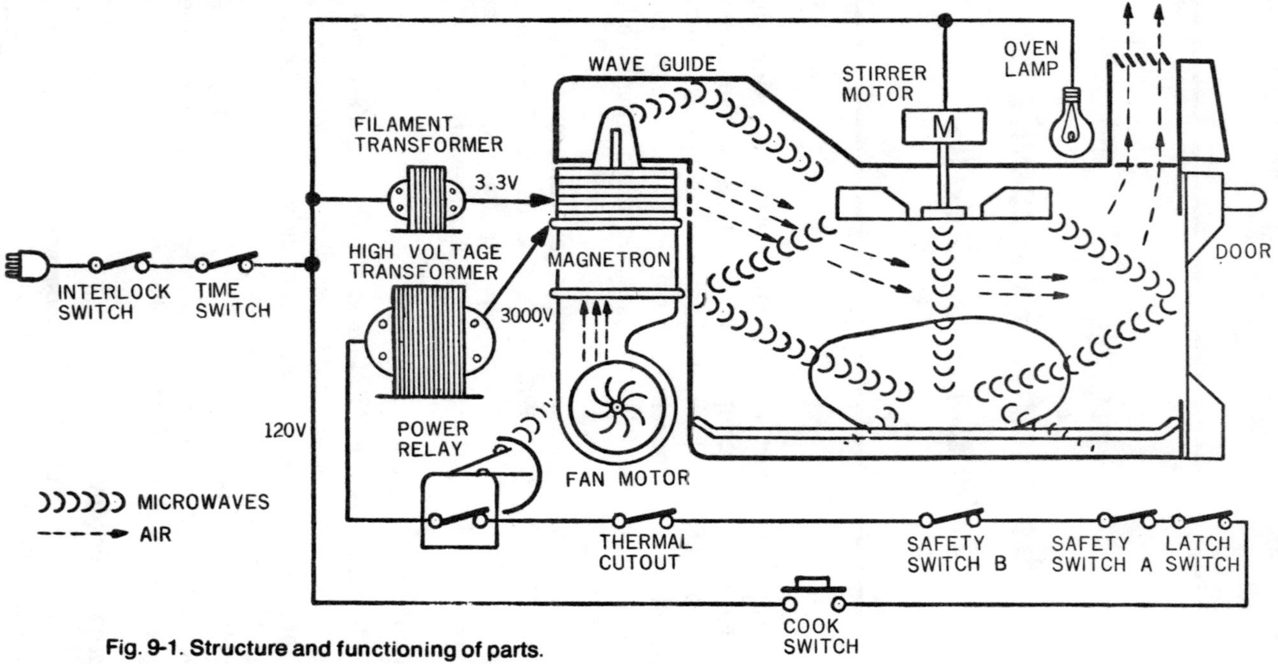

Fig. 9-1. Structure and functioning of parts.

cooling fins attached to the magnetron. If the magnetron is not cooled sufficiently—as if the fan motor stops functioning or an account of the ambient condition—the thermal cutout will be energized to stop oscillation of the magnetron. Under normal conditions, the contact of the thermal cutout is kept closed. The air, after cooling the magnetron, is taken into the oven, forcing out the vapor and odor in the oven from the front of the upper panel. The stirrer motor distributes the microwaves for more uniform cooking by rotating the aluminum stirrer blade. The oven lamp glows to illuminate the inside of the oven. At the same time, voltage is applied to the filament transformer, and from the secondary side, 3.3V is generated and supplied to the magnetron filament to heat it.

Pressing the COOK Switch. The power relay is energized as voltage is applied through the COOK switch, latch switch, safety switch A, safety switch B, and thermal cutout; it relays the voltage to the HV transformer. At the secondary side of the HV transformer, about 3000V is generated and transferred to the magnetron to cause it to oscillate at 2.45 GHz. The microwaves are led into the oven by the waveguide and then distributed by the stirrer blade for uniform cooking. The magnetron is a type of diode with an added magnet for producing microwaves.

Door Is Opened During Cooking. The microwave oscillation is stopped by the latch switch before the door is opened. The microwave-generating circuit is interrupted two more times by safety switch A and safety switch B when the door is opened.

NE-6300 Self-Maintaining Circuit

When the COOK switch is pressed, 100V is applied to the power-relay coil through the thermal protector, safety switch A, COOK switch, thermal cutout, and time switch, as shown by solid line in Fig. 9-2, thereby energizing the power relay and closing contacts 2 and 4. Then 120V is supplied to the inrush-reducing relay through contact 2 and 4, thereby energizing the inrush-reducing relay and closing contacts 3 and 4 plus 1 and 2. Contacts 1 and 2 transfer the voltage to the HV transformer bypassing resistor (15Ω). Contacts 3 and 4 are a self-maintaining circuit that transfers the voltage to the

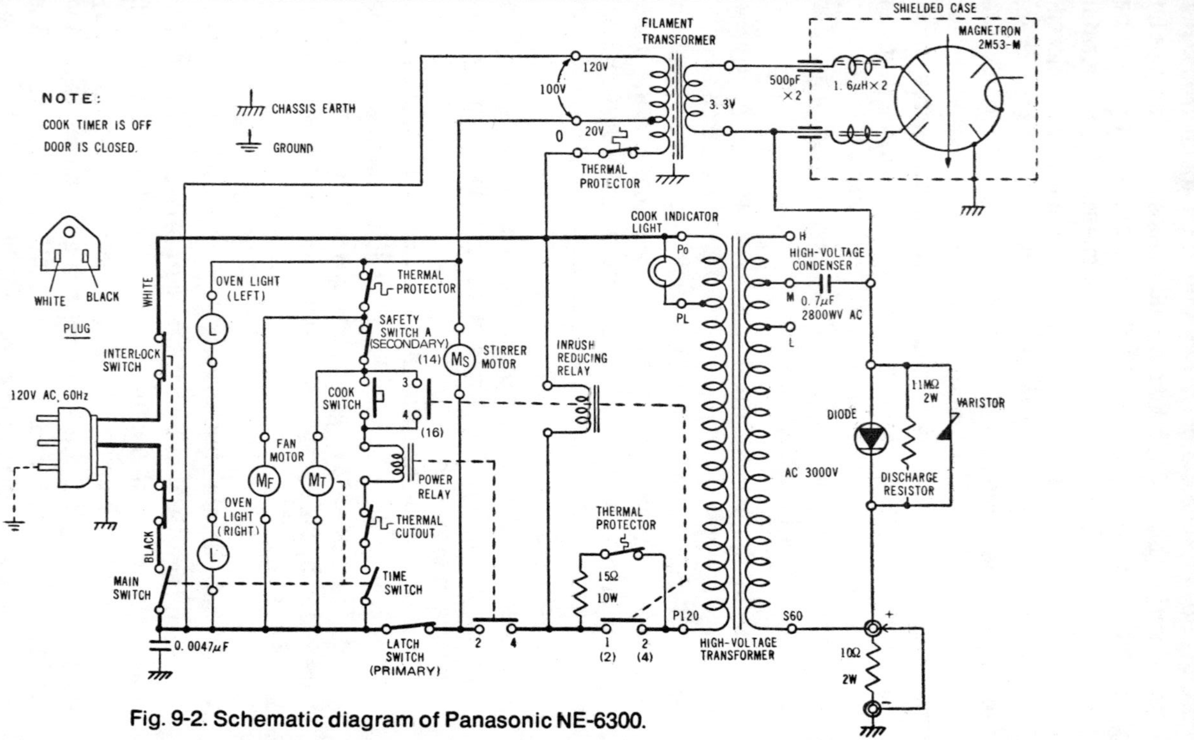

Fig. 9-2. Schematic diagram of Panasonic NE-6300.

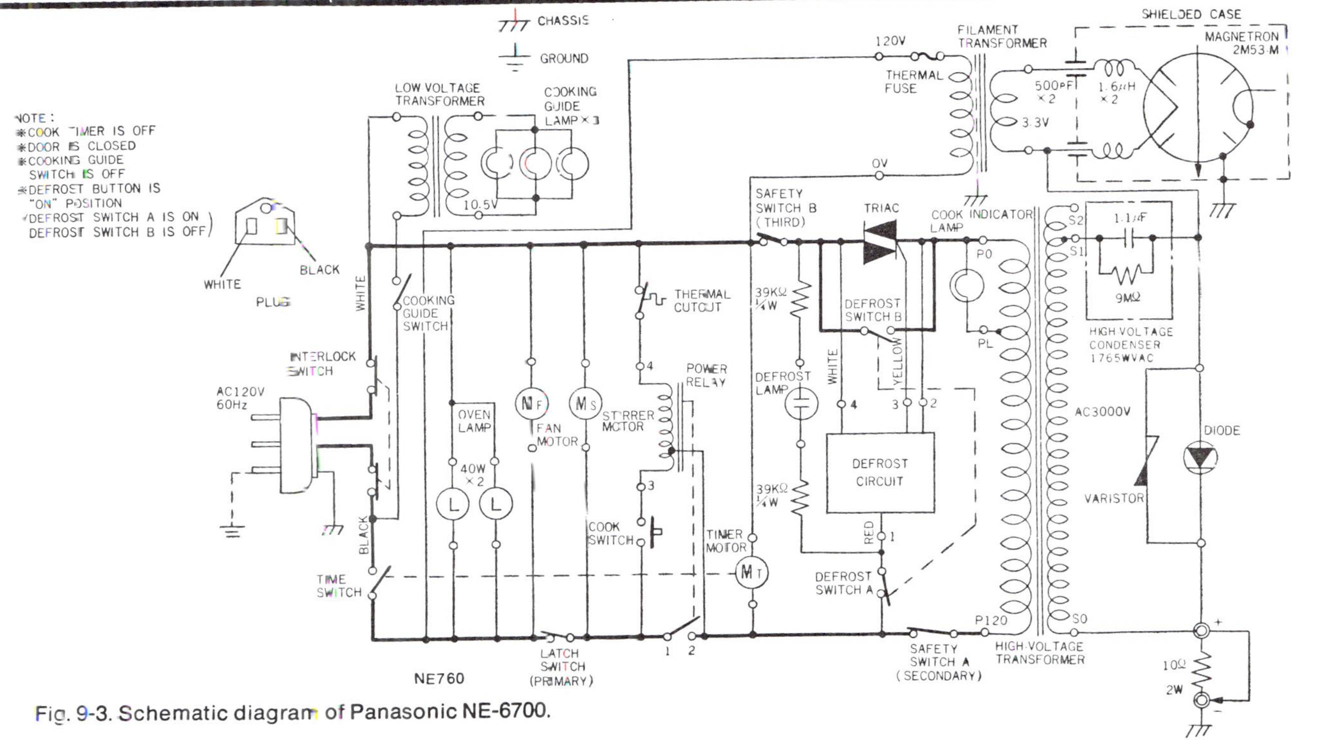

Fig. 9-3. Schematic diagram of Panasonic NE-6700.

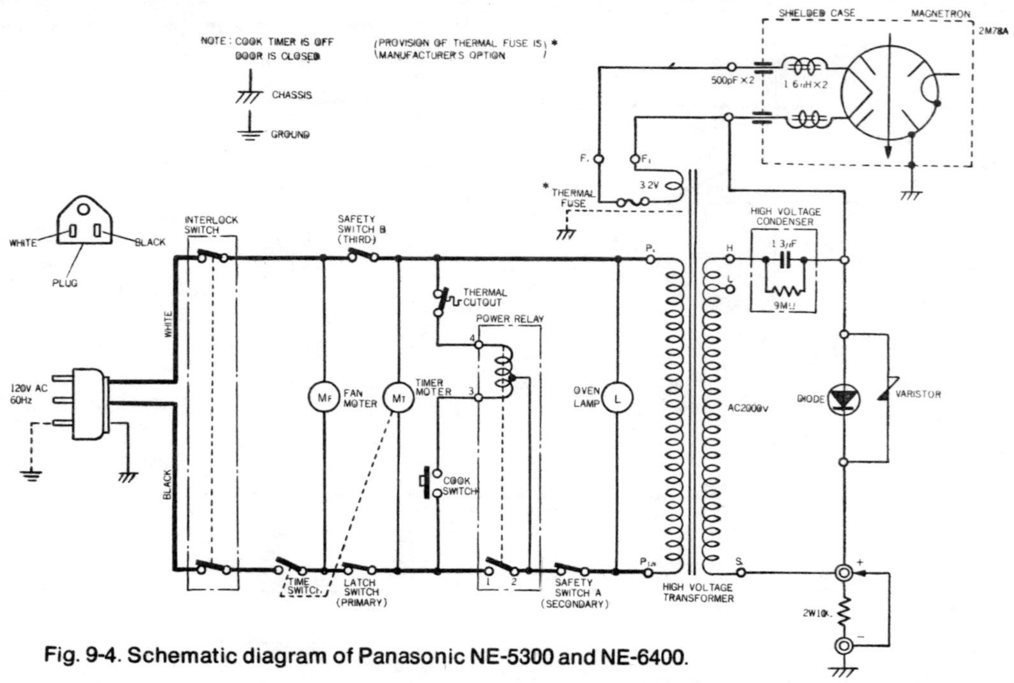

Fig. 9-4. Schematic diagram of Panasonic NE-5300 and NE-6400.

power-relay coil continuously after the COOK switch is released.

NE-5300, -6400 & -6700 Self-Maintaining Circuit

After the pressing of the COOK switch, voltage is supplied to the power relay through the thermal cutout, COOK switch, and latch switch (see appropriate schematic), thereby energizing the power relay and closing contacts 1 and 2. After the COOK switch is released, voltage is supplied to the power relay through self-maintaining contacts 1 and 2 of the power relay. Then the power relay is continuously operated even when the COOK switch is released.

NE-6300 Surge-Suppression Circuit

After the power relay is energized and contacts 2 and 4 are closed, the voltage is first applied to the HV transformer through the 15Ω resistor and thermal protector. The 15Ω resistor reduces the inrush current that occurs momentarily upon pressing the COOK switch, which prevents the circuit breaker from opening. At the same time, the voltage is supplied to the inrush-reducing relay, closing contacts 3 and 4 plus 1 and 2. Then the voltage is applied to the HV transformer through contacts 1 and 2, bypassing the 15Ω resistor. The thermal protector attached to the 15-ohm resistor works when the resistor's temperature increases unusually due to a malfunction of contacts 1 and 2 of the inrush-reducing relay. This is to prevent the 15Ω resistor from burning.

NE-5300, -6400 & -6700 Surge-Suppression Circuit

After the COOK switch is pressed, voltage is applied to the power relay to operate it, and separately, voltage is applied to the HV transformer through part of the power-relay coil and safety switch A. At this moment, part of the power-relay coil functions as a resistor to decrease the inrush current and prevent the circuit breaker from opening. After the power relay is energized and contacts 1 and 2 are closed, the voltage is applied to the HV transformer through contacts 1 and 2 and safety switch A, bypassing the power-relay coil.

HV Circuit

The 3.3V generated from the secondary side of the filament transformer is applied to the magnetron filament

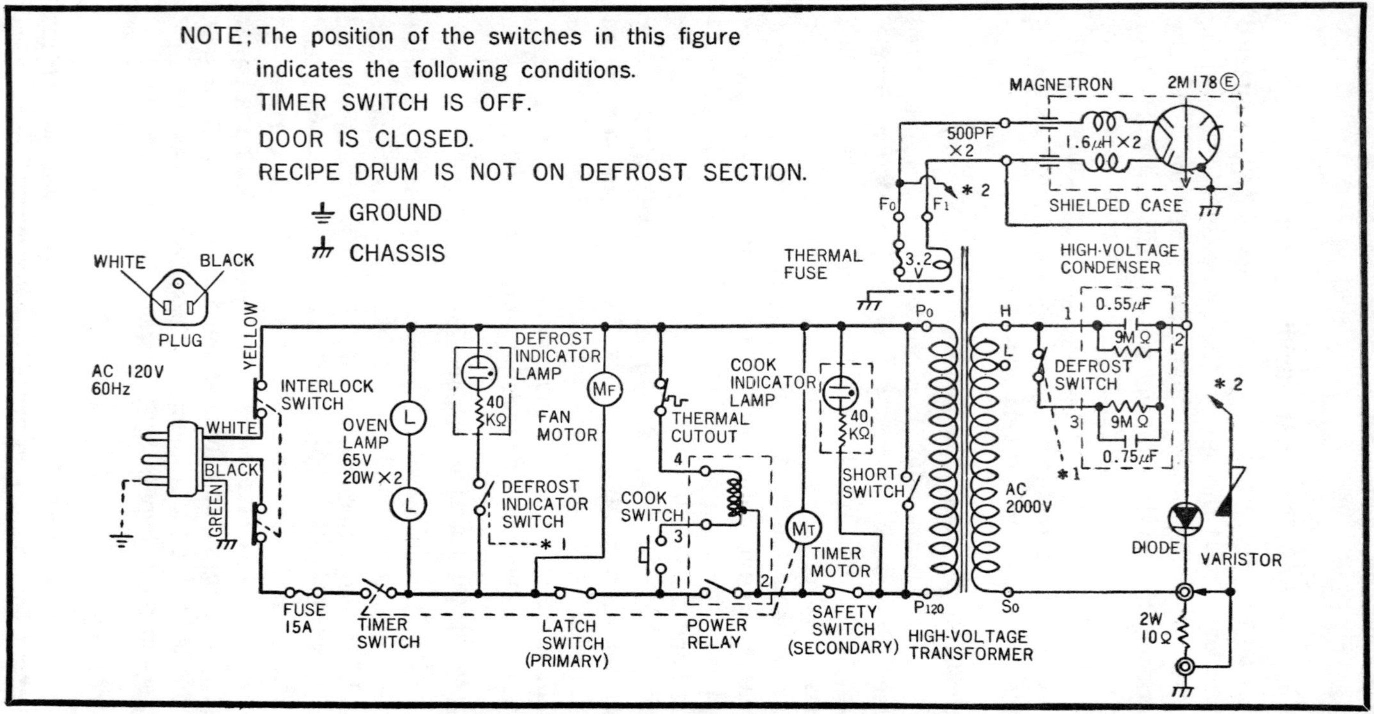

Fig. 9-5. Schematic diagram of Panasonic NE-6600.

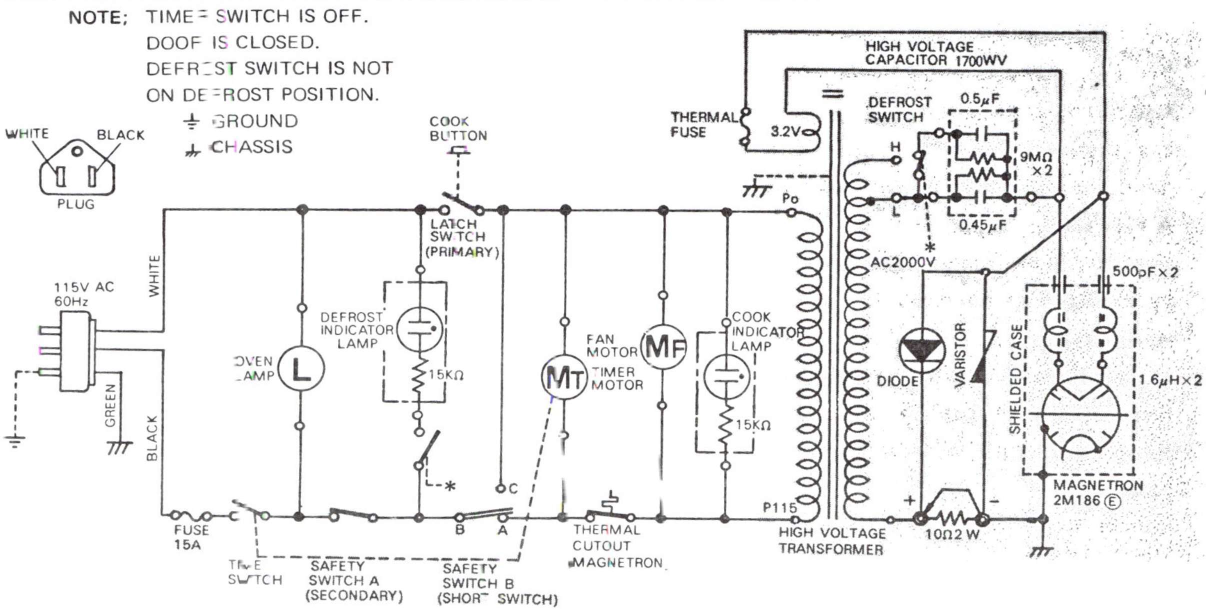

IMPORTANT SAFETY NOTE; The shaded area on this schematic diagram incorporates special features important for protection from microwave radiation, fire and electrical shock hazards. When servicing it is essential that only manufacturer's specified parts be used for the critical components in the shaded areas of the schematic.

Fig. 9-6. Schematic diagram of Panasonic NE-5610.

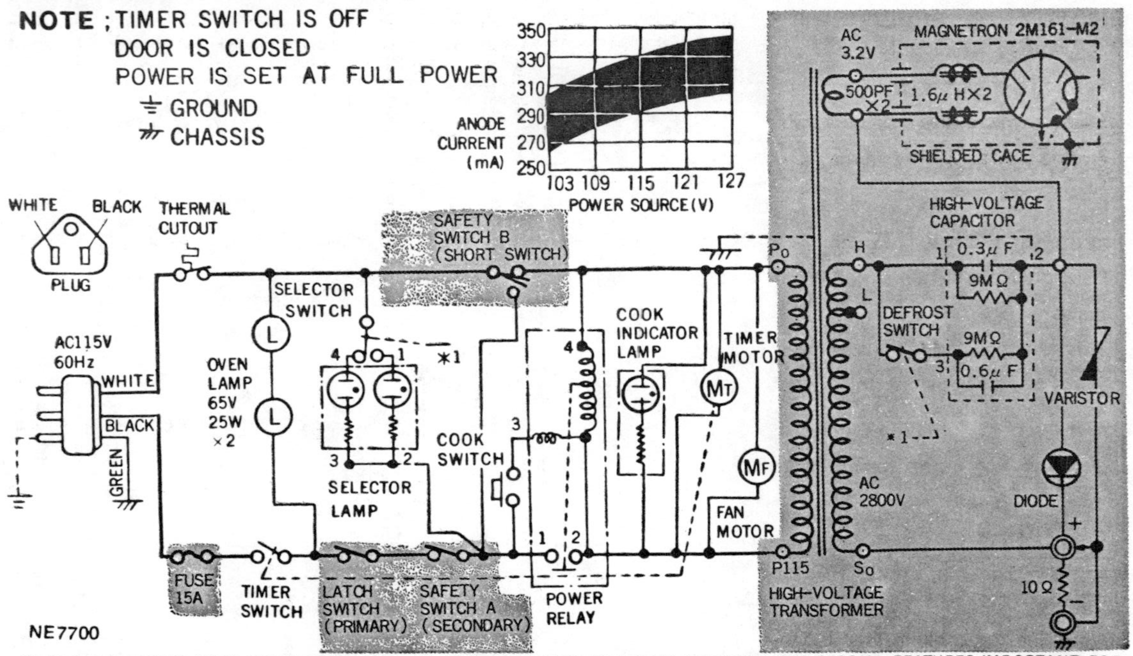

IMPORTANT SAFETY NOTE; THE SHADED AREA ON THIS SCHEMATIC DIAGRAM INCORPORATES SPECIAL FEATURES IMPORTANT FOR PROTECTION FROM MICROWAVE RADIATION, FIRE AND ELECTRICAL SHOCK HAZARDS. WHEN SERVICING IT IS ESSENTIAL THAT ONLY MANUFACTURER'S SPECIFIED PARTS BE USED FOR THE CRITICAL COMPONENTS IN THE SHADED AREAS OF THE SCHEMATIC.

Fig. 9-7. Schematic diagram of Panasonic NE-7700.

through two noise-preventing choke coils of 1.6 μH. These two coils and two noise-preventing capacitors prevent disturbance waves generated by the magnetron from affecting TV and radio reception. After energization of the power relay, the 120V is applied to the HV transformer to then, at the secondary side, generate 3000V. This 3000V is increased by the action of the diode and the charging of the HV capacitor. When forward voltage to the diode is generated, current flows and 3000V is stored in the HV capacitor. During the second half-cycle, a backward voltage is generated, and the electric charge remaining in the HV capacitor and the HV generated from the secondary side are combined and applied to the magnetron. The magnetron oscillates at 2.45 GHz. After the HV action ceases, the electricity remaining in the HV capacitor is discharged by the discharge resistor (9M), which is contained in the HV capacitor. The time required for complete discharge is about 30 seconds. *When servicing the unit, be sure to short both terminals of the HV capacitor, and start repair work only after it is completely discharged.*

The varistor ordinarily has almost infinitely high resistance. However, when a surge voltage is applied, the internal resistance decreases and current is allowed to flow. It may sometimes happen that a surge voltage of more than 10 kV is generated at the magnetron. The varistor prevents breakdown of circuits by this surge voltage. It has the characteristic that when the applied voltage exceeds a certain value, the internal resistance is decreased so that the varistor absorbs the abnormal surge voltage to protect the HV circuit.

NE-6700 Defrost Circuit

A 5-second-on and 5-second-off signal is conducted between the cathode (terminal 2 of the defrost circuit) and the gate (terminal 3 of the defrost circuit) of the triac, as shown in Fig. 9-9, due to the function of the defrost circuit (multivibrator actions). Then the triac operates current flows between the anode and the cathode alternating 5 seconds on and 5 seconds off so that the magnetron produces microwaves intermittently (5 seconds on and 5 seconds off) for defrosting.

After the operating of the power relay, 120V is applied between terminals 1 and 4 of the defrost circuit. When a

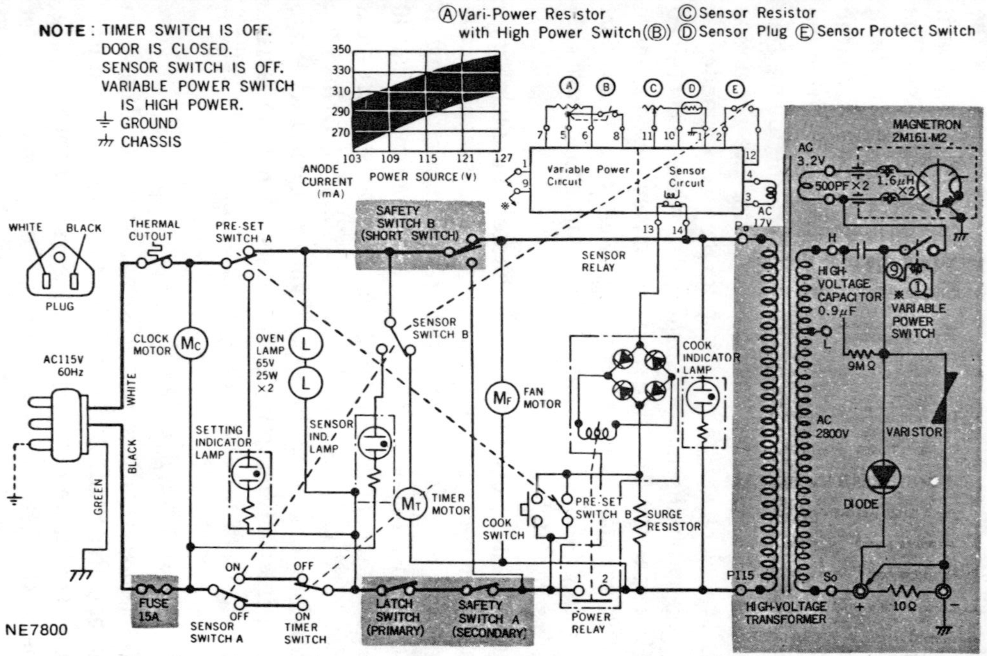

IMPORTANT SAFETY NOTE: THE SHADED AREA ON THIS SCHMATIC DIAGRAM INCORPORATES SPECIAL FEATURES IMPORTANT FOR PROTECTION FROM MICROWAVE LEAKAGE, FIRE AND ELECTRICAL SHOCK HAZARDS WHEN SERVICING IT IS ESSENTIAL THAT ONLY MANUFACTURERS SPECIFIED PARTS BE USED FOR THE CRITICAL COMPONENTS IN THE SHADED AREAS OF THE SCHEMATIC DIAGRAM.

Fig. 9-8. Schematic diagram of Panasonic NE-7800.

forward voltage is applied to diode D1, voltage is applied to both capacitor C1 and the *multivibrator* circuit. In the opposite case, when a reverse voltage is applied, the electric charge in capacitor C1 is discharged to the multivibrator circuit since diode D1 cannot pass the current. The multivibrator circuit is supplied with a mean DC voltage of 156V as shown in Fig. 9-10.

This 156V DC is divided by R1-R2 combination and the R5-R6 combination. About 20V DC is applied to both of the thyristor gates of T1 and T2. But due to the noise in the multivibrator circuit, the anode voltage of T2 becomes more than 20V DC (the gate voltage), T2 conducts, and the 20V DC on the gate of T2 becomes approximately 0V. The 156V DC is allowed to pass as shown in Fig. 9-11A, so that capacitor C2 is charged. After 5 seconds, the electric charge in capacitor C2 becomes more than 20V so that T1 conducts and T2 does not, as shown in Fig. 9-11B.

The thyristor conducts current only when its anode voltage is more than 20V (the gate voltage). For these 5 seconds, the gate voltage of thyristor T2 is applied to the base of the transistor through resistor R6, so that the transistor conducts,

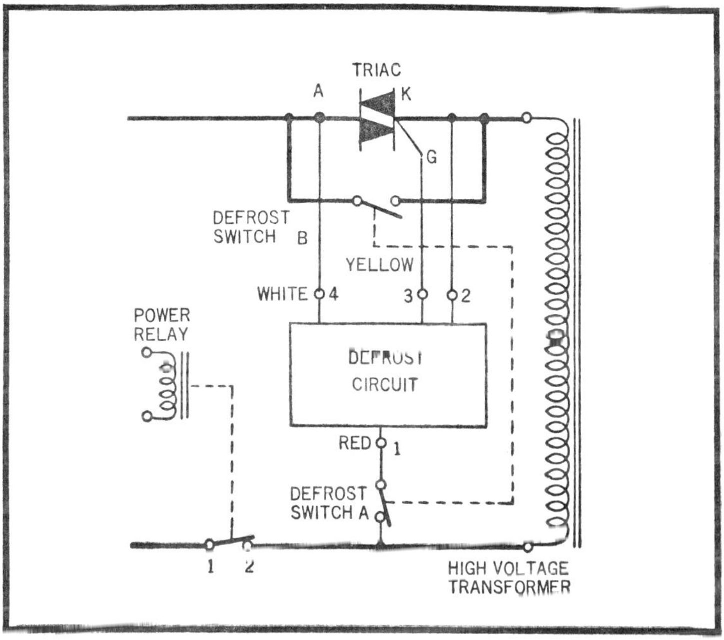

Fig. 9-9. Defrost circuit—Model NE-6700.

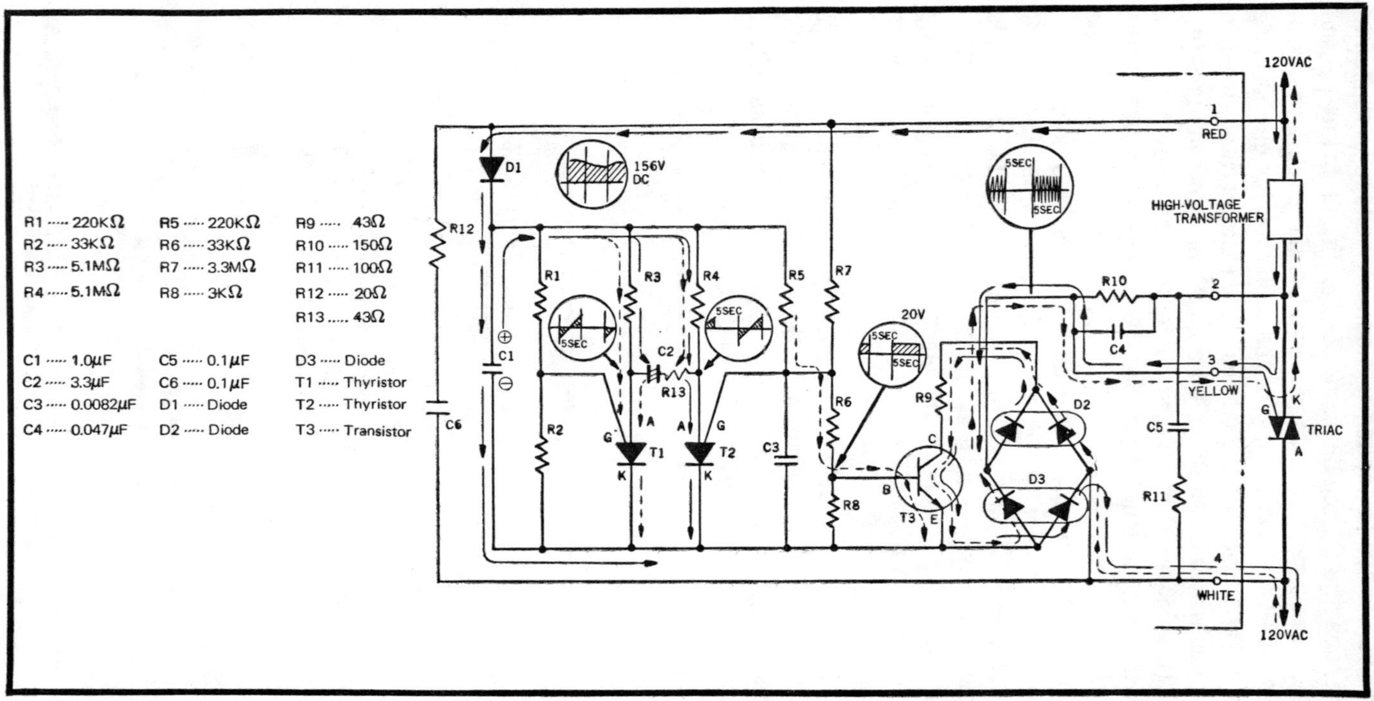

Fig. 9-10. Defrost circuit board.

allowing 120V AC to flow between its collector and emitter. Due to this current flowing between terminals 3 and 4 of the defrost circuit, a current flows between the cathode and gate of the triac, so that between the cathode and the anode of the triac a current also flows, supplying voltage to the HV transformer, causing the magnetron to oscillate. (Resistors

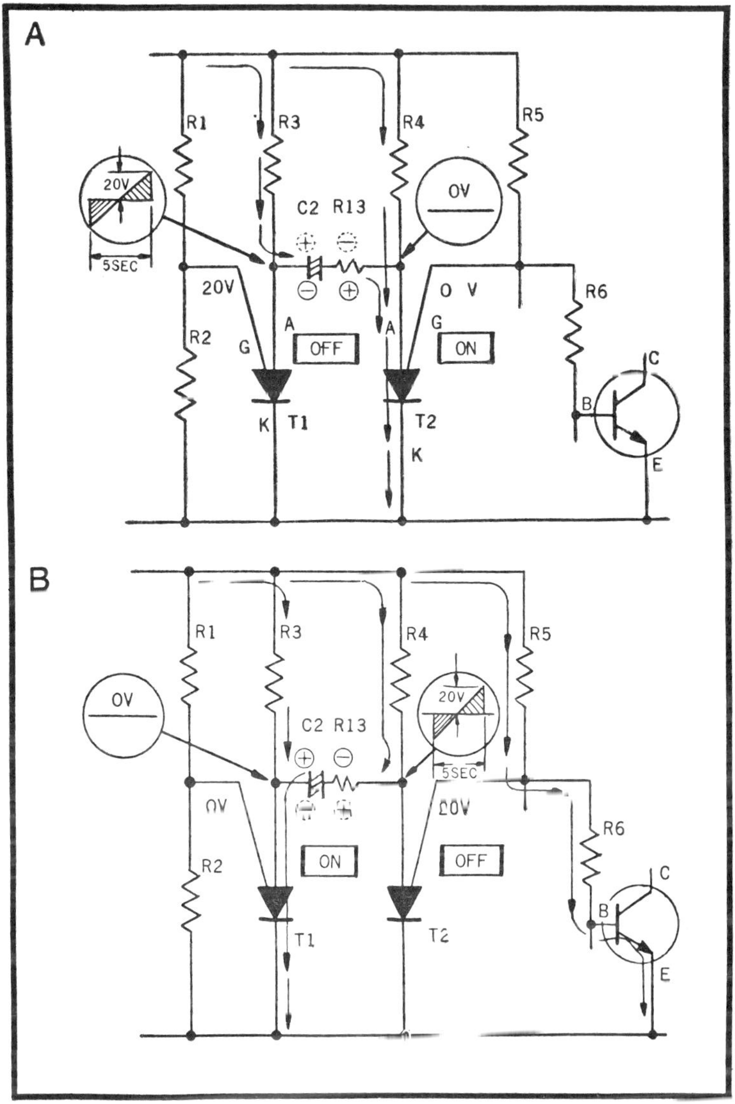

Fig. 9-11 Operation of multivibrator circuit.

R10, R11, and R12, along with capacitors C4, C5, and C6 are noise-preventing components.)

In this way, the anode voltage of the thyristors becomes more than 20V due to the time constant between R3-C2-R13 and R4-R13-C2, causing T1 and T2 to conduct current intermittently. Then this signal is applied to the triac, so that the magnetron operates 5 seconds on and 5 seconds off for defrosting.

NE-6300 Oscillation-Interrupting Circuit

When the door handle is grasped, the latch switch is activated to stop microwave oscillation before the door is opened. Besides this, as soon as the door begins opening, safety switch A is operated to stop the power relay from functioning, opening contacts 2 and 4 of the power relay.

NE-5300, -6400 & -6700 Oscillation-Interrupting Circuit

The latch switch is operated to stop microwave oscillation by pressing the press-to-open button or handle lock before the door is opened. Besides this, two safety switches are provided to insure that oscillation is stopped when the door begins to open.

TROUBLESHOOTING & TESTING

When the upper panel of the oven is removed for servicing, an interlock switch will be opened so that turning the cook timer will not operate the oven. The interlock can be temporarily disabled during servicing by pressing the interlock switch by hand. Be very careful to avoid shock from the HV section when the interlock is defeated.

Specific checks to make for the various Panasonic models are given in Table 9-1. The following are some general checks to be made in troubleshooting Panasonic microwave ovens:

1. Check whether the oven operates (fan motor and oven lamp) when the cook timer is set.
2. Check whether the power relay operates when the COOK switch is pressed.
3. Check whether the power relay is energized after the COOK switch is released.
4. Check whether the COOK indicator lamp glows.

Table 9-1. Troubleshooting Chart for Panasonic Microwave Ovens

OVEN DOES NOT OPERATE THOUGH THE COOK TIMER IS SET—OVEN LAMP & FAN MOTOR DO NOT OPERATE.

CONDITION	NE-6300	NE-5300, 64-00 & 6700
Household outlet has proper volt-age.	1. Malfunction of main switch contact of cook timer. 2. Disconnection of pow-er-supply cord or mal-function of interlock switch. 3. Disconnection of fil-ament-transformer pri-mary. Defective ther-mal protector of fila-ment transformer. The filament-transformer secondary is presumably shorted if the thermal protector has functioned. Check it and replace filament transformer if necessary.	1. Malfunction of timer switch contact. 2. Disconnection of pow-er supply cord or mal-function of interlock switch.

CONDITION	NE-5300, -6300, -6400 & -6700
House circuit-breaker trips	1. Insufficient electric power. 2. Defective HV capacitor. 3. Defective HV transformer. 4. **NE-6700 only**. Defective defrost circuit assembly.

HV CAPACITOR & TRANSFORMER CHECK

After making sure that the electric power is sufficient, remove one side of the lead wire connected to the HV capacitor and operate the unit. If the home circuit breaker is not interrupted under this condition, it is con-ceivable that the HV capacitor is defective. If the home circuit breaker still functions, the HV transformer must be defective.

POWER RELAY DOES NOT OPERATE WHEN THE COOK SWITCH IS PRESSED

CONDITION	NE-6300	NE-6700	NE-5300 & -6400
Power relay does not op-erate.	1. Defective ther-mal protector. 2. Malfunction of safety switch A. 3. Malfunction of cook switch contact. 4. Disconnected power relay coil. 5. Defective ther-mal cutout. 6. Malfunction of time switch.	1. Malfunction of the latch switch. 2. Malfunction of the cook switch. 3. Disconnection of the power relay coil. † 4. Defective ther-mal circuit. †	1. Malfunction of the latch switch. 2. Malfunction of the cook switch. 3. Disconnection of the power relay coil. † 4. Defective thermal cutout. † 5. Malfunction of the safety switch B.

† —In this case, the power relay may chatter only when the cook switch is pressed.

Table 9-1 cont.

MALFUNCTION OF THE SELF-MAINTAINING CONTACT OF THE POWER RELAY.		
CONDITION	NE-6300	NE-5300, -6400 & -6700
Malfunction of the self-maintaining contact of the power relay.††	1. Malfunction of latch. 2. Incomplete contact (3-4) of inrush reducing relay. 3. Disconnected inrush reducing relay coil. 4. Incomplete contact (2-4) of power relay.	Incomplete contact (1-2) of the power relay.
††—The power relay operates when the cook switch is pressed, but does not operate when the cook switch is released.		
COOK INDICATOR LAMP DOES NOT GLOW AND UNIT DOES NOT COOK		
CONDITION	NE-6700	NE-5300 & -6400
Cook indicator lamp does not glow.	1. Malfunction of the safety switch A. 2. Malfunction of the safety switch B.	Malfunction of safety switch A.

MAGNETRON FILAMENT DOES NOT HEAT WHEN THE COOK TIMER IS SET	
CONDITION	NE-5300, -6300, -6400 & -6700
Magnetron filament does not heat.	1. Disconnection of the magnetron filament. 2. Disconnection of the filament tranformer or of the filament side of the HV transformer.
NO MAGNETRON ANODE CURRENT	
CONDITION	NE-S3000, -6300, -6400 & 6700
No magnetron anode current.	1. Short circuit of the stack diode. 2. Faulty magnetron or open HV circuit.
ANODE CURRENT IS ALMOST HALF OR SLIGHTLY LESS THAN RATED CURRENT	
CONDITION	NE-5300, -6300, -6400 & -6700
Anode current is slightly less than rated current or nearly half.	1. Faulty insulation of the filament transformer (or filament side of the HV transformer) or inside of the shielded case. 2. Faulty magnetron (anode current sometimes is less than half).
OUTPUT POWER IS LOW	
CONDITION	NE-5300, -6300, -6400 & -6700
Output is low.	1. Faulty magnetron. 2. Aging deterioration of the magnetron or decrease in power source voltage.

Table 9-1 cont.

OVEN DOES NOT OPERATE NORMALLY WHEN DEFROSTING— OVEN OPERATES NORMALLY WHEN COOKING	
CONDITION	**NE-6700**
Circuit breaker interrupts.	1. Diode D1 shorted. 2. Diode D2 (A) shorted. 3. Diode D3 (D) shorted. 4. Diode D2 (B) shorted. 5. Diode D3 (C) shorted.
Time of both on operation and off operation when defrosting shortens.	1. Thyristor T1 between A-G shorted. 2. Thyristor T1 between A-K shorted.* 3. Thyristor T1 between G-K shorted. 4. Thyristor T2 between A-G shorted.*
Magnetron does not oscillate microwaves.	1. Capacitor C2 shorted. 2. Thyristor T2 between G-K shorted. 3. Capacitor C3 shorted. 4. Transistor T3 between B-E shorted. 5. Triac between G-K shorted. 6. Capacitor C4 shorted.
Magnetron oscillates microwaves continuously instead of alternating ON-OFF oscillation pattern.	1. Thyristor T2 between A-K shorted. 2. Transistor T3 between C-B shorted. 3. Transistor T3 between C-E shorted. 4. Triac between A-K shorted. 5. Triac between A-G shorted. 6. Capacitor C1 shorted.**

*—In this case, only the OFF-operation time is shortened. Defrosting cycle approximates to continuous cooking time.

**—In this case, Diode D3(C), D3(D) and D1 might become defective. Check them before mounting new parts.

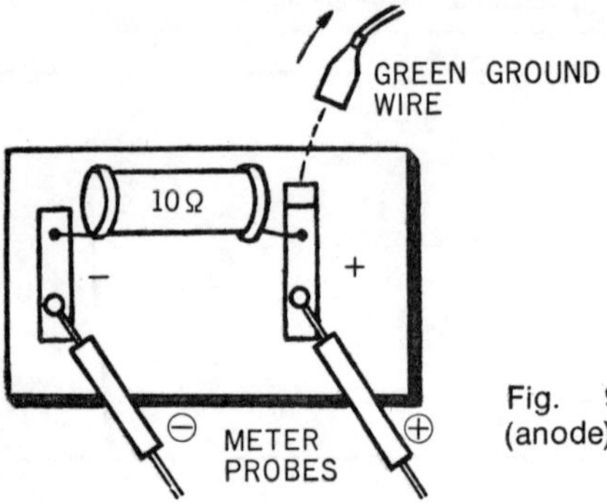

Fig. 9-12. Measuring plate (anode) current.

5. Check whether the magnatron filament heats.
6. Measure the tube's anode current.

Anode Current Measurement

Whether the magnetron operates properly or not can be determined by measuring the anode current. After replacement of the HV transformer, HV capacitor, and magnetron, measure the anode current of the magnetron according to the following procedure. To measure the anode current of the magnetron, do not use an ammeter; use the 5V DC or 10V DC range of a voltmeter.

As shown in Fig. 9-12, remove the green ground wire connected to the positive (+) side of the 10Ω resistor, insert the meter probes into both resistor terminal holes observing the correct polarity, and press the COOK button. After taking the measurement, connect the green ground wire to the positive (P) side of the 10Ω resistor. To determine the anode current on the 10V scale, use this formula:

$$\text{Anode current} = \frac{\text{meter indicated voltage}}{10}$$

If the meter pointer indicates 3.0V, for instance, the anode current can be calculated as follows:

$$\text{Anode current} = 3.0 \div 10 = 0.3 = 300\ \text{mA}$$

The anode current of each model, under normal conditions, is as shown below; but the anode current may change, depending on the power source voltage.

MODEL	RATED ANODE CURRENT
NE-6300	250 ± 25 mA
NE-6700	290 ± 25 mA
NE-5300	315 ± 25 mA
NE-6400	315 ± 25 mA

If the anode current is outside the range of rated current after replacing the magnetron, HV transformer or HV capacitor, change the position of the HV transformer secondary tap. In changing from *H* to *L*, or S2 to S1, the anode current is decreased. In the reverse case of changing from *L* to *H*, or S1 to S2, the anode current is increased.

LATCH SWITCH ADJUSTMENT

This is a very important step in assuring safety from the radiation leakage. Therefore, after replacing the latch switch, adjust the new latch switch according to the following procedures:

NE-5300, -6300 & -6400. With the latch switch held to the door hook, the distance *l* when the latch switch is on and off is specified as shown in Fig. 9-13. Move the rear part of the latch switch to adjust.

When mounting the *door hook* to the oven assembly, adjust the door hook. The gap between the outside surface of the oven and the inside surface of the door is specified as shown in Fig. 9-14. Make this adjustment while pulling on the door handle. At this time do not operate the handle lock. Tighten the screws completely.

NE-6700. With the door in the closed position ,the clearance between the latch-switch lever and the latch-switch surface is specified as shown in Fig. 9-15. For adjustment, loosen the left-hand nut of the latch switch and adjust the position as shown in the figure. While adjusting, do not bend the lever. After adjusting, tighten the nut completely to prevent the latch switch from moving.

SAFETY SWITCHES A & B ADJUSTMENT

Opening the door slowly will turn off safety switch A and safety switch B. The movement distance between the closed-door position and the operating position when safety

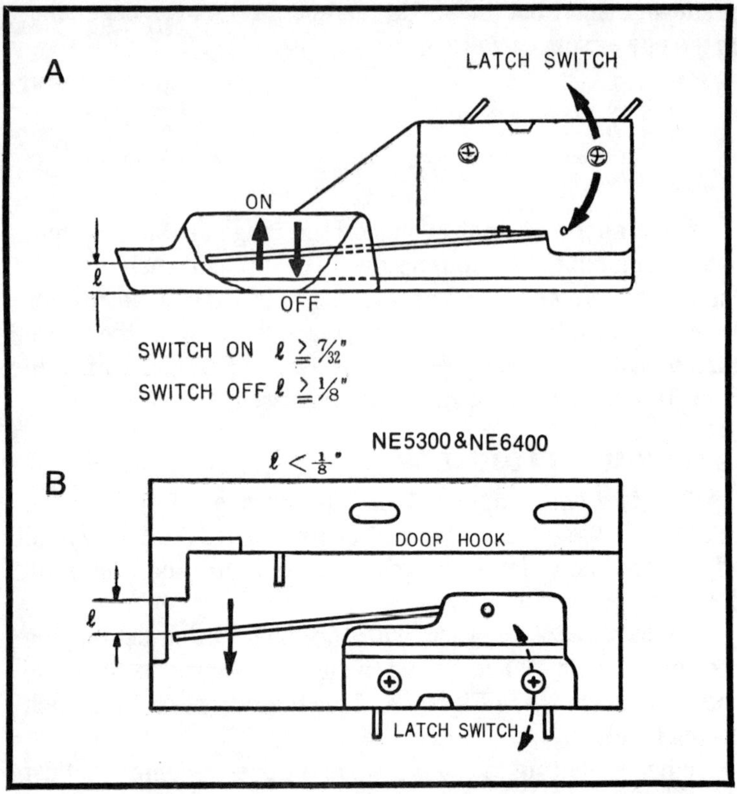

Fig. 9-13. Latch-switch adjustment. (A) NE-6300, (B) NE-5300 and NE-6400.

switches A and B are turned off is specified as shown below and in Fig. 9-16.

MODEL	SAFETY SWITCH A*	SAFETY SWITCH B*
NE-6300	3/32–5/32	–
NE-6700	3/32–5/32	5/16–15/32
NE-5300	1/16–5/32	1/4–13/32
NE-6400	1/16–5/32	1/4–13/32

*—dimensions in inches.

For the switch adjustment, loosen the switch screws or nuts without bending the lever; then move the switch. Bending the lever may later cause faulty switch operation.

Door Assembly Adjustment

The door assembly must be mounted parallel to the oven surface and ferrite rubber. If the door is not mounted parallel, microwaves might leak from the clearance between the oven and the door. When mounting the door, be sure to confirm the clearance to be parallel as shown in Fig. 9-17. For the adjustment, move both the upper and lower (or right and left sides) of the door hinges in the direction shown in Fig. 9-18.

NE-5300 DISASSEMBLY & PARTS REPLACEMENT

Outer Panel Removal. First remove the upper panel. The oven is so designed that the back panel cannot be removed before the upper panel. Remove the 20 screws from the 4 sides

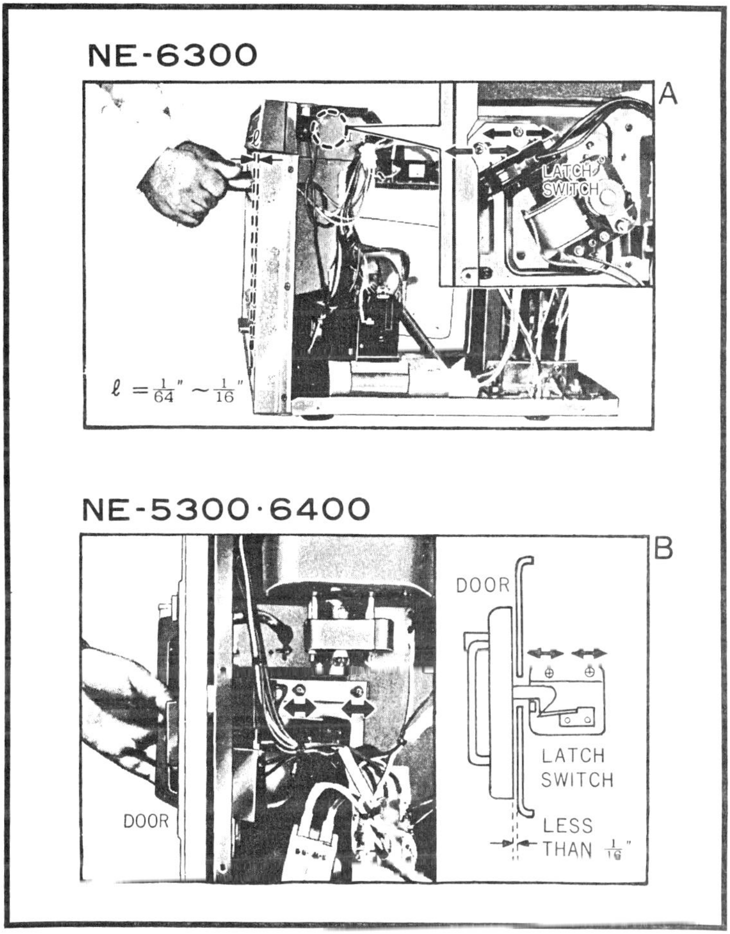

Fig. 9-14. Adjustment of door hook to oven assembly. (A) NE-6300, (B) NE-5300 and NE-6400.

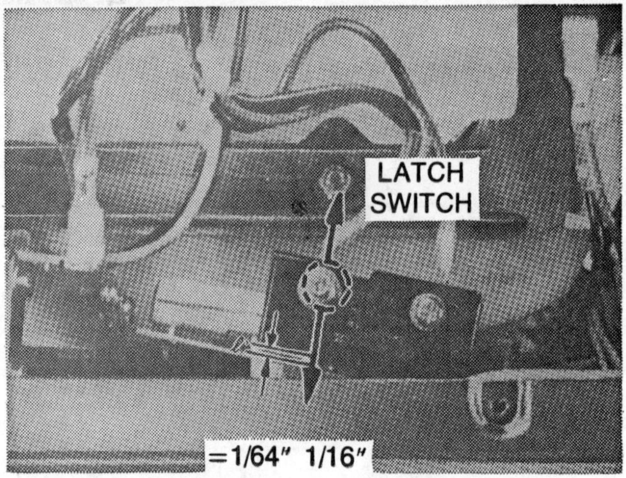

Fig. 9-15. Latch-switch adjustment—NE-6700.

of the upper panel and raise top side until the end of the upper panel is separated from the back panel and sash. One of the screws which mounts the upper panel is different from all other screws. Always use this to install the upper panel.

After removing the upper panel, remove the four back screws. Then the back panel can be removed.

Magnetron Replacement. *Caution:* Remove your watch whenever working close to or repairing the magnetron. Remove the magnetron with the shielded case, permanent magnet, choke coils, and 500 pF capacitors because they are one assembly. Using caution, proceed as follows:

1. Discharge the electric charge remaining in the HV capacitor.
2. Remove the thermal cutout from the magnetron by pulling toward you.
3. Remove the belt from the pulley connected to the fan motor.
4. Remove the fan motor assembly by removing three screws and pulling toward you.
5. Disconnect the magnetron-filament wires connected to the terminal-board assembly.
6. Remove the four nuts that hold the magnetron assembly, while supporting the magnetron from the lower side with your hand.

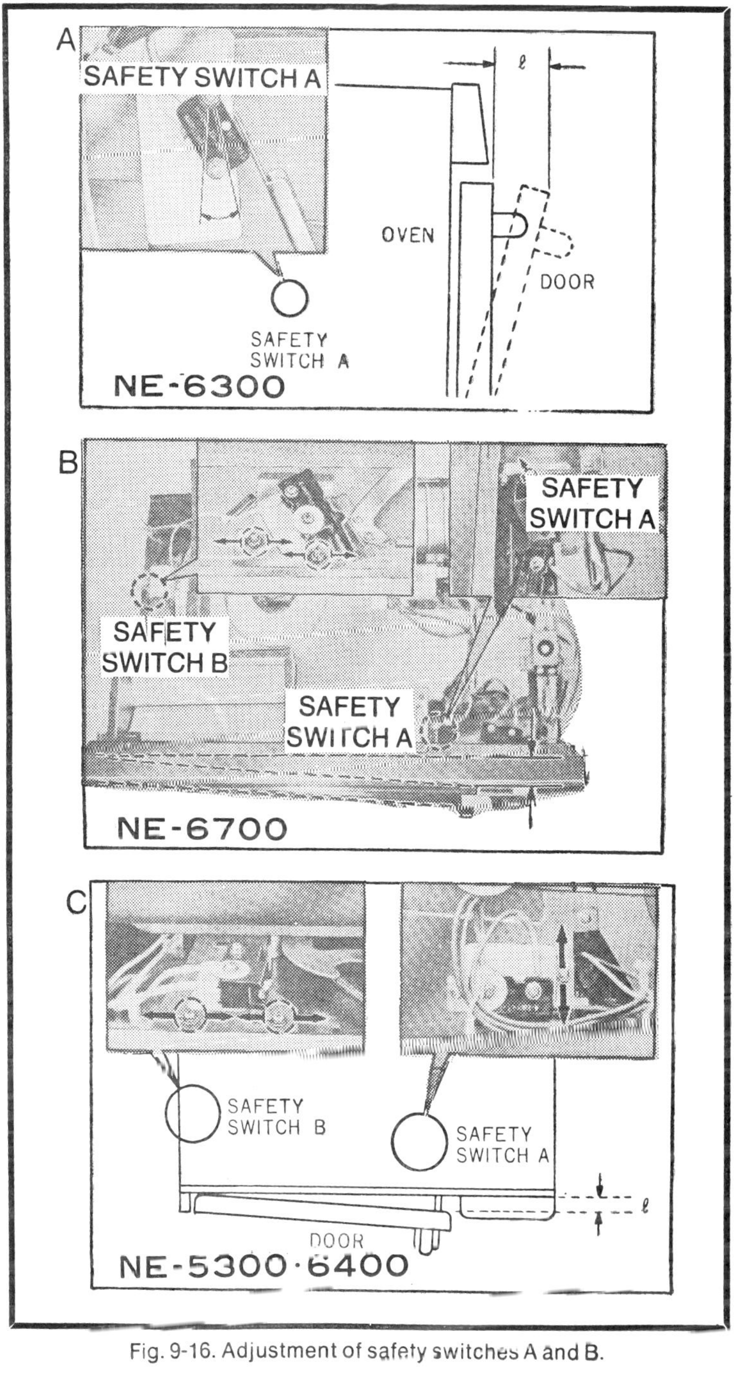

Fig. 9-16. Adjustment of safety switches A and B.

Fig. 9-17. Door mounted parallel to oven surface and ferrite rubber.

7. After removing the magnetron, install a new magnetron by following the above procedure in reverse.

When installing the magnetron, use care not to break the glass section. To prevent microwave leakage, tighten the mounting nuts properly, making sure there is no gap between the waveguide and the magnetron. Make sure that the embossed ring is rigidly attached on the magnetron anode.

HV Capacitor Replacement. Using caution, proceed as follows:

1. Discharge the electric charge remaining in the HV capacitor.
2. Disconnect the lead wires of the primary from the terminal-board assembly.
3. Disconnect the four lead wires from the HV transformer secondary and filament side from the terminal-board assembly and HV capacitor.
4. Remove the two nuts to remove the cable holders.
5. Remove two nuts holding the HV transformer. Remove the HV transformer by pulling it toward you.

Fan Motor Replacement. Remove the belt from the pulley connected to the fan-motor shaft. Disconnect the two lead wires of fan motor from the terminal-board assembly. Remove the three screws holding the fan motor to the oven assembly and pull it toward you. Then the fan motor assembly can be removed.

Power Relay Replacement. Remove the two screws to remove the terminal-board assembly. Disconnect the connector from the power-relay terminal. Remove the two screws to remove the terminal-board bracket. Then the power relay can be removed.

Cook Timer Replacement. Using caution, proceed as follows:

1. Remove the three screws to remove the escutcheon.
2. Remove the three screws to remove the sash bracket and right sash.
3. Remove the cook-timer knob by pulling it toward you.
4. Disconnect two lead wires from the timer switch contact of the cook timer.
5. Disconnect the connector of the cook timer motor.
6. Remove the two screws which have been holding the escutcheon base of the cook timer to the oven assembly.
7. Remove the three screws to remove the cook timer.

Cook Switch Replacement. Remove the two screws to remove the cook switch bracket from the oven assembly. Disconnect the two lead wires connected to the cook switch by using the soldering iron. Remove the two nuts to remove the cook switch from its bracket.

Fig 9-18. Adjuctmcnt of dooı hinges.

HV Transformer Replacement. Using caution, proceed as follows:

1. Discharge the electric charge remaining in the HV capacitor.
2. Disconnect the three lead wires from the capacitor terminals.
3. Remove the nut from capacitor bracket.
4. When mounting a new HV capacitor, make sure that the HV capacitor sheet is situated under the capacitor.

Door Assembly Replacement. Using caution, proceed as follows:

1. Remove the escutcheon by removing the three screws.
2. Spread one side of the door spring from the door arm.
3. Loosen the four nuts of hinge A and hinge B. Then the door assembly can be removed by spreading both hinges to the outside.
4. When mounting the door assembly to the oven assembly, be sure to adjust the door assembly parallel to the oven and the ferrite rubber by moving hinge A and hinge B. If the door assembly is not mounted parallel, microwaves may leak from the clearance between the door and the oven.
5. As shown in Fig. 9-19, hinge spacer C is attached between the hinge A and the hinge bracket to adjust the distance between the hinge and door assembly so that the door assembly will be tightly attached. If the clearance l is more than a sixteenth-inch, remove hinge spacer C.

Latch Switch Replacement. Using caution, proceed as follows:

1. Disconnect the two lead wires from the latch switch.
2. Remove the two screws which mount the door hook to the oven assembly.
3. Pull the door hook with the latch switch to the back side.
4. Remove two nuts to remove the latch from the door hook.
5. When mounting this switch, be sure to adjust it according to steps given earlier.

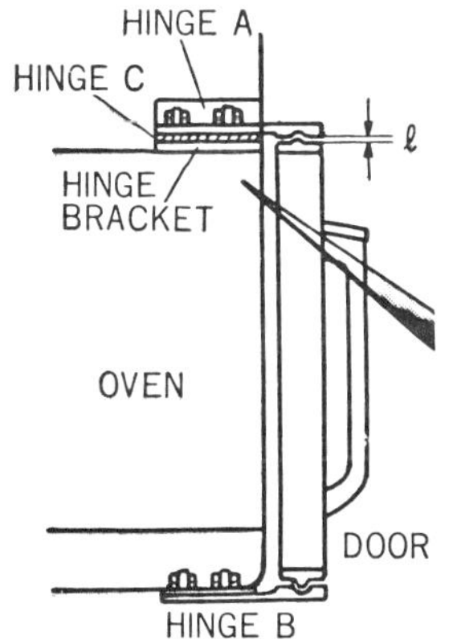

Fig. 9-19. Adjusting distance between hinge and door assembly.

NE-6300 DISASSEMBLY & PARTS REPLACEMENT

Outer Panel Removal. First remove the upper panel. The oven is so designed that the back, right, and left panels cannot be removed before the upper panel. Remove the three screws from the upper-front section and the three screws from the upper-rear section and raise the front part. Then the upper panel can be removed.

After removing the upper panel, remove the eight screws on three sides. Then the back panel can be removed.

Magnetron Replacement. Remove the magnetron with the shielded case, permanent magnet, choke coil, and 500 pF capacitor as one unit. Remove your wrist watch when setting to work because the powerful magnet in the magnetron may damage the watch. Proceed as follows:

1. Short the HV circuit to chassis, or short both ends of the discharge resistor, by using a screwdriver.
2. Disconnect the connector of the thermal-cutout lead wire.
3. Remove the thermal-cutout lead wire and HV lead wire from the lead-wire holders.
4. Remove the lead wires which have been connected to the terminal on the secondary side of the filament transformer.
5. Remove the four screws holding the air guide located around the magnetron antenna.

6. Remove the air guide assembly from between the magnetron and fan motor.
7. Remove the thermal cutout.
8. Remove four nuts holding the magnetron.
9. After taking out the magnetron, mount a new magnetron by following the steps in reverse.

When mounting the magnetron, use care not to break the glass section. Tighten the holding nuts completely. Otherwise, radiation may leak. Be sure to confirm that the emboss ring is rigidly attached, check whether dirt or dust has adhered around the emboss rings. If dust adheres, radiation may leak.

Cook Timer Replacement. Using caution, proceed as follows:

1. Disconnect the lead wires and connectors attached to the cook timer, cook indicator light, and switch.
2. Remove the right-hand sash by removing the two screws.
3. Remove the cook timer together with the escutcheon base, cook switch, and cook indicator light by removing the four screws.
4. Remove the four screws holding the cook timer to the escutcheon base.
5. Remove the cook timer knob by pulling it toward you.
6. Remove the cook timer pointer by pulling it toward you.
7. Remove the cook timer dial by removing two screws.
8. After taking out the cook timer, mount a new cook timer by following the steps in reverse. When mounting the cook timer knob and cook timer pointer, use care not to disturb the zero point.

Cook Timer Zero Adjustment. Mount the cook timer knob with the triangular mark facing upward. Turn the cook timer knob about one-fourth of a turn clockwise and return it. Two clicks will be heard. Adjust it so that the cook timer pointer is at zero when the first click is heard, and at OFF when the second is heard. Push the cook-timer knob and cook-timer pointer in completely so that they will not fall out.

Stirrer Motor Replacement. Using caution, proceed as follows:

1. Disconnect the connector of the stirrer-motor lead wire.
2. Remove the ceiling.
3. While supporting the stirrer blade by hand from the oven side, remove the stirrer blade by hand from the oven side, holding wire pin with long-nose pliers.
4. Remove three screws holding the stirrer motor.

NE-6600 DISASSEMBLY & PARTS REPLACEMENT

Outer Panel Removal. First remove the upper panel. The oven is so designed that left and right side panels cannot be removed before the upper panel. Remove the three screws from the upper front section and the three screws from the rear section to raise the front part. Then remove the two screws from the front sections of each side and the three screws from the rear sections of each side.

Magnetron Replacement. *Caution*: Remove your watch whenever working close to or repairing magnetron.

Remove the magnetron with the shielded case, permanent magnet, choke coils, and 500 pF capacitors as one assembly. Proceed as follows:

1. Discharge the electric charge remaining in HV capacitor.
2. Disconnect the magnetron leads connected to the terminal-board assembly.
3. Pull the thermal cutout from the magnetron assembly.
4. Remove the two nuts from the upper and the two nuts from the lower sections of the magnetron assembly while supporting the screws with the screwdriver inside the oven. Pull the magnetron assembly toward you.
5. After removing the magnetron, install a new assembly by following the above steps in reverse.

When installing the magnetron, use care not to break the glass section. To prevent microwave leakage, tighten the mounting nuts and screws properly, making sure there is no gap between the oven wall and the magnetron. Make sure that the embossed ring is rigidly attached on the magnetron anode.

HV Transformer Replacement. Using caution, proceed as follows:

1. Discharge the electric charge remaining in the HV capacitor.
2. Remove the two nuts from the terminal board to remove the two filament-side lead wires of the HV transformer.
3. Disconnect the connector from the HV capacitor.
4. Disconnect the stack-diode lead wire from the terminal board and remove the two-lead wire holder.
5. Remove the screw holding the green lead wire to the interlock-switch bracket.
6. Disconnect the connector of the lead-wire harness from the terminal board.
7. Disconnect the ground lead wire of the HV transformer.
8. Remove the two nuts holding the HV transformer to the base assembly and pull it toward you.
9. Disconnect the two lead wires of the primary from the terminal board.
10. Remove the two screws holding the terminal board onto the HV transformer.

Fan Motor Replacement. Remove the motor assembly with gear, gear case, and pulley as one part. Remove the belt from the pulley. Disconnect the connector of motor assembly. Remove the screw to remove the air guide. Remove the three screws holding the motor assembly to the oven assembly and pull it upward.

Cook Timer Replacement. Using caution, proceed as follows:

1. Disconnect five connectors of the cook switch, cook indicator lamp, defrost lamp, timer motor, and defrost indicator switch.
2. Disconnect the two lead wires connected to the cook switch.
3. Disconnect the two lead wires of the defrost switch from the HV capacitor.
4. Pull the cook timer knob in your direction.

5. Remove the five screws holding the escutcheon base to the oven assembly.
6. Remove four screws holding the drum-cover assembly to the escutcheon base.
7. Extend the tension spring by carefully pulling the dial thread to remove the dial pulley.
8. Pull the dial pulley in your direction from the axis of the cook timer.
9. Remove the three screws holding the cook timer to the drum cover assembly.

Dial Thread Replacement. Using caution, proceed as follows:

1. Form the two loops as shown in Fig. 9-20.
2. Turn the dial pulley counterclockwise until you hear a click at the end.
3. Attach a loop *A* of the dial thread to the dial pulley.
4. Install the long part (l_1) of the dial thread as shown by the dotted line in Fig. 9-21.
5. While holding the long part (l_1) of the dial thread, install the short part (l_2) of the dial thread as shown by the solid line in Fig. 9-21.
6. Attach a loop *B* (Fig. 9-21) of the dial thread to one hook of the tension spring and pull the end of the long part of the dial thread as shown, while attaching it to the other hook of the tension spring. Adjust the length, of the tension spring as follows:

$$l = 1\frac{19}{32} \text{ to } 1\frac{11}{16} \text{ inches}$$

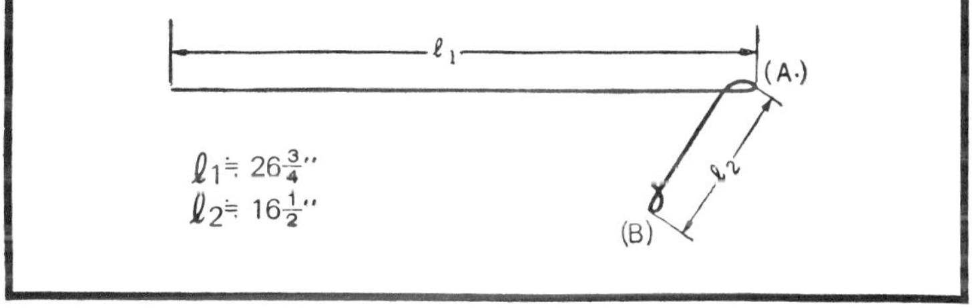

Fig. 9-20. Sketch of NE-6600 dial cord, showing loops mentioned in text.

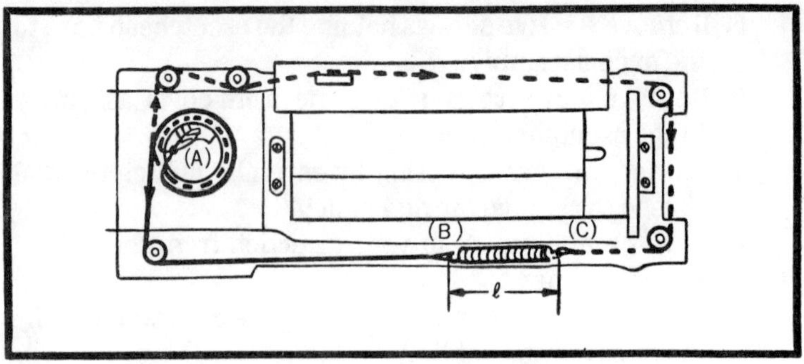

Fig. 9-21. Dial-cord diagram.

When the tension spring becomes this length, form a loop at point *C* of the long part of the dial thread.

7. Attach the pointer to the dial thread.
8. After installation of the pointer, adjust the OFF point of the cook timer. Turn the cook-timer knob clockwise about one-fourth of a turn; then return it. A click will be heard. Adjust it so that the cook-timer pointer is at OFF point when the click is heard. Then cement the pointer to dial thread so that it will not move.

Door Assembly Replacement. Using caution, proceed as follows:

1. Remove the wire pin to remove the door arm from the shutter-lever assembly.
2. Detach the end of door spring from door arm.
3. Remove the two screws holding the upper sash to the oven assembly.
4. Remove the two screws holding the upper hinge to the oven assembly. Pull the door assembly upward to remove the door assembly from the lower hinge then pull it toward you.
5. When mounting the door assembly to the oven assembly, be sure to adjust the door assembly parallel to the sash by moving the upper and lower hinges. The gap between the outside surface of the oven and the door must be less than a sixteenth-inch. If the door assembly is not mounted parallel, microwaves may leak from the clearance between the door and the oven.

Outer Panel Removal. First remove the upper panel. The oven is so designed that the back, right, and left panels cannot be removed before the upper panel. Remove the three screws from the upper-front section and the three screws from the upper-rear section to raise the front part. Then the upper panel can be removed.

After removing the upper panel, remove the eight screws on three sides. Then the back panel can be removed. Remove the three screws from the lower section of each side and swing the rear part outward. Then the side panels can be removed.

Magnetron Replacement. *Caution*: Remove your watch whenever working close to or repairing the magnetron.

Remove the magnetron with the shielded case, permanent magnet, choke coils, and 500 pF capacitor because they are all one assembly.

1. Discharge the electric charge remaining in the HV capacitor.
2. Remove the insulation cover from the filament transformer by removing one screw.
3. Disconnect the two filament lead wires from the filament-transformer secondary by removing the two nuts.
4. Disconnect the connector of the thermal-cutout lead wire.
5. Remove the air guide around the antenna of the magnetron by removing the two screws.
6. Remove the air guide assembly from between the magnetron and fan by removing one screw.
7. Disconnect the connector of the fan motor.
8. Remove the motor assembly by removing one nut.
9. Remove the four nuts holding the magnetron to the waveguide to remove the magnetron, supporting the magnetron from the lower side with your hand.
10. Remove the thermal-cutout assembly.
11. Disconnect the lead wire from the magnetron filament.
12. After removing the magnetron, install a new one by following the above procedure in reverse.

When installing the magnetron, use care not to break the glass section. To prevent microwave leakage, tighten the mounting nuts properly, making sure there is no gap between the waveguide and the magnetron. Make sure that the emboss ring is rigidly attached to the other emboss ring which has already been attached to the magnetron anode.

HV Capacitor Replacement. Discharge the electric charge remaining in the HV capacitor. Disconnect the three lead wires from the HV capacitor. Remove the screw from the HV capacitor bracket. Then the HV capacitor can be removed.

Filament Transformer Replacement. Using caution, proceed as follows:

1. Discharge the electric charge remaining in the HV capacitor.
2. Remove the insulation cover by removing one screw.
3. Disconnect the three lead wires which have been connected to the filament-transformer secondary.
4. Disconnect the connector from the filament-transformer primary.
5. Remove the screw from the bracket.

Rectifier Assembly Replacement. Discharge the electric charge remaining in the HV capacitor. Remove the insulation cover. Disconnect the lead wires from the rectifier terminals. Remove the two screws which have been holding the rectifier assembly to the filament-transformer bracket.

Stirrer Motor Replacement. Disconnect the connector of the stirrer-motor lead wire. Remove the ceiling inside the oven by removing four screws and pushing it to the rear. Remove the stirrer blade by pulling its center part to the bottom with your hand. Remove the two screws holding the stirrer motor on the oven.

Cooking-Guide Lamps & Switch Replacement. Using caution, proceed as follows:

1. Remove the exhaust-guide assembly by removing the three screws.
2. Disconnect the three connectors from the cooking-guide lamps and the switch.
3. Remove the escutcheon with cooking-guide lamps and switch by removing three screws.

4. For replacement of the three cooking-guide lamps, remove one screw of each; then pull the lamp toward you.
5. For replacement of the cooking-guide switch, remove the four screws and take out the switch with the lamp-bracket assembly. After that, pull the cooking-guide switch button and remove two screws. Then the cooking-guide switch can be removed.

Door Assembly Replacement. Using caution, proceed as follows:

1. Spread one side of the door spring from the door arm.
2. Loosen the two nuts mounting the lower hinge to the base assembly.
3. Loosen the two nuts mounting the upper hinge assembly to the oven assembly.
4. Hold the door assembly with your hand and pull it forward. Then the door assembly can be removed from the unit.
5. When mounting the door assembly, adjust the door assembly parallel to the oven and the ferrite rubber. If the door assembly is not mounted parallel, microwaves may leak from the clearance between the door and the oven. Adjust by moving both the upper and lower side of each door hinge.

Defrost Circuit Replacement. Disconnect the lead wires connected to the defrost-circuit assembly. Remove the two screws holding the defrost-circuit assembly to the heat-radiating metal. Then the defrost-circuit assembly can be removed.

HV Transformer Replacement. Using caution, proceed as follows:

1. Discharge the electric charge remaining in the HV capacitor.
2. Remove the filament transformer according to preceding steps.
3. Remove the filament-transformer bracket with defrost-circuit assembly, the heat-radiating metal, and the rectifier assembly by removing two screws.

When removing the two screws holding filament-transformer bracket, it is advisable to remove the HV capacitor before removing filament-transformer bracket.

4. Disconnect the three lead wires connected to the HV transformer primary.
5. Disconnect two lead wires from secondary. One is connected to the HV capacitor, and other is connected to the rectifier assembly.
6. Remove the two nuts. Then the HV transformer can be removed.
7. Disconnect two lead wires from the secondary with soldering iron.

Chapter 10

Servicing Roper Microwave Ovens

This chapter provides servicing information for pratically all of the microwave ovens offered by Roper Sales[1]. It includes complete wiring diagrams for five models. It also includes step-by-step servicing instructions for a typical Roper model, and this information can, with certain adaptations, be used in servicing any Roper model.

Energy-emission and microwave-power measurements follow the general procedures given in Chapter 3. Be sure to make these measurements carefully and accurately and follow all of the safety precautions given earlier.

HOW THEY WORK

Refer to the appropriate wiring diagram for the model of interest. See Figs. 10-1 through 10-5 for schematics of Models 2095, 2904, 2905, 2915, and 2926.

Anytime the door is unlatched or open, the magnetron's blower motor and the oven-cavity light will be on. The auxiliary blower motor on the 2095 will also be on. When the oven door is closed and latched and the timer is set to OFF, the magnetron-blower motor—as well as the 2095 auxiliary blower motor—and oven light are off.

---

1. All of the illustrations in this chapter are courtesy of Roper Sales.

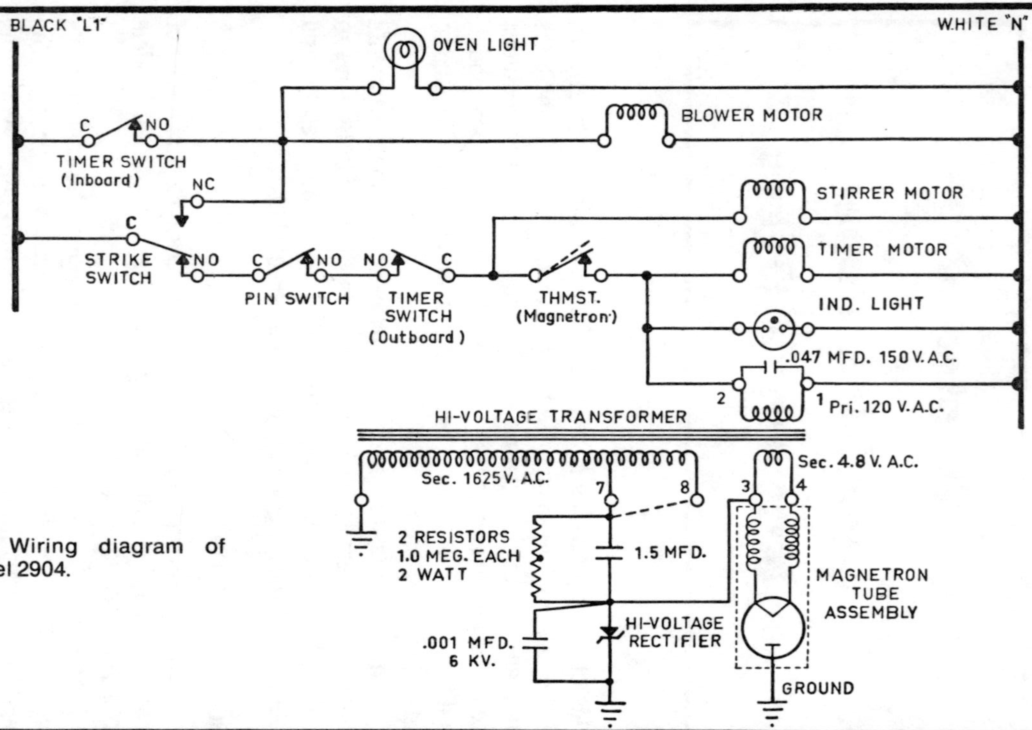

Fig. 10-1. Wiring diagram of Roper Model 2904.

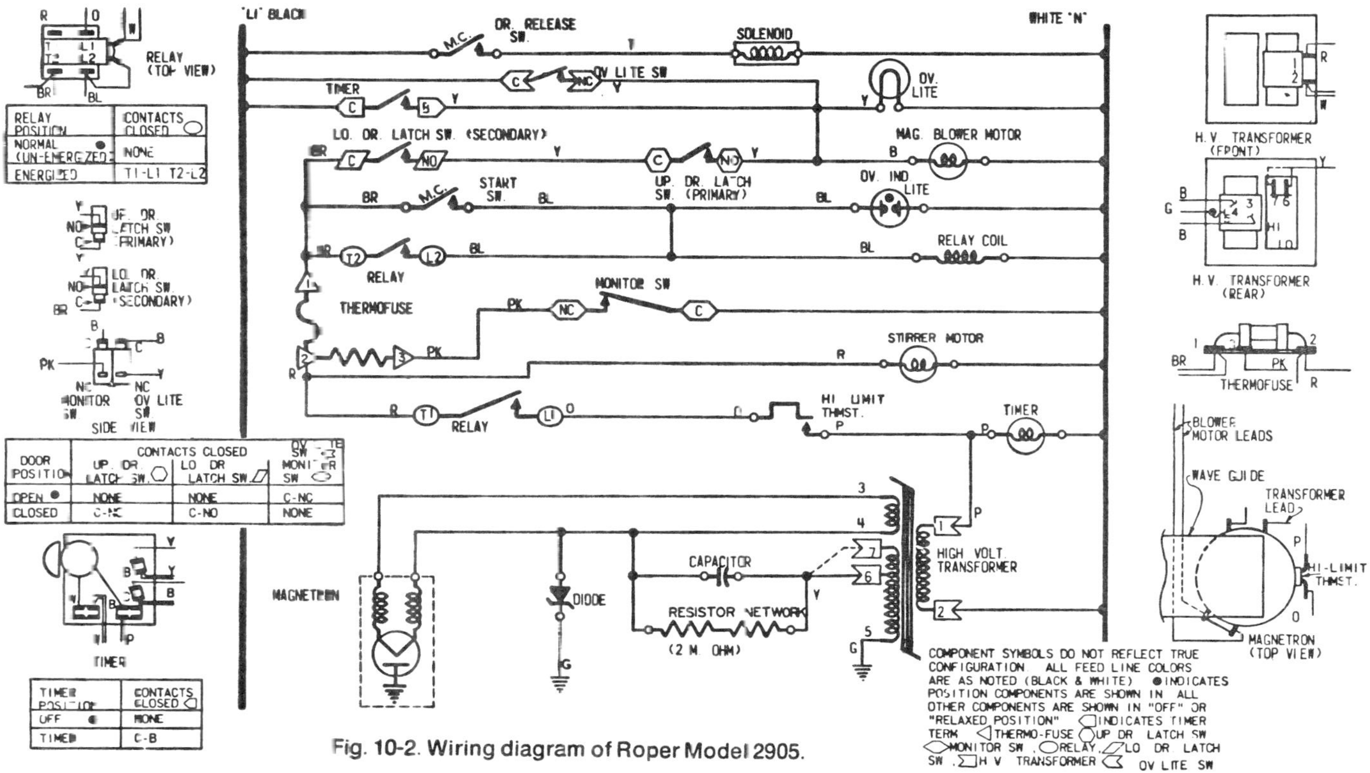

Fig. 10-2. Wiring diagram of Roper Model 2905.

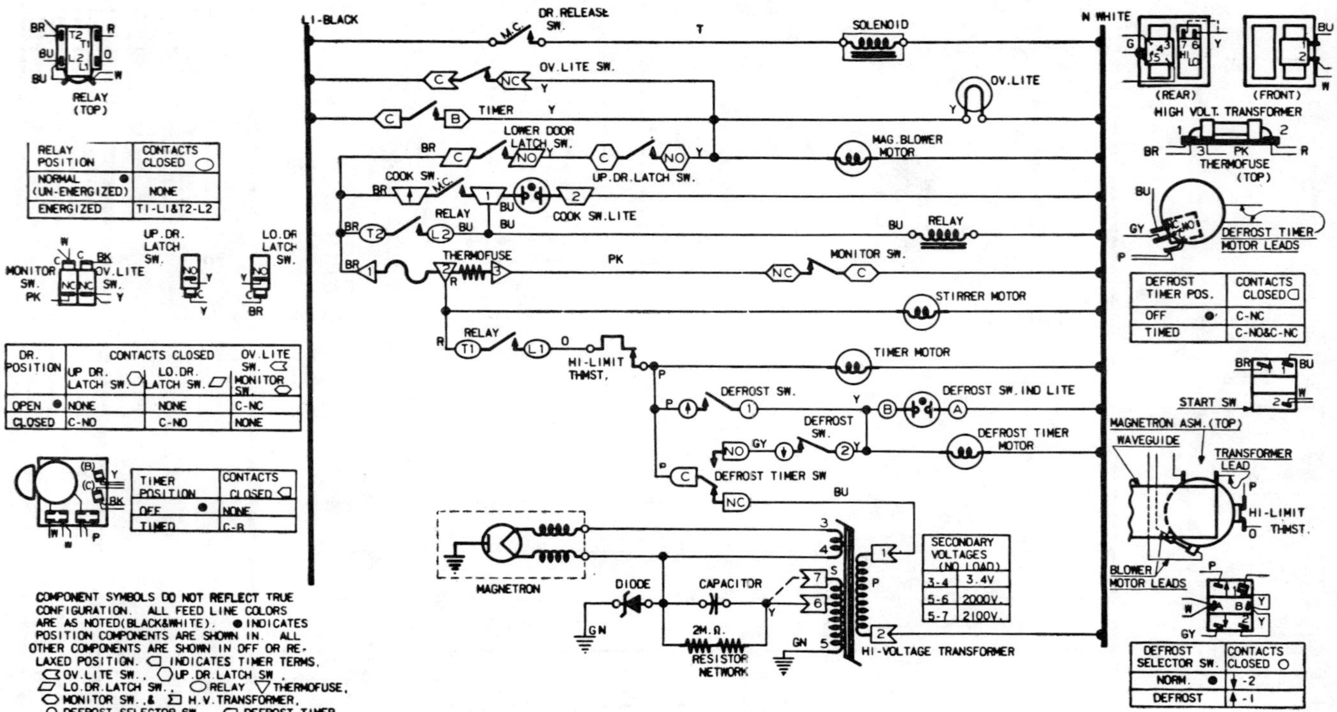

Fig. 10-3. Wiring diagram of Roper Model 2915.

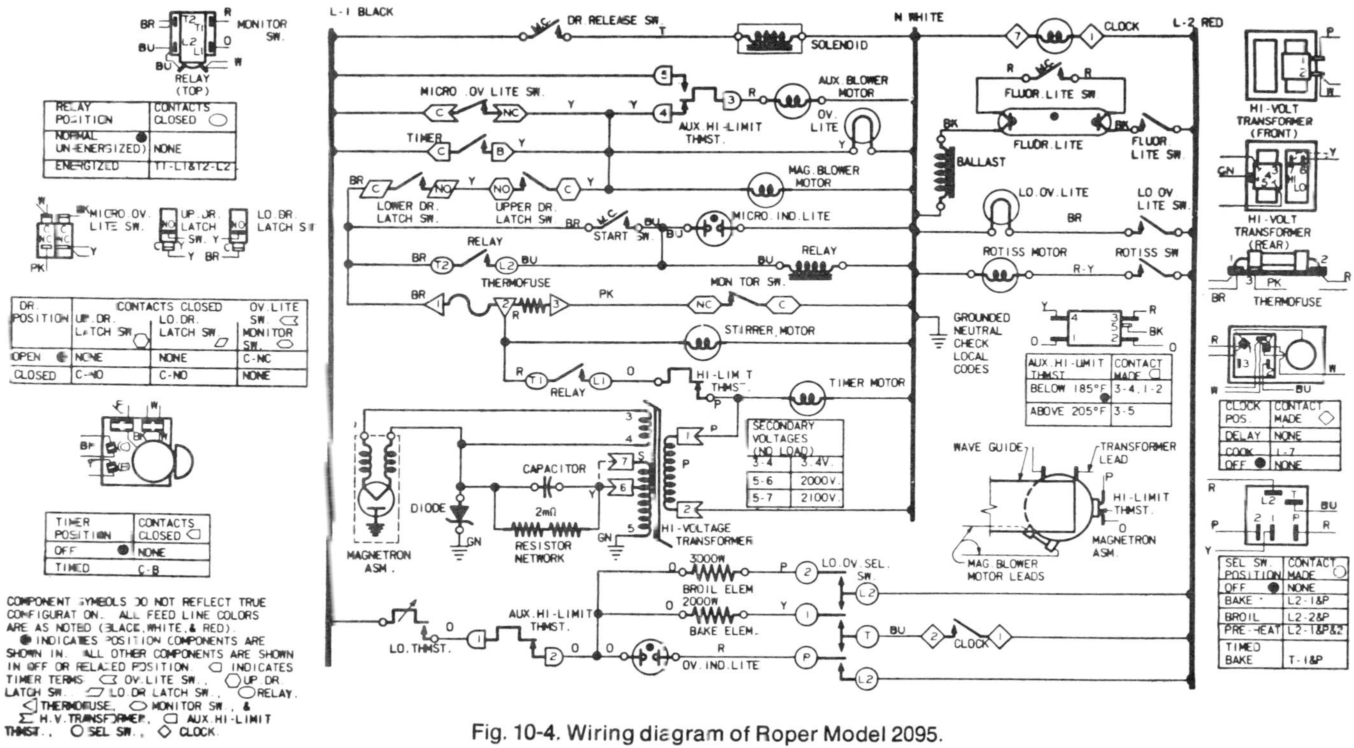

Fig. 10-4. Wiring diagram of Roper Model 2095.

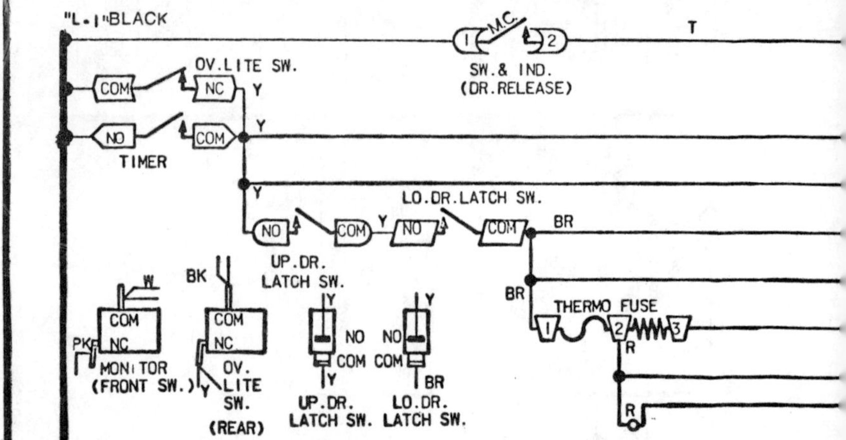

DR.POSITION	CONTACTS CLOSED MONITOR SW. / OV.LITE SW.	UP.DR.LATCH SW. / LO.DR.LATCH SW.
OPEN ●	COM-NC	NONE
CLOSED	NONE	COM-NO

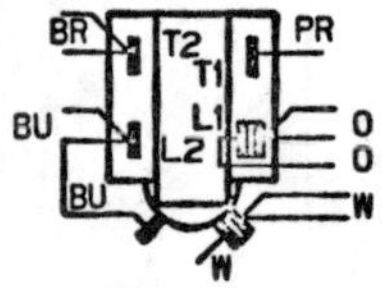

RELAY POSITION	CONTACTS CLOSED
UNENERGIZED ●	NONE
ENERGIZED	T2-L2,T1-L1

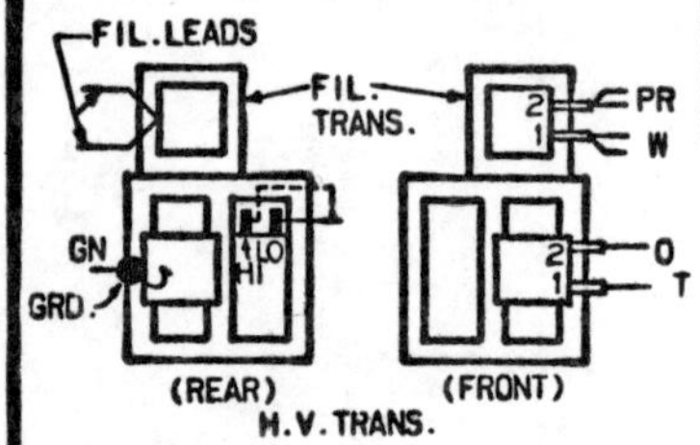

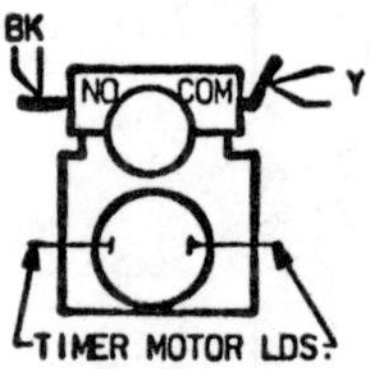

TIMER POSITION	CONTACTS CLOSED
OFF ●	NONE
TIMED	COM-NO

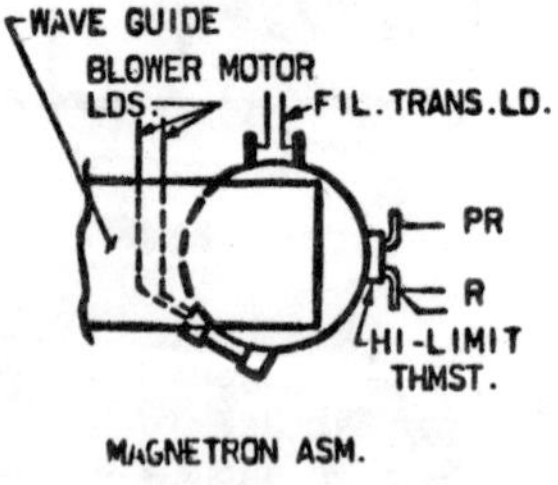

SW. & IND. POSITION (STATION DEPRESSED & HELD)	CIRCUITS CLOSED
NONE ●	NONE
DR.RELEASE	1-2
COOK START	8-9
RESISTOR SPECIFICATIONS	
R3-18KΩ ±20%,0.25W	

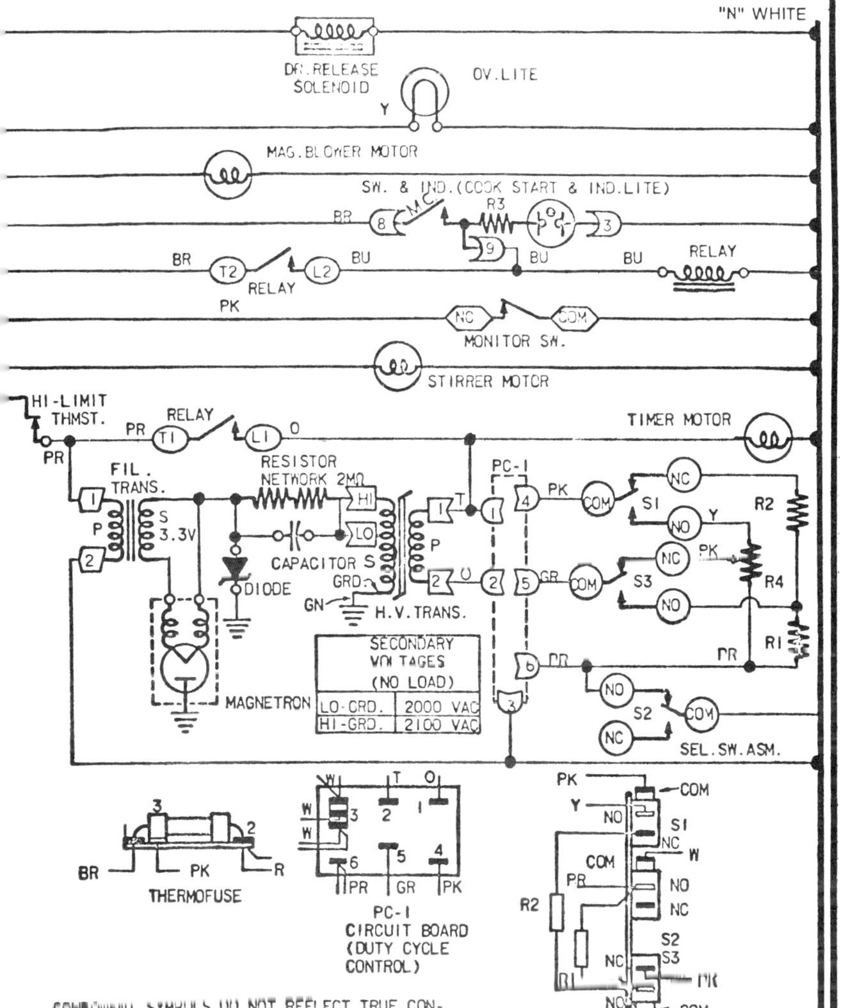

COMPONENT SYMBOLS DO NOT REFLECT TRUE CONFIGURATION. ALL FEED LINE COLORS ARE AS NOTED(BLACK & WHITE). ● INDICATES POSITION COMPONENTS ARE SHOWN IN. ALL OTHER COMPONENTS ARE SHOWN IN "OFF" OR "RELAXED" POSITION. ◁ INDICATES TIMER TERMS. ▭ OV.LITE SW., ◻ UP.DR.LATCH SW., ▱ LO.DR.LATCH SW., ○ RELAY ▽ THERMOFUSE ⬭ MONITOR SW., ▭ H.V.TRANSFORMER, ⊂⊂ SW. & IND., ○ SELECTOR SW.ASM. ▱ FIL TRANS.

SELECTOR SW.ASM. POSITION	CONTACTS CLOSED S1	S2	S3
MICRO SELECT	COM-NO	COM-NC	COM-NC
NORMAL COOK ●	COM-NC	COM-NO	COM-NC
DEFROST	COM-NC	COM-NC	COM-NO

RESISTOR SPECIFICATIONS
R1-330KΩ ±20%,0 5W
R2 220KΩ ±20%,0.5W
R4-VARIABLE,0 TO 540KΩ ±10%,0.125W

Fig. 10-5. Wiring diagram of Roper Model 2926.

Closing and latching the oven door closes the primary and secondary interlock switches. When the timer is set for cooking and the COOK/START switch is depressed, the relay is energized. A circuit now exists from L1 (black line) through the timer switch to the oven light and magnetron blower motor, and auxiliary blower motor on the 2095. Simultaneously, another circuit exists from L1 through the timer and both interlock switches, branching to: 1) the latching contacts of the relay to the relay coil and indicator light, 2) the thermal fuse and stirrer motor, 3) the power contacts of the relay, the normally closed magnetron high-limit thermostat, the main timer, the defrost timer switch on the 2915, and the transformer's primary winding.

When the timer is set for cooking and the COOK/START and DEFROST switches are depressed (2915 only), the same circuits exist as stated previously. However, an additional circuit will also exist through the DEFROST switch and defrost timer motor.

The secondary of the transformer has two windings. One is a 3.7V filament winding (terminals 3 and 4), and the other is a 2000V AC winding from ground (chassis) to terminal 6 or 7 (terminal 7 will be approximately 100V more). The oven is normally operated with the HV lead connected to terminal 6. The function of the 3.7V winding is to heat the magnetron filament.

The HV winding of the transformer, in conjunction with the 1.05 μF capacitor and HV rectifier, supplies the high DC voltage necessary to operate the magnetron in order to convert 120V, 60 Hz power into microwave energy.

On model 2095 only, if the microwave oven is not in use but the lower conventional oven is, a circuit will exist from L1 (black) through the auxiliary high-limit switch to the auxiliary blower motor if the temperature in the microwave oven areas rises to 205°F. The auxiliary blower motor will remain on until the temperature in the oven falls to 185°F. This condition is generally caused from long periods of preheat or broiling in the conventional lower oven.

Model 2926 Exceptions

Two types of transformers are used in the 2926. The smaller of the two is a filament transformer. When the timer is set for cooking and the door is closed, a circuit exists which provides the primary winding of the filament transformer with 120V AC. This, in turn, is converted to 3.3V AC from the secondary winding of the filament transformer. The 3.3V AC is used to heat the filament of the magnetron. The heating of the magnetron filament by the filament transformer enables the magnetron tube to immediately emit microwave energy when the COOK/START switch is depressed and the secondary winding of the HV transformer is energized.

The circuit of Fig. 10-6 has been isolated from the regular wiring diagram so that the functions of the interlock-and-monitor switch system may be better understood. This branch of the system must be understood to make basic adjustments on the interlock and monitor switches.

The power supply is controlled by the timer switch. When the timer is set for a cooking period, power is supplied to the interlock switches, both primary and secondary, which must make when the door is closed. When this circuit is closed, there is power to terminal 1 of the thermal fuse. The key to the entire system is that at this point the monitor switch must be open (it is normally closed). (Refer to Fig. 10-6.) Note that if the monitor switch is closed at the same time as the interlock switches, a direct short is created with only the limiting resistor in series. The resistor limits the current, although the current will destroy the fuse. The fuse being destroyed is a

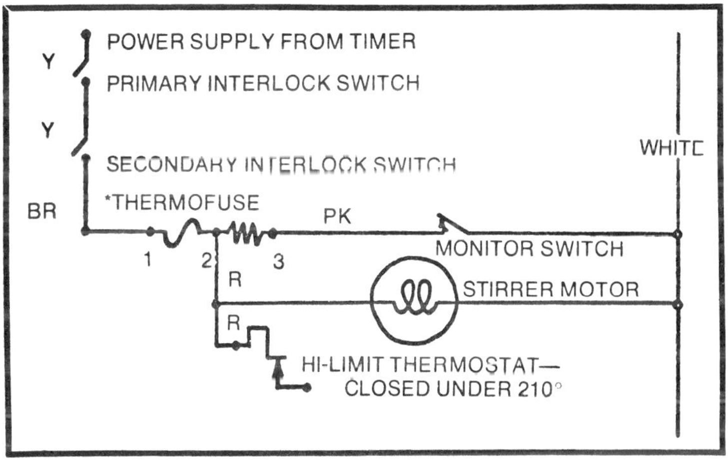

Fig. 10-6. Interlock switches in Roper models.

signal that the switches must be checked for proper adjustment. The cure is not to merely change the fuse, as it is conceivable that the same condition could reoccur and blow the fuse a second time. One quick yet sure check to determine if the fuse is good is to inspect the stirrer motor. If this motor is operating normally, the fuse is intact.

TROUBLESHOOTING & REPAIR

To diagnose the trouble with one of these ranges, refer to the listing of troubles and probable causes given in Table 10-1. Then refer to the specific servicing information for the component or section that seems to be at fault. Specific servicing information for the various components and sections is given in the following pages.

PREPARATION FOR SERVICING

On portable models, disconnect the unit from its power source and remove the top cover to gain access to all components. Remove the cover by taking out the seven screws in the back and the four on the bottom of the unit.

On the double-cavity built-in unit, the microwave module must be removed for all servicing. Disconnect the entire oven from the power line. Remove the two screws and remove the trim piece. Pull the module out a maximum of 4 inches. This should allow enough room to reach in to disconnect the module from the main wiring harness. Pull the module all the way out and set it on a table or work area. An auxiliary power cord must be used to supply power to the module during servicing. It must be a 3-wire grounded cord for use with a 110V outlet.

MODELS 2905, 2904 & 2915 SERVICE PROCEDURES

Never operate or permit a customer to operate a faulty or damaged microwave oven until the unit has been properly repaired, tested, and checked to determine that no microwave emission greater than 5 mW/cm^2 exists.

Excessive microwave energy emission in an undamaged, properly assembled oven is limited to the door area. Make an inspection of the door area before the unit is turned on.

If the door or door frame is warped, components must be replaced. If improper latch adjustment is evident, proper

Table 10-1. Troubleshooting Chart for Roper Models

TROUBLE	CAUSE
Nothing operates, door won't open.	1. Household fuse. 2. Power cord not connected.
Nothing operates, door will open.	1. Timer switch. 2. Interlocks. 3. Relay. 4. Push-to-start switch.
Oven light doesn't come on.	1. Bulb. 2. Loose wire.
Blower won't run.	1. Blower motor. 2. Loose wire.
Food not heating (oven ind. light out, stirrer not running).	1. Door-interlock switch (primary). 2. Door-interlock switch (secondary). 3. Timer switch. 4. Monitor switch and thermofuse. 5. Relay.
Timer won't advance (stirrer running, ind. light on)	1. Timer motor. 2. Magnetron thermostat.
Ind. light out (food heats).	Indicator light.
Stirrer not turning (cook light on, timer advancing).	Stirrer assembly.
Food not cooking (oven and ind. lights on, blower motor running).	1. Transformer. 2. Rectifier. 3. Magnetron. 4. Capacitor. 5. Magnetron thermostat. 6. Thermofuse.
Arcing in cavity	1. Stirrer alignment. 2. Carbonized dust in oven cavity. 3. Metal utensils.
Microwave leakage	Door adjustment
Microwave turns off during cooking cycle.	1. Load too small. 2. Blower assembly. 3. Stirrer. 4. Magnetron thermostat.
Poor heating pattern	1. Stirrer assembly. 2. Improper placement or rotation of food.

adjustment must be made prior to operating the microwave oven.

Door Removal, Adjustment & Replacement

1. Disconnect power to oven.
2. Refer to Fig. 10-7. Remove the two screws (B) from the hinge-adjustment bracket. Lift the oven door from

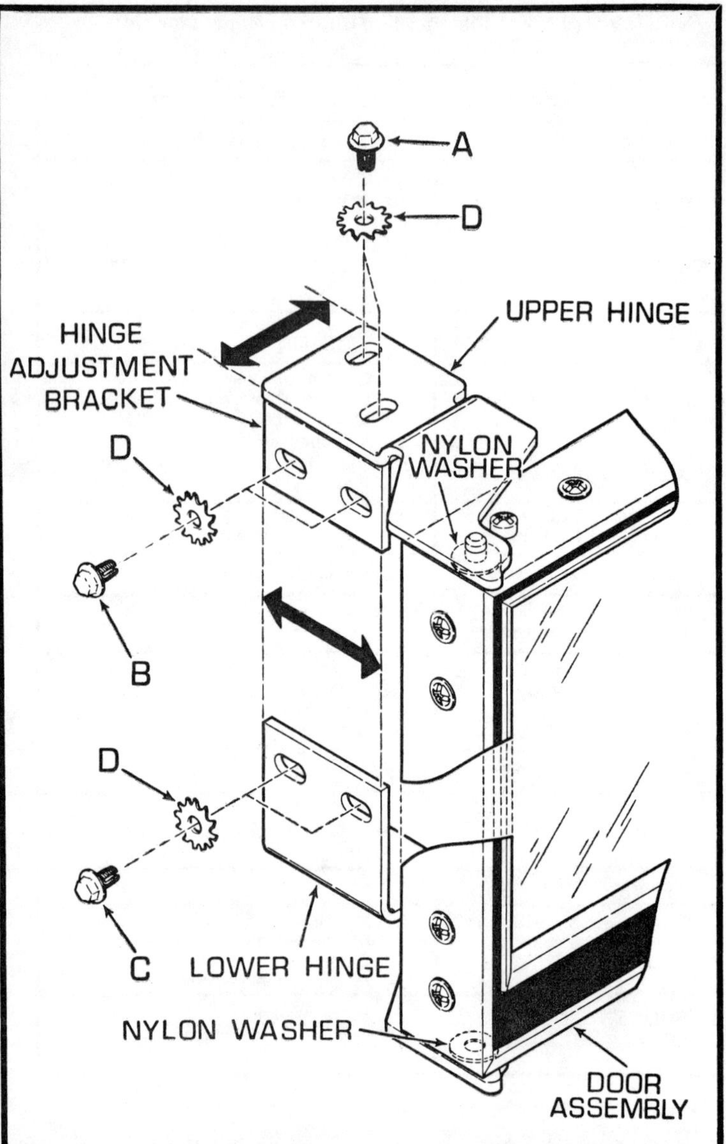

Fig. 10-7. Door removal and adjustment.

the lower hinge. Retain the two nylon washers from the upper and lower hinges.

3. Position door assembly so door strikes are evenly spaced in the front frame openings. This is done by

loosening the two screws (C) on the lower hinges and the four screws (A and B) on the upper hinges and hinge-adjustment bracket. Make sure the washer is in place on the lower hinge.

4. With door positioned, hold the lower hinge in place by pulling up on the hinge and tightening the two screws (C) while maintaining compression on the door seal. The lower hinge should be tight against and parallel with the bottom of the front frame.
5. Place the washer over the upper-hinge pin of door. Secure the upper hinge to upper-hinge adjustment bracket using two screws (A) and two lockwashers (D).
6. Place the upper-hinge assembly on the upper-hinge pin of the door. Secure the upper hinge while applying pressure to the door and seal when tightening the two screws (B) on the hinge-adjustment bracket.
7. Close the door and recheck the door strike spacing in the front frame openings. Reposition, if necessary, by loosening the two screws (A) on the upper hinge and sliding the hinge left or right to obtain the correct positioning. Tighten the screws.

SAFETY INTERLOCK SWITCHES

The microwave oven has two concealed snap-action safety switches (Fig. 10-8). Both operate independently of each other. Both break the main power before the door begins to open. Both switches are actuated by the door strike attached to the door.

Testing Electrical Operation

1. Open the oven door and place a load in the oven.
2. Close and latch the door.
3. Set the cook time on timer. The oven light and blower motor should come on. Push the COOK/START switch. The cook indicator light should come on.
4. If the cook indicator does not come on, one or both of the interlock switches may be defective or misaligned.
5. Open the oven door. The COOK indicator light will go out.

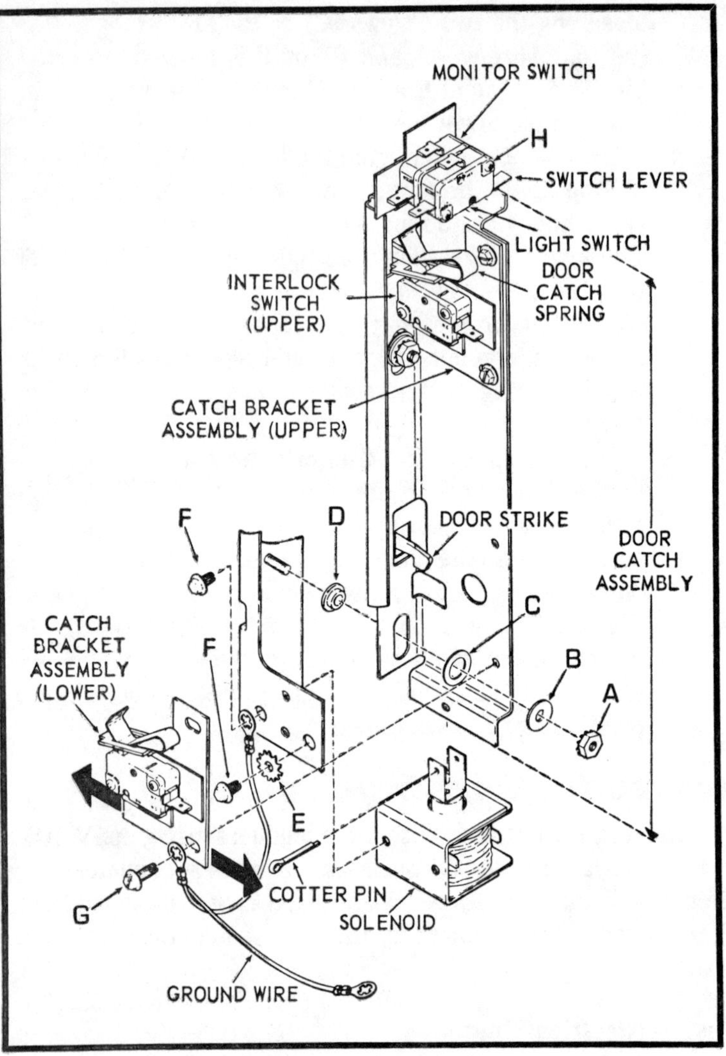

Fig. 10-8. Switches and door-latch assembly on Roper models.

6. Wait 10 seconds (minimum). Push the COOK/START switch. The indicator light should remain off. If the COOK indicator light comes on when the door is open, the fuse, as well as interlock switches or wiring, must be replaced.
7. Close and latch the door. Turn the timer to OFF. The oven-cavity light and blower motor will turn off.

Interlock Actuation

1. Push the door-release switch to open door.
2. Disconnect the oven from its electrical supply.
3. Close the door slowly. Observe the door strikes to see that they pass freely through the front frame slots and enter the slots in the catch-bracket assembly (Fig. 10-8).
4. If door strikes do not enter the front frame freely, make a door-hinge adjustment.
5. Each door strike should enter and engage its catch-bracket assembly. Each door-catch spring should ride up on the door strike, pulling the door-catch assembly upward when both strikes are over the catch-bracket-assembly slots.
6. If the interlock switches do not *make* (audible click), check for;
   a. Improper door-catch-spring tension.
   b. The solenoid binding.
   c. The monitor lever arm bottoming on the switch when the door is closed.
   d. The catch assembly binding.

   Correction for these problems is as follows:
   a. Replace the spring.
   b. Remove the solenoid from the oven and actuate it on a bench test using 120V AC. Replace it if the solenoid does not actuate.
   c. Loosen the top screw on the monitor and light switches (H of Fig. 10-8) and slide the switches upward.
   d. Check the tension on nut *A* and washer *B* (Fig. 10-8). The latch should float freely. If binding or rubbing is evident, grease may be applied to the bushing (D).

Catch-Bracket Assembly Adjustment

If the strike fails to enter the slot in the catch-bracket assembly, adjustments can be made as follows:

1. Loosen the catch-bracket screws (G of Fig. 10-8).
2. Apply pressure to the outside of the door in the strike area.

3. Slide the catch-bracket assembly so that the door strike just enters the catch-bracket slot.
4. Tighten the screws (G), making sure the catch bracket is parallel with the back edge of the door-catch assembly.
5. Repeat the procedure for the other catch bracket, if necessary.
6. Plug the oven into a 120V AC power supply. Push the door release. The door should open smoothly.
7. If the tension on the strikes seems too great, repeat steps 1 through 5 above.
8. Run a radiation-emission test to determine if the level is less than 5 mW/cm^2. If the emission level is greater than 5 mW/cm^2, steps 1 through 5 above must be repeated until the emission level is below 5 mW/cm^2.

Monitor Switch Adjustment

If interlock switches are not making, the monitor switches may require adjustments. Proceed as follows:

1. Loosen the screws (H of Fig. 10-8).
2. Close the door, if not closed, noting the actuation of the interlock switches.
3. The monitor switch is properly set when actuation of the monitor switch occurs just prior to the actuation of the interlock switches.
4. Tighten both screws.

Solenoid

There is no means of adjusting the solenoid. It can only be replaced. To remove, proceed as follows:

1. Remove the cotter pin.
2. Disconnect the coil wires.
3. Remove the two screws (F of Fig. 10-8), and slide the solenoid out.

To replace it, proceed as follows:

1. Position and pull up on the solenoid while securing to the bracket with the screws (F).
2. Position the holes in the plunger with the hole in the latch assembly. Insert the cotter pin.
3. Connect the coil wires.

Thermofuse

As the name implies, the thermofuse is a fuse link which will become interrupted provided the temperature surrounding the fuse link is raised to the melting point. This temperature is rapidly reached when a high current is passed through the resistor heater.

As shown in the wiring section, the fuse link is in series with the primary and secondary interlock switches and also the oven light and timer switches. The heater winding is in series with the monitor switch, which, in turn is connected to the neutral bus. With the door open, the oven light and monitor switches are closed and the primary and secondary interlock switches are open. If the primary and secondary interlock switches become defective in an unsafe mode (welded closed) a circuit will be completed through the oven-light switch and timer, the primary and secondary interlock switches, the fuse link and heater, and the monitor switch. This circuit will cause the thermofuse to open in approximately 2.0 to 8.0 seconds.

Warning: Do not replace thermofuse before replacing defective interlock switches.

Replace the fuse as follows:

1. Unplug the unit from the power source.
2. Remove the three wires from the thermofuse.
3. Remove the two screws holding thermofuse and replace with fuse of the *same part number* as indicated on the warning label (Fig. 10-9).
4. Replace the wires per the wiring diagram.
5. Replace the upper and lower interlock switches as previously outlined.
6. Replace the wrapper after making operational check.

Operational Check. Use a Simpson 260 meter or equivalent for the following test:

1. Set the meter to the R × 10 scale.
2. Measure the resistance of the heater. It should be approximately 100Ω.
3. Connect one lead to L1 and the other side to neutral.
4. With the door open, defeat each interlock switch individually. The meter reading should not change.

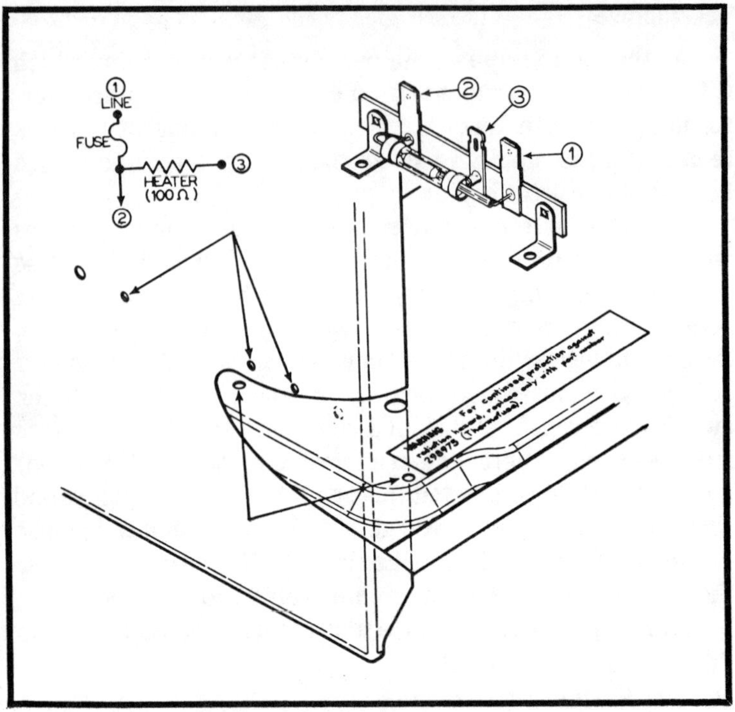

Fig. 10-9. Thermofuse schematic and physical appearance.

5. Defeat both interlock switches simultaneously. The meter should deflect to a lower reading.
6. Check the condition of monitor switch by measuring the resistance between common to the N.C. contacts. Acutate the switch, noting the meter deflection (make sure one wire is removed) before making tests.

MAIN TIMER

Each timer is a single-speed, 20-minute interval timer with an SPST (single-pole, single-throw) switch. Each has a bell that will signal the approach of zero time. The timer must be set before any microwave energy is generated. To check timer:

1. Plug the unit into a 120V AC source.
2. With the door closed, set the timer. The oven light and blower motor(s) will come on if the timer contacts are closing.

3. Turn the timer to OFF. The oven light and blower motor will turn off. If they remain on, check the timer and wiring.

To remove the timer refer to Figs. 10-10 and 10-11 and proceed as follows:

1. Disconnect the oven from the outlet.
2. Remove the timer knob and control-panel screws *A* and lockwashers *C*. Pull the panel assembly away and lay it to one side.
   Note: The wiring will need to be disconnected from the door-release switch on Model 2915 to swing the panel assembly far enough to expose the main-timer screws. *Do not* remove any wiring from the other models.
3. Remove the timer skirt from the 2905 and 2915.
4. Remove main-timer screws *B* and lockwashers *C*. Pull the timer from its position.
5. Disconnect the main-timer wiring.

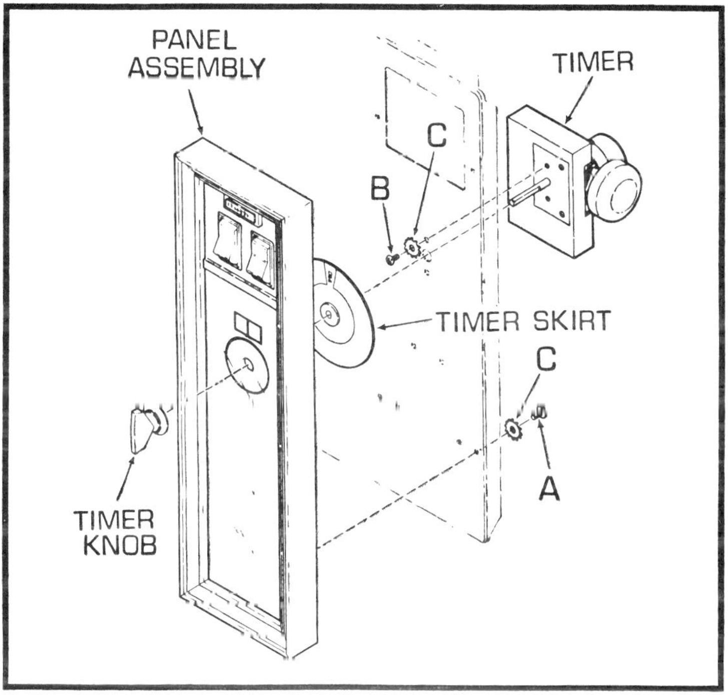

Fig. 10-10. Panel assembly and timer of Roper Models 2905 and 2915.

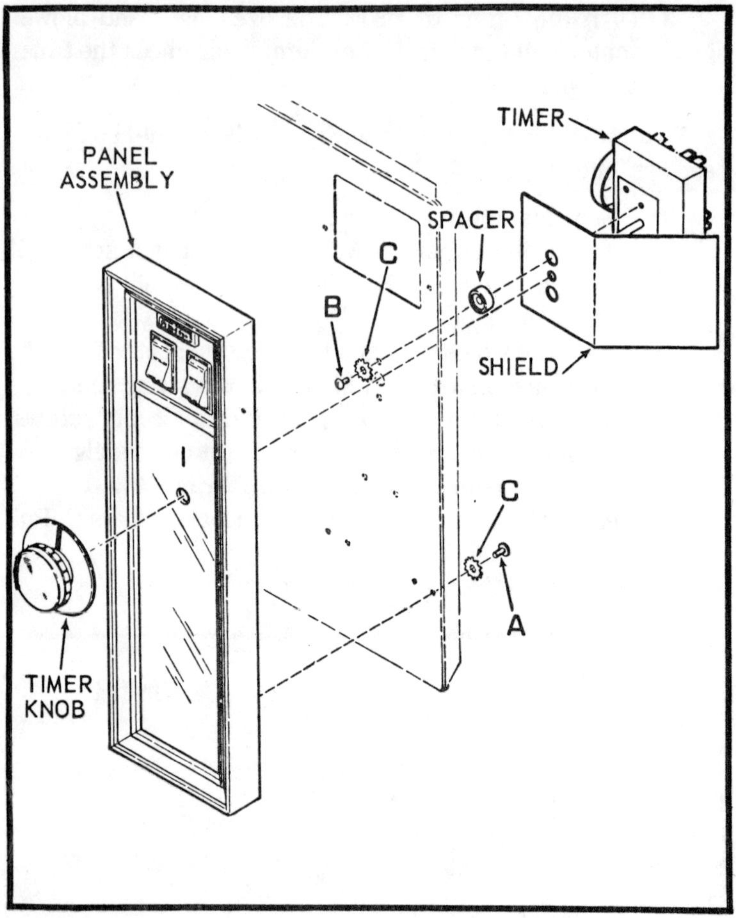

Fig. 10-11. Panel assembly and timer of Roper Model 2095.

To replace the main timer:

1. Secure the timer to the front frame with screws *B* and lockwashers *C*.
2. Replace the timer skirt on the 2905 and 2915.
3. Position the control-panel assembly on the front frame and secure with screws *A* and lockwashers *C*. Remember to connect the door-release-switch wiring on the 2915 model.
4. Connect the timer wiring as shown in Fig. 10-12.
5. Replace the timer knob.

Defrost Timer

Model 2915 has a defrost timer (Fig. 10-13). This timer is an electromechanical-pulsating timer made up of a drive motor and cam operating a microswitch. This microswitch makes and breaks the transformer primary circuit at a rate of 30 seconds *on* and 30 seconds *off*. The defrost circuit is before the primary of the transformer.

When the unit is set for defrosting (main timer set, DEFROST and COOK/START switches made), the defrost timer motor drives the cam to make and break the defrost timer switch.

If the unit is turned off and COOK is selected, a period of up to 30 seconds may elapse before microwave energy is generated. The DEFROST light will be on. This will happen due to the defrost timer being in the OFF postion when the unit was turned off. It will finish its off cycle and the defrost light will go out before going into a cook cycle. To check defrost timer:

1. Disconnect the unit from its power source.
2. Remove the wires from the microswitch.

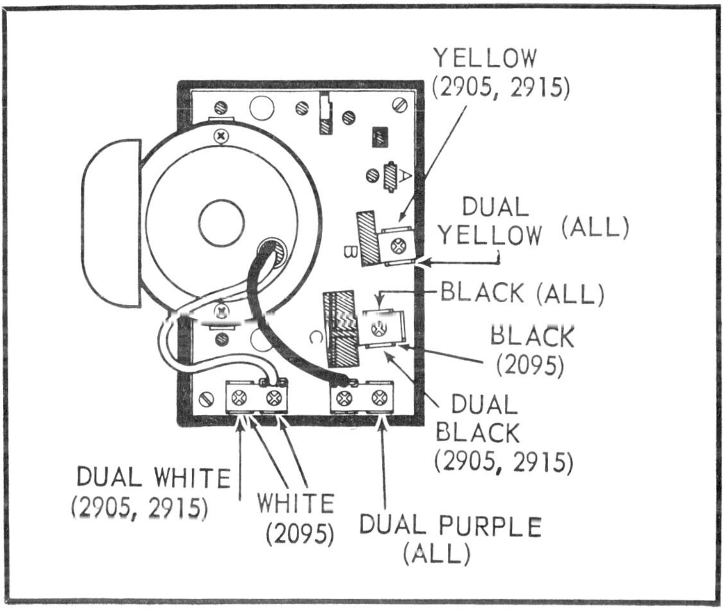

Fig. 10-12. Timer wiring.

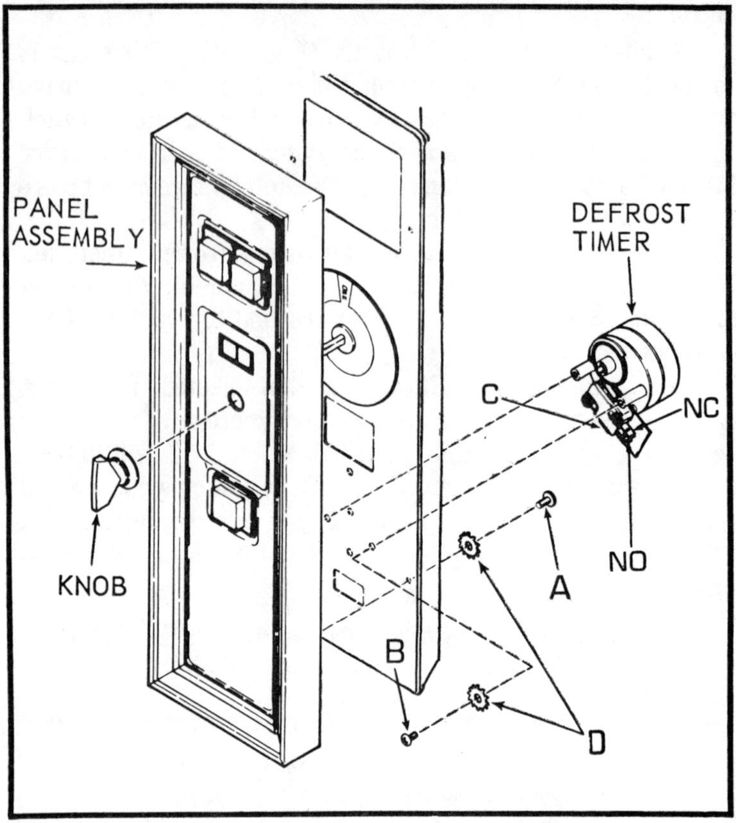

Fig. 10-13. Sketch of control panel and defrost timer—Model 2915.

3. Using a Simpson 260 or equivalent meter, check for continuity between the C and N.C. terminals (Fig. 10-13). Make sure the microswitch lever is fully depressed. Continuity should exist between C and N.O. No continuity should exist between C and N.C. If these conditions do not exist, the timer/ microswitch should be replaced.

To replace timer/microswitch (Fig. 10-13):

1. Disconnect the unit from its electrical supply.
2. Remove the control knob.
3. Remove control-panel screws *A* and lockwashers *D*.
4. Disconnect the door-release switch wires and tilt control-panel assembly, exposing timer screws *B*.

5. Remove timer screws *B* and lockwashers *D*.
6. Disconnect the timer wires.
7. Mount the replacement timer/microswitch, using screws *B* and lockwashers *D*.
8. Connect the door-release switch wires. Position the control panel and secure with screws *A* and lockwashers *D*.
9. Wire the timer with the blue lead attached to the N.C. terminal, the dual purple lead attached to the C terminal, and the gray lead to the N.O. terminal. (See Fig. 10-3.)
10. Replace the control knob.

BLOWER

The fan must be on whenever the oven door is open or the oven is in operation. Inadequate air movement will cause fast cycling of the magnetron's thermostat. Continued faulty fan operation can reduce the expected life of the magnetron.

If the oven cavity light is on and fan motor is not operating, check for defective wiring. If wiring checks out, fan motor is defective and needs to be replaced.

Fan & Motor Replacement

1. Disconnect the oven from its power source.
2. Rotate the fan by hand to ensure free movement of blades.
3. Disconnect the two fan motor leads (Fig. 10-14).
4. Remove screws *A* to disconnect the ground wire and separate the fan assembly from the magnetron tube. It may be necessary to remove the magnetron from its mounted position before removing the fan screws.
5. Pull the fan from the motor.
6. Loosen screws *B* (2) and pinch the motor support to remove from motor.
7. Loosen the new motor's screws and position the motor support. Tighten the two screws.
8. Press the fan onto the motor shaft.
9. Position the motor assembly on the magnetron. Make sure the motor coil is towards the oven front. Secure it

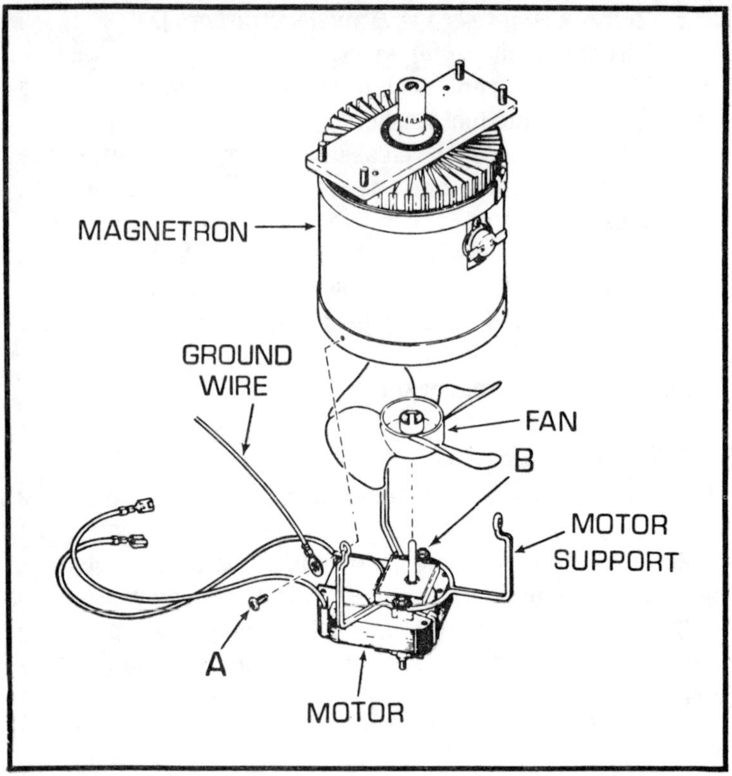

Fig. 10-14. Illustration for fan and motor replacement.

with screws *A*, making sure the ground wire is in place.

10. Rotate the fan by hand to ensure free movement. Check the distance between the fan blades and magnetron; there must be a quarter-inch between the blades and magnetron (Fig. 10-15).
11. Attach the motor leads. Make sure that they are clear of the fan blade.
12. Replace the top cover.
13. Plug the unit into a 120V AC power supply.
14. Open the oven door. The blower should come on when door is open.

STIRRER

If stirrer assembly fails to rotate during timer operation with door closed and timer set, the stirrer assembly must be inspected for tightness on the shaft and proper blade clearances.

Stirrer Test & Repair Procedures

Disconnect the unit from its power source. Remove the top cover and stirrer cover (Fig. 10-16).

1. Check the blade clearance (Fig. 10-17).
2. The stirrer motor can be checked by connecting a voltmeter across the motor leads. Make certain the meter is set to the 250V AC scale.
3. Connect the power cord to a 120V AC supply.
4. Close and latch the door.
5. Set the timer for 2 minutes. The line voltage should read 120V AC on the meter. Replace the defective stirrer motor if it fails to rotate.

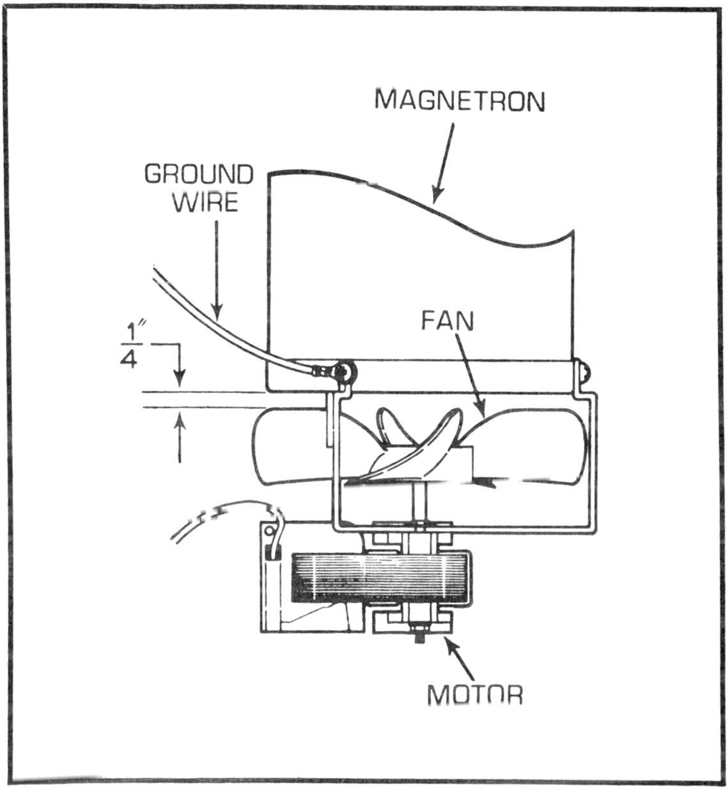

Fig. 10-15. Illustration of clearance between fan and magnetron.

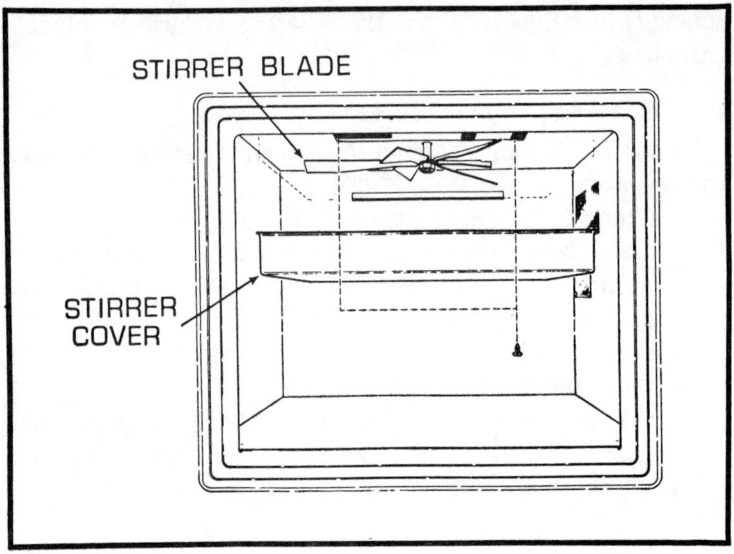

Fig. 10-16. Removal of top cover and stirrer cover.

Stirrer Motor Replacement

1. Disconnect the unit from its power source.
2. Remove the retainer clip from under the motor. This detaches the stirrer shaft from motor shaft. Hold the stirrer blade to prevent it from falling.
3. Disconnect the motor leads and remove screws *A* of Fig. 10-17.
4. Position the replacement motor and secure with screws *A*.

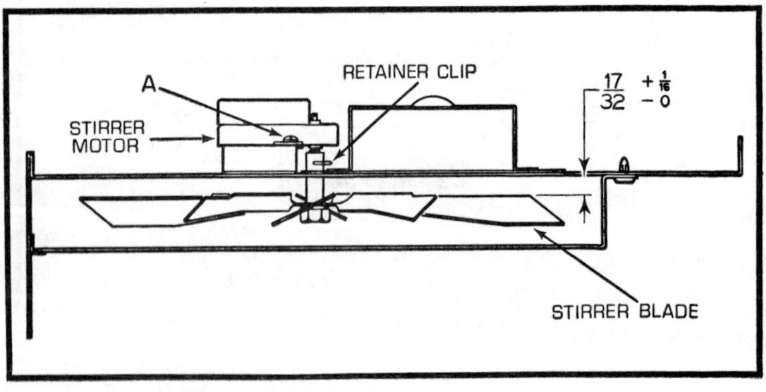

Fig. 10-17. Stirrer-motor replacement.

5. Position the stirrer blade on motor shaft and insert retainer clip.
6. Check the clearance between the blade and cavity top. This clearance must be maintained, or arcing may result. Blades may be bent slightly at the hub to attain the proper clearance.
7. Connect the motor leads.
8. Replace the stirrer cover.
9. Connect the unit to a power source and test it for proper operation.

POWER SUPPLY COMPONENTS

Caution: Make sure the power cord is disconnected before making the following tests. Wait at least 15 seconds, then remove the wire leads from terminals 1 and 2 of the transformer.

HV Transformer Check

1. Attach the ohmmeter leads to terminals 1 and 2 of the transformer (Fig. 10-18).
2. Observe a meter deflection, stabilizing at approximately 2.0Ω.
3. Reverse the meter leads and repeat the check. The meter should indicate an identical reaction.
4. If no continuity is present, a slight meter deflection may be observed prior to infinite resistance. This indicates that the primary is open and transformer must be replaced.
5. If continuity exists between terminals 1 and 2, remove leads 3 and 4 from the magnetron and capacitor. Check the continuity between leads 3 and 4.
6. If no continuity exists between 3 and 4, the transformer's filament secondary is open and transformer must be replaced.
7. If continuity exists between transformer leads 3 and 4, remove the lead from transformer terminal 6 to check for continuity between the ground lead and this terminal.

   Note: It is possible that no continuity between ground and terminal 7 exists, while continuity will exist between ground and terminal 6.

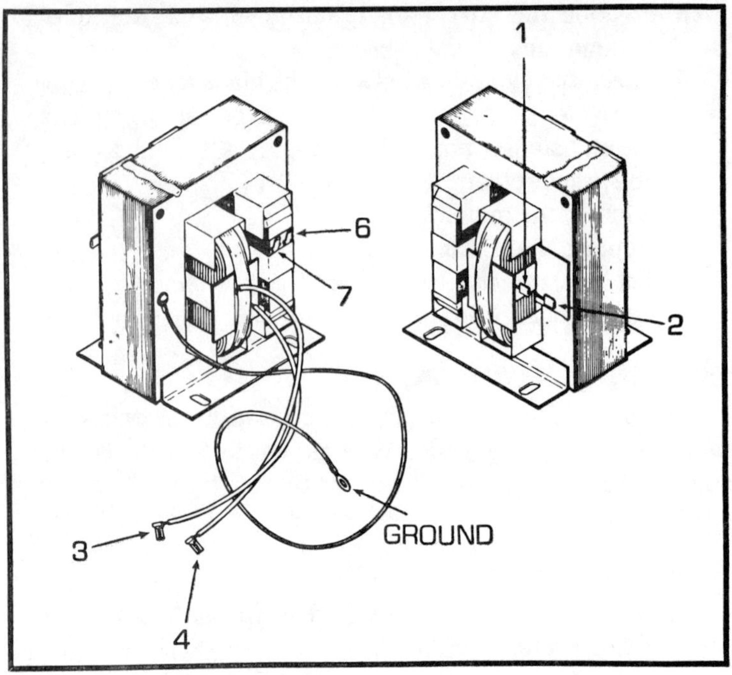

Fig. 10-18. Terminals on high-voltage transformer, referred to in text.

Filament Voltage Check

Caution: Make sure the unit has been disconnected from the power source. Then proceed as follows:

1. Remove the wire lead from terminal 6 or 7 of the transformer.
2. Set the meter to the 10V AC scale. Attach the meter leads to the transformer leads connected to the magnetron's filament terminals.
3. Place a load in the oven, close and latch the door, connect the power cord, set the timer, and push the COOK/START button. The meter should read 3.2 ± 0.3V AC with line voltage at 120V AC. If the meter does not read within the above range, the transformer is defective and must be replaced.

HV Check

Caution: Disconnect the oven from its power source. Then proceed as follows:

1. Remove the wire lead from terminal 6 or 7 of the transformer (Fig. 10-18).
2. Set the meter to the 5000V AC scale. Attach the black meter lead to the oven chassis (ground) and the red meter lead to terminal 6 or 7. The meter should be on an insulated surface beside the oven.
3. Place a load in the oven, close and latch the door, connect the power cord, set the timer, and push the COOK/STARTER button. The meter should read 1850–2150V AC with line voltage at 120V AC. If the meter does not read within the above range, the transformer must be replaced.

HV Transformer Replacement

1. Disconnect the oven from its electrical supply source.
2. Locate the oven so that the control panel and power-components section is overhanging a table or bench.
3. Locate the transformer and remove its mounting screws from under the oven.
4. Disconnect all transformer wiring.
5. Remove the old transformer from the chassis. Remove the speed nuts from the old transformer and position them on the replacement transformer.
6. Position the replacement transformer and secure with the four mounting screws.
7. Connect the wire leads to the replacement transformer, making sure the leads are connected to the proper terminals (refer to wiring diagram).

HV Capacitor Check

1. Disconnect the oven from its electrical supply course.
2. Set the ohmmeter on the R×10K scale.
3. Remove the magnetron, transformer, and resistor assembly leads from the capacitor.
4. Attach the meter leads to the outside terminals of the capacitor.
5. Reverse the polarity of the leads. This will cause a meter deflection or kick.

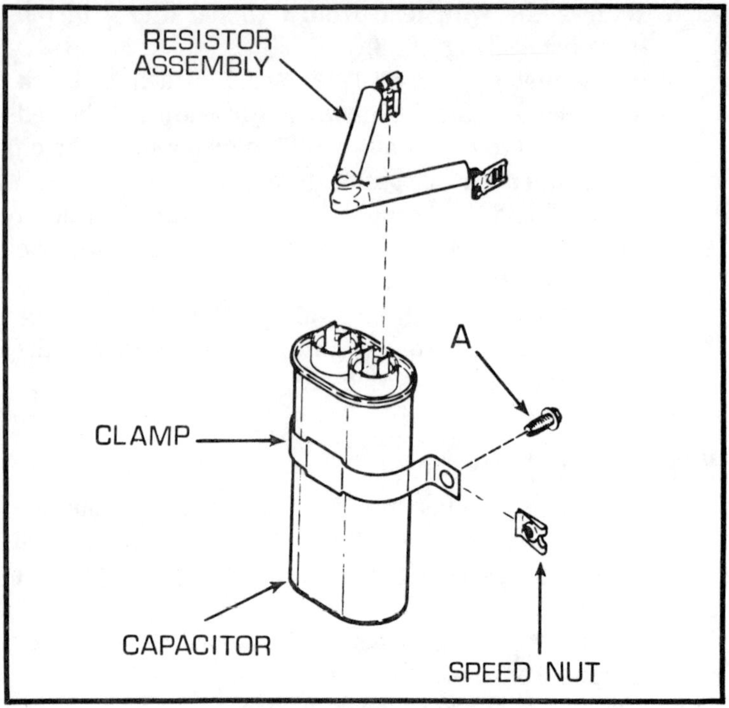

Fig. 10-19. Capacitor and mounting hardware.

6. Check for continuity between each terminal and capacitor housing. If no continuity, and deflection or kick was noted, capacitor is good.
7. Check the resistor assembly. A typical reading should be approximately 2M. If the reading shows an open (infinite), replace the resistor assembly.

HV Capacitor Replacement

1. Disconnect the oven from its electrical source and allow 15 seconds to elapse. Remove the oven top.
2. Remove capacitor clamp screws *A* of Fig. 10-19 from the chassis rear.
3. Disconnect the wires and resistor assembly.
4. Position the replacement capacitor and secure with the clamp and screws *A*.
5. Connect the wire leads and resistor assembly. Make sure leads are attached to the correct terminals. Refer to the wiring diagram.

Rectifier Check

1. Disconnect the oven from its electrical supply source.
2. Remove the top cover.
3. Remove the diode leads. Set the ohmmeter to the R × 10K scale. Attach the meter leads to the diode terminals. Note the reading.
4. Reverse the leads or polarity of the leads and note the reading. One reading must be of infinite resistance, and the other approximately 100K to 300K. If it is otherwise, replace the diode.

Rectifier Replacement

1. Disconnect oven from its outlet. Remove the top cover.
2. Remove the rectifier mounting screws and hex nuts.
3. Disconnect the wire leads from the diode.
4. Mount the new diode and secure with screws and hex nuts.
5. Attach the wire leads, making sure the green flag wire (to transformer ground) is attached to the *negative* side of the replacement diode.
   Caution: Power cord must be disconnected.

MAGNETRON TESTING

Set the ohmmeter to the R × 1 scale. Check the magnetron and filament leads for continuity.

Remove both leads from the transformer. Attach the meter leads to the transformer leads. If no continuity is observed, check for continuity at the magnetron filter choke connections. If no continuity is observed, replace the defective magnetron. If continuity is observed, replace the defective HV wiring harness. *Do not* attempt to isolate the defective filter choke on models 2095, 2905, or 2915. It is a part of the magnetron.

Magnetron Replacement

1. Disconnect the unit from its power source.
2. Discharge the HV capacitor by shorting the terminals with a screwdriver.
3. Disconnect the magnetron filament, thermostat, and blower leads.

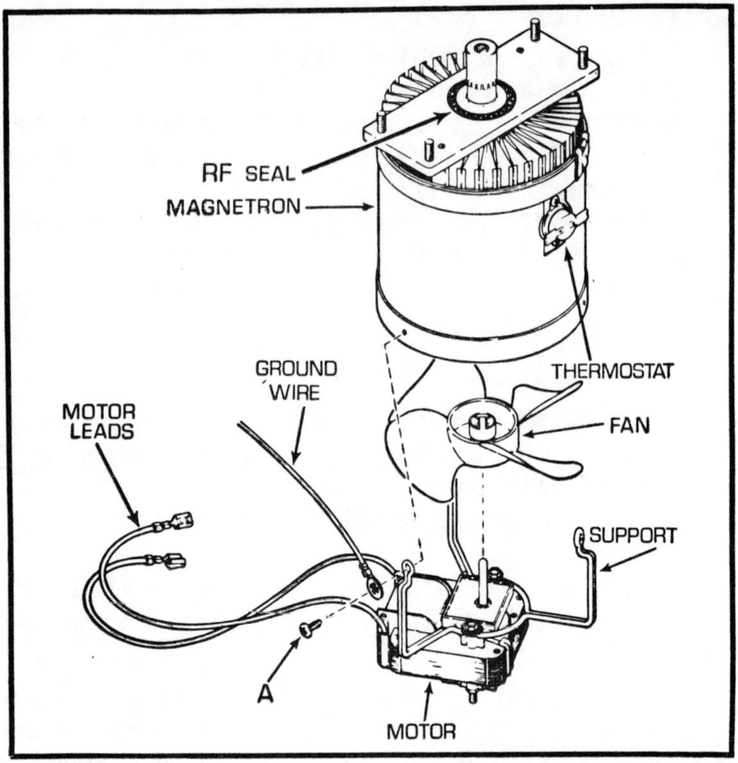

Fig. 10-20. Magnetron and blower motor.

4. Remove the air barrier from around the magnetron by removing the wire clip and pulling out and up.
5. Remove the four magnetron hex nuts and lower magnetron down and away.
6. Remove the blower assembly from the tube by removing the screws from the support bracket (Fig. 10-20).
7. Position the new tube on the support bracket and secure with screws.
8. Make sure the RF seal is in place on the new magnetron tube (Fig. 10-20). If it is not excessive energy emission and magnetron damage will occur.
9. Position tube assembly. Secure it with the four hex nuts.
10. Replace the air barrier.
11. Attach the filament, thermostat, and blower leads.

Chapter 11

Servicing Sharp Electronic Ovens

Now one of the multinational giants of the electronics field, this company started life in a rather suprising way—as a maker of mechanical pencils. If you're old enough, you may remember when the company's trade name Ever Sharp was synonymous with mechanical pencils. Now the name Sharp[1] is synonymous with the latest, most sophisticated electronic products, including microwave ovens. In the first dozen years since it produced its first microwave oven back in 1962, Sharp produced over a million such units.

Four countertop Sharp models are covered in this chapter. Models 5000 and 6300 are economy models. These two models have single-speed timers of 15 and 25 minutes and produce power outputs of 550W and 650W.

Model 6500 is a 650W unit with a 2-speed 25-minute timer and a defrosting feature. Model 8200 adds to this a browning and variable-cooking-speed capability plus a carousel turntable. In most ovens the microwaves are agitated by a stirrer blade to promote even cooking throughout the food load; but in Sharp models equipped with the carousel feature, the food is rotated through the microwave field on a turntable, and this is said to give superior results.

1. All of the illustrations in this chapter are courtesy of Sharp Electronics Corporation.

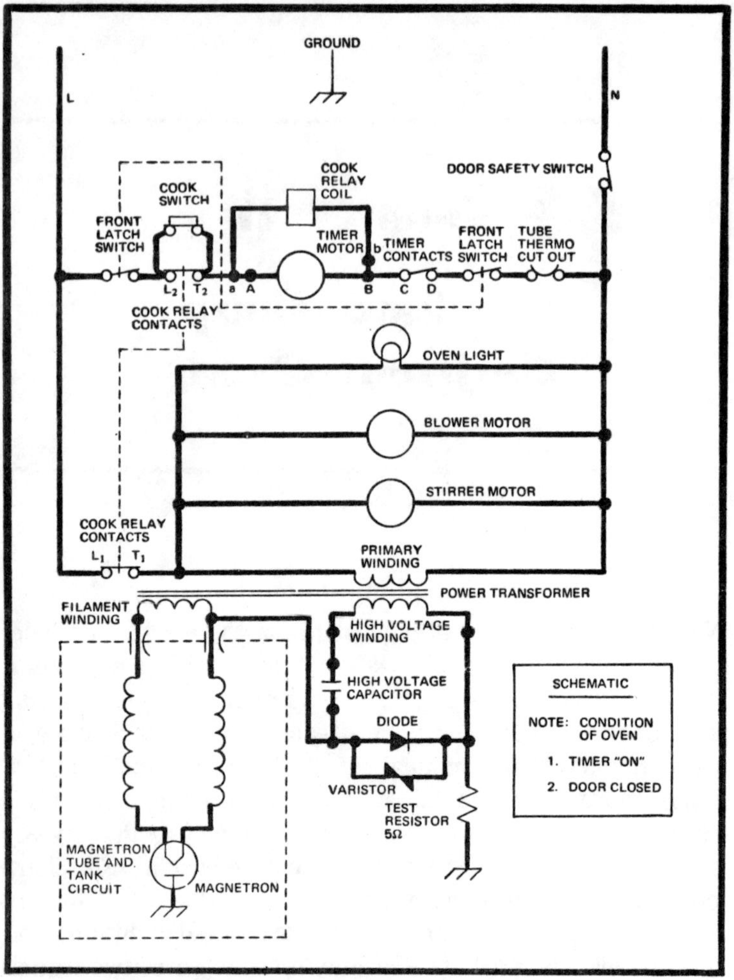

Fig. 11-1. Circuit of Sharp R-5000—cooking condition.

HOW MODELS 5000 & 6300 WORK

The schematic and wiring diagrams of these models are given in Figs. 11-1 through 11-4. Refer to the schematic as you read the following description of the circuit operation. In case of a circuit malfunction, an understanding of normal circuit operation will help you to quickly localize the source of trouble.

Off Condition

With the pointer of timer knob at zero, no components in the oven will operate.

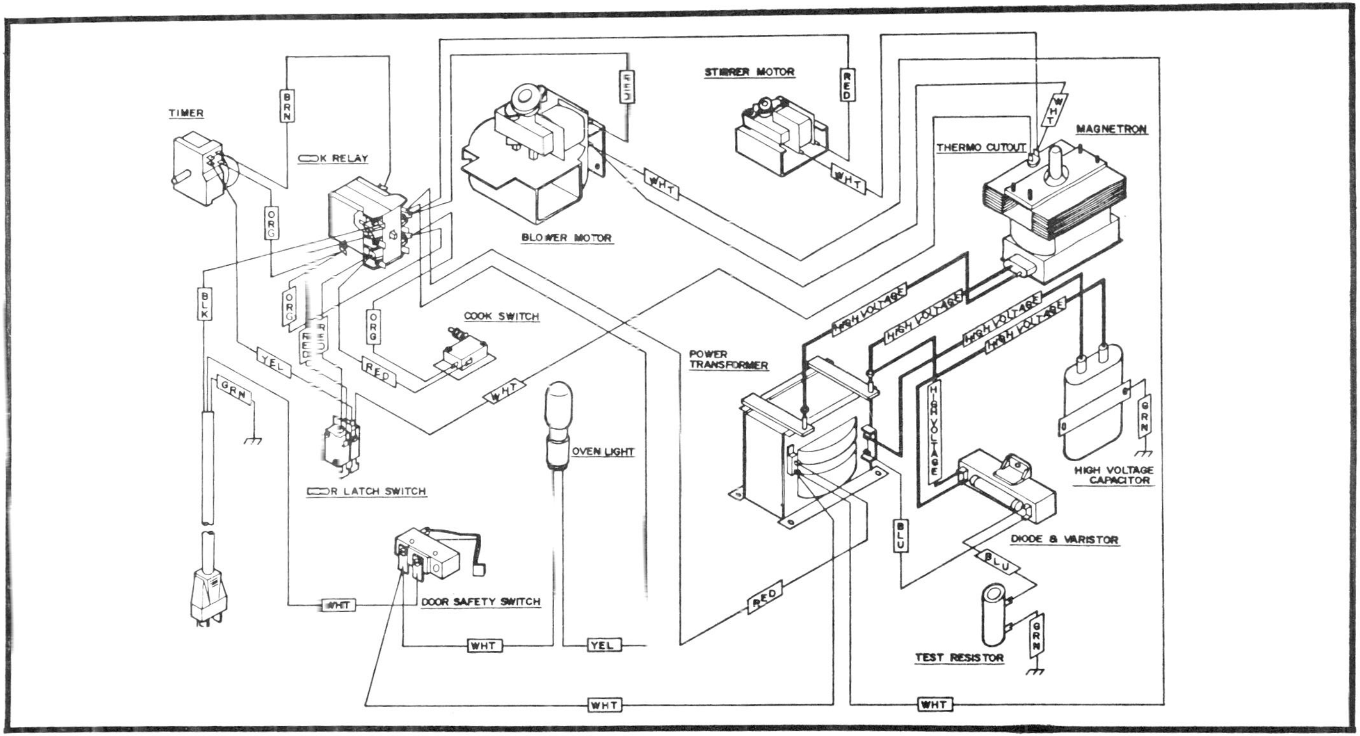

Fig. 11-2. Wiring diagram cf Sharp R-5000.

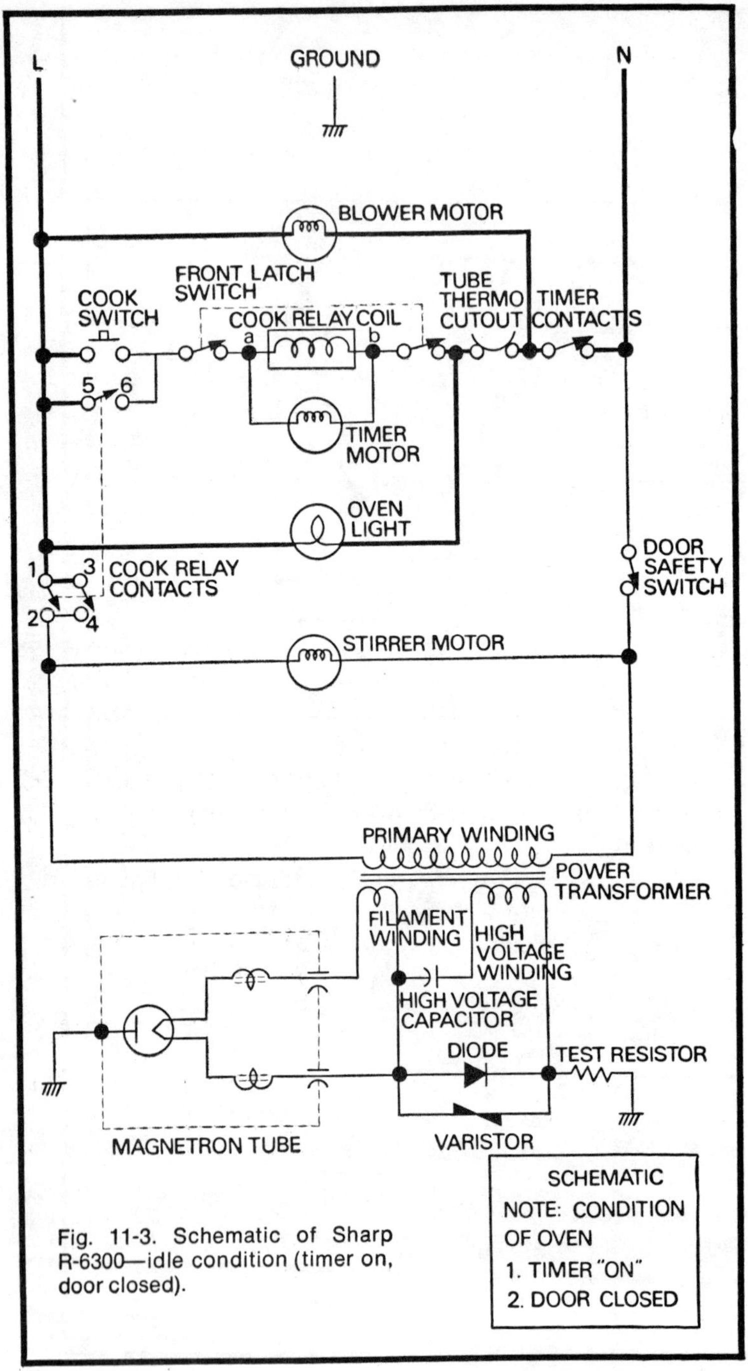

Fig. 11-3. Schematic of Sharp R-6300—idle condition (timer on, door closed).

Idle Condition

Closing the door activates both door-interlock switches—the front-latch switch and door-safety switch. Selecting a cooking time closes the timer switch contacts leading to the cook relay and timer motor; the blower motor and oven light activate on Model 6300.

Cooking Condition

When the COOK switch is depressed, the coil of the cook relay is energized. Two relay contacts close to provide a current path to the timer motor, which begins to return to the OFF position. This also provides the holding circuit for energizing the coil of the cook relay when the START switch is released. In addition, two relay contacts close to provide a current path to the stirrer motor and the power transformer.

The 120V AC input to the primary winding of the power transformer is converted to 3.2V AC output on the filament winding and approximately 1900V AC output on the secondary winding. The 3.2V AC output heats the magnetron filament. The 1900V AC output from the HV winding is sent to a voltage-doubler circuit consisting of a capacitor and diode. This 1900V AC is converted to approximately −3800V DC (peak) by the voltage doubler circuit and sent to the magnetron.

The −3800V DC applied to the cathode of the magnetron tube causes it to oscillate at the 2.45 GHz cooking frequency. The RF energy produced by the magnetron tube is channeled through a waveguide into the cavity feedbox, past the stirrer blade, and then into the cavity, where the food is placed to be heated.

When the cooking time expires, the timer switch opens, and the coil of the cook relay is de-energized. (The lamp will turn off and the blower motor will cease to operate on the 6300.) The contacts of the cook relay open with the following results: 1) HV is removed from the magnetron so that no more RF energy is produced. 2) The timer bell rings to indicate the end of the cook cycle. The oven has reverted to the off condition.

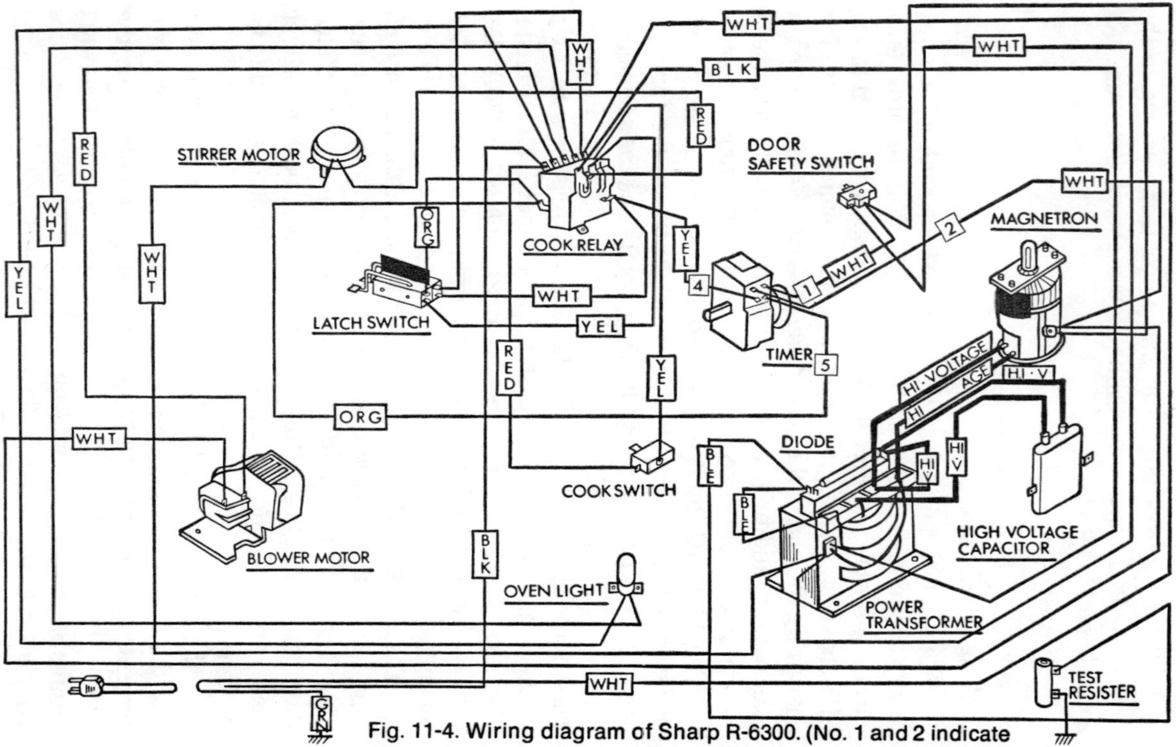

Fig. 11-4. Wiring diagram of Sharp R-6300. (No. 1 and 2 indicate terminals of switch contacts; 4 and 5 indicate terminals of motor coil.)

HOW MODEL 6560 WORKS

The following is complete description of component function during the three stages of oven operation. Refer to Figs. 11-5 and 11-6.

Off Condition

With the pointer of timer knob at zero , no components in the oven will operate.

Idle Condition

Closing the door activates both door-interlock switches—the primary-interlock switch and secondary-interlock switch. In this condition, the monitor switch is opened. Selecting a cooking time closes the timer contacts leading to the blower motor, oven lamp, cook relay, and timer motor; and the blower motor and oven lamp turn on.

Cooking Condition

When the COOK switch is depressed, the coil of the cook relay is energized. Relay contacts 10 and 12 close to provide a current path to the timer motor, which begins to return to the OFF position. This also provides the holding circuit for energizing the coil of the cook relay when the COOK switch is released. Relay contacts 2 and 4 close to provide a current path to the stirrer motor, the power transformer, and the indicator. The magnetron is then powered as with the previous models.

In the defrost cycle with the selector switch in the depressed position, the DEFROST lamp turns on and the defrost motor rotates at 1 RPM. The 30-seconds-*on* and 30-second-*off* repetition cycle of the defrost switch, which works by means of a cam provided on the defrost motor shaft, applied power to the power transformer intermittently. By this, RF energy is applied for 30-second intervals.

When the cooking time expires, the timer switch opens, and the coil of the cook relay is de-energized. Also the lamp will turn off and the blower motor will cease operating. The contacts of the cook relay open with the following results: 1) HV is removed from the magnetron so no more RF energy is produced. 2) The timer bell rings to indicate the end of the cook cycle. The oven has reverted to the off condition.

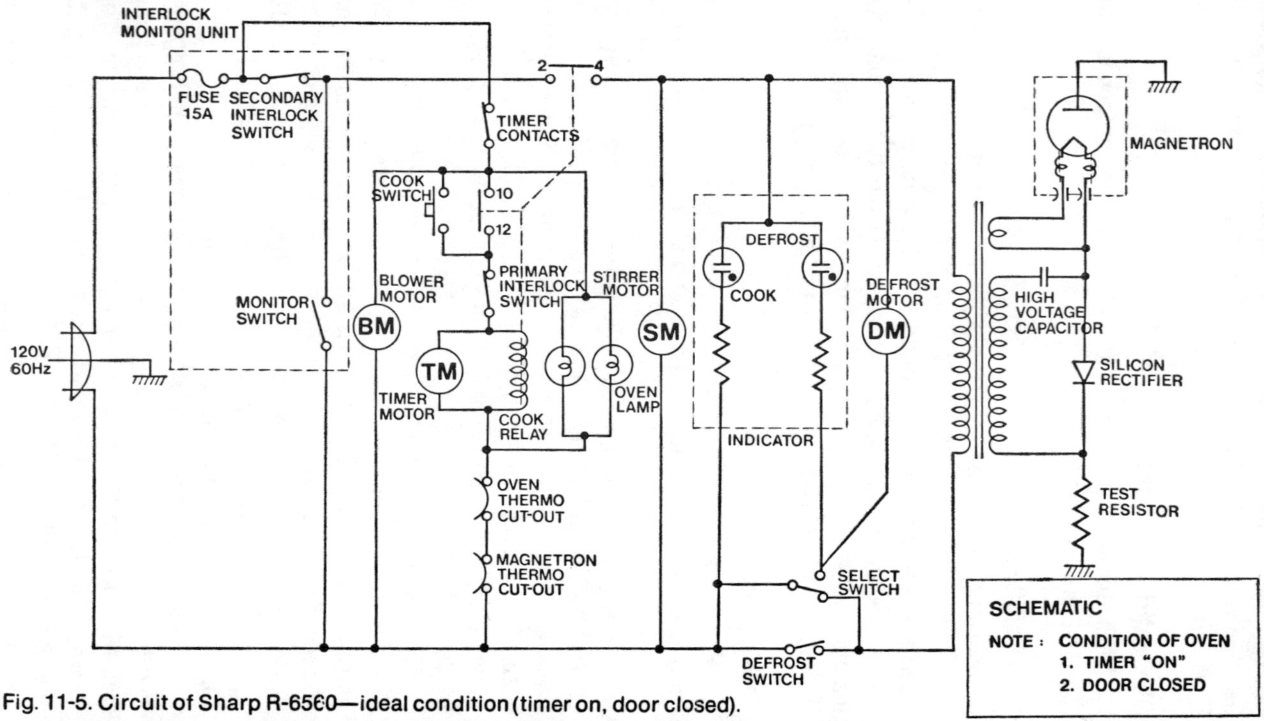

Fig. 11-5. Circuit of Sharp R-6560—ideal condition (timer on, door closed).

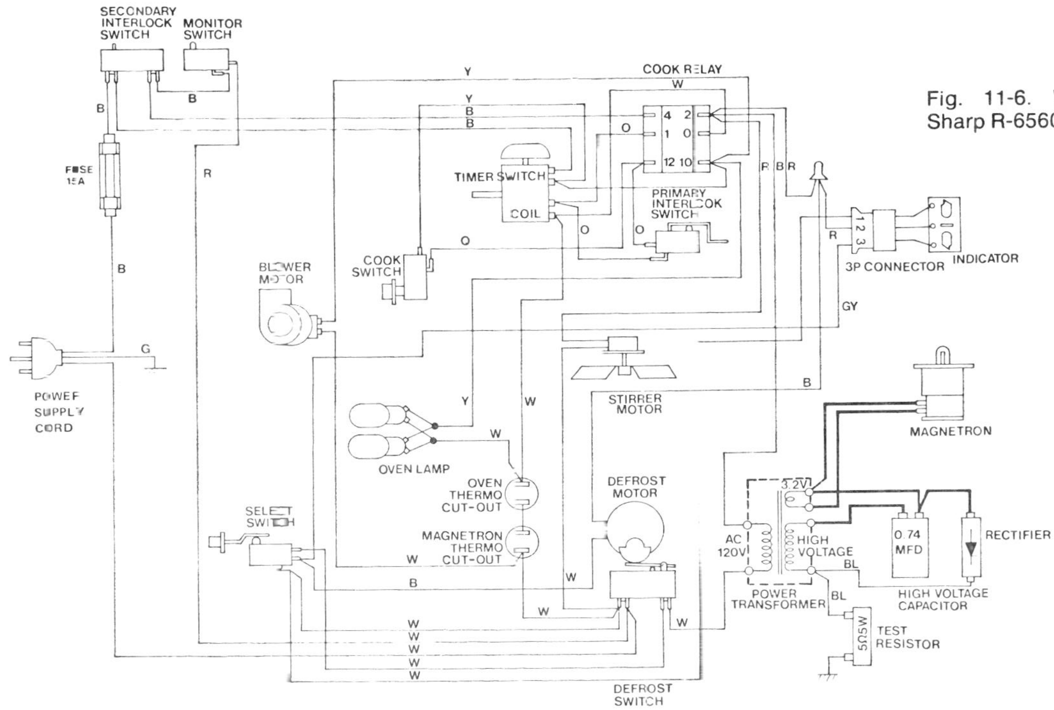

Fig. 11-6. Wiring diagram of Sharp R-6560.

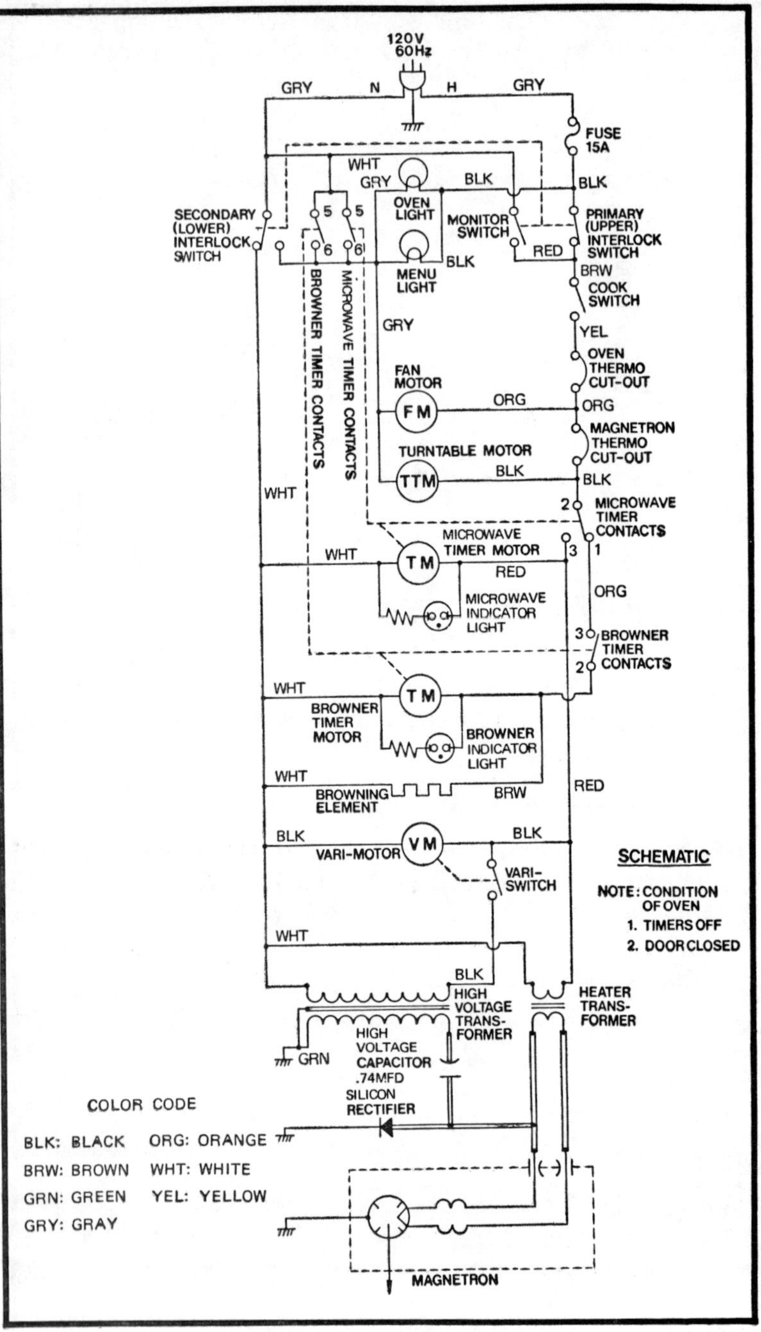

Fig. 11-7. Circuit of Sharp R-8200—timers off, door closed.

When the door opens during the cooking cycle, the primary-interlock switch opens, and the coil of the cook relay opens. The oven reverts to the idle condition.

The monitor switch, which electrically monitors the operation of the secondary-interlock switch, is mechanically associated with the door so that it may function in the following sequence.

1. When the door opens from the closed position, the primary- and secondary-interlock switches first operate to open the contacts; then the monitor switch closes.
2. When the door is closed from the open position, the monitor switch first opens the contacts, then the primary and secondary switches close.

In case the secondary-interlock switch has failed with the contacts closed, the monitor switch creates a short circuit through the fuse and the secondary switch when the door opens; the fuse will blow.

HOW MODEL 8200 WORKS

The following is a complete description of component functions during the three stages of oven operation. Refer to Figs. 11-7 and 11-8.

Off Condition

With the timer at zero and the door closed, no component in the oven will operate. The oven and menu lights will turn on when the door is opened.

Idle Condition

Closing the door activates both interlock switches—primary (upper) interlock switch and secondary (lower) interlock switch. In this condition the monitor switch is opened. Selecting a cooking time closes the microwave or browner timer contacts leading to the oven and menu lights, therefore both oven and menu lights turn on.

Cooking Condition

Full-Power Condition. When the COOK switch is depressed and the variable cooking mode selector is turned fully right (FULL-POWER position), the following operation occurs:

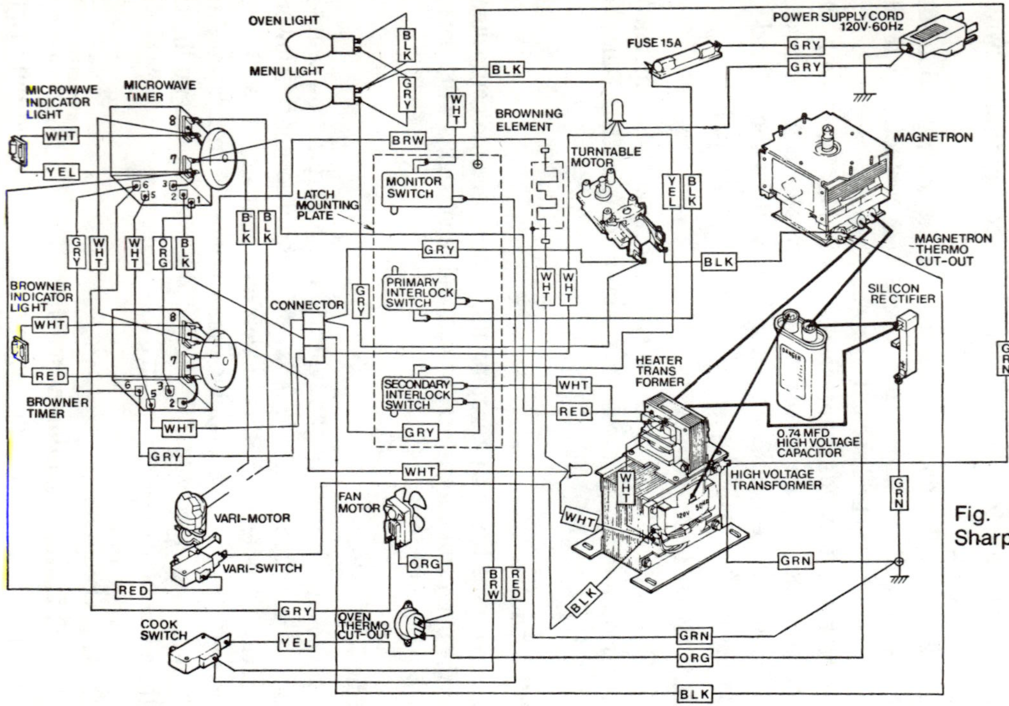

Fig. 11-8. Wiring diagram of Sharp R-8200.

1. The fan motor, turntable motor, microwave timer motor, microwave indicator light, and vari-motor are energized.
2. The 120V AC input to the primary winding of the heater transformer is converted to a 3.2V AC output on the secondary winding.
3. The 3.2V AC output from the transformer heats the magnetron filament.
4. The 120V AC is supplied to the primary winding of the HV transformer through the vari-switch.
5. The 2400V AC output from the secondary winding of the HV transformer is sent to a voltage-doubler circuit consisting of a HV capacitor and a silicon rectifier. The 2400V AC is converted to approximately − 4000V DC voltage (peak) by the voltage-doubler circuit and sent to the magnetron assembly.
6. The − 4000V DC is applied to the cathode of the magnetron tube. This causes it to oscillate at a 2.45 GHz cooking frequency.
7. The RF energy produced by the magnetron tube is channeled through a waveguide into the oven-cavity feedbox; then it is fed into the cavity, where the food has been placed to be heated.
8. Upon completion of the selected cooking time, the timer bell rings once, and the timer switch is deactivated. The oven has reverted to the off condition.

Defrost, Simmer, and Roast Condition. When the COOK switch is depressed and the variable cooking-mode selector is set to DEFROST, SIMMER, or ROAST, the following operation occurs:

1. Steps 1, 2, and 3 of the *Full-Power Condition* occur.
2. The vari-motor rotates at 2 RPM. If the mode selector is set at DEFROST, 120V AC is supplied to the HV transformer intermittently through the vari-switch, which is operated 9 seconds *on* and 21 seconds *off* by means of the cam provided on the vari-motor shaft. The repetition rates for other modes are as follows:

   Simmer: 15 seconds on, 15 seconds off
   Roast: 21 seconds on, 9 seconds off

3. The 2400V AC output is converted to − 4000V while the 120V AC is supplied to the HV tansformer.
4. Then steps 5, 6, 7, and 8 of the *Full-Power Condition* occur.

Browning Condition

When the COOK switch is depressed and the browner timer set, the following operations occur: The fan motor, turntable motor, browner timer motor, and browner indicator light are energized. The 120V AC is supplied to the browning element. Upon completion of the selected cooking time, the timer bell rings once as the timer switch is deactivated, and the oven reverts to the off condition.

When both microwave and browner timers are set, microwave energy is produced first. In this case, the browning element will be activated when the microwave timer returns to zero.

TROUBLESHOOTING & TESTING

When troubleshooting the microwave oven, it is helpful to follow a logical sequence of operation in performing the checks. Such sequences are given in Tables 11-1 and 11-2. Many of the possible causes of trouble will require that a specific test be performed. These tests are given in the following paragraphs.

Important: If the oven becomes inoperative because a fuse blows, the whole interlock-monitor unit must be replaced with a new one.

HV Component Tests

The following test is required if little or no heat is produced in a cooking load but oven apperars to be operating normally (120V AC is supplied to the primary winding of HV transformer and magnetron filament glows).

HV is present during the cook cycle, so extreme caution should be observed. Discharge the HV capacitor before touching any oven components or wiring.

Magentron Assembly. To test for an open filament, isolate the magnetron from the HV circuit. A continuity check across the filament leads should indicate less then 1Ω.

To test for a shorted magnetron, connect the ohmmeter leads between the filament leads to the tube and chassis ground. This test should indicate an infinite resistance. If there is little or no resistance, the magnetron is grounded and must be replaced.

Power output of the magnetron can be measured by performing a water temperature-rise test as described in Chapter 3.

Silicon Rectifier. Isolate the rectifier from the circuit. Using the highest ohms scale of the meter, read the resistance across the rectifier terminals. Reverse the meter leads to the rectifier terminals. A normal rectifier should read infinite resistance in one direction, and very high resistance when the meter leads are reversed. If a short is indicated in both

Table 11-1. Troubleshooting Guide—Sharp Models 5000, 6300, and 6560

OFF CONDITION		
PROBLEM	**POSSIBLE CAUSE**	**TEST OR CORRECTION**
Line fuse blows when power cord is plugged into wall receptacle.	Shorted wire in power cord or wire harness.	Replace cord or check wiring
IDLE CONDITION		
PROBLEM	**POSSIBLE CAUSE**	**TEST OR CORRECTION**
Oven completely inoperative with timer set.	1. No power from service entrance.	Check wall outlet.
	2. Open wire in power cord or wiring harness.	Replace same or repair wiring.
	3. Defective timer.	Replace.
	4. Blown fuse in secondary-interlock monitor.	Replace whole secondary-interlock monitor unit.
Blower motor inoperative.	1. Defective blower motor.	Test.
	2. Open wiring in circuit to blower.	Check wiring.
Oven cavity light does not come on when timer is set.	1. Defective bulb.	Replace.
	2. Open wiring in circuit to light.	Check wiring.
	3. Defective timer contacts.	Test.
	4. Defective thermal cutout.	Test
	5. Open wiring in circuit to thermal cutout.	Check wiring.

Table 11-1. cont.

COOKING CONDITION		
PROBLEM	**POSSIBLE CAUSE**	**TEST OR CORRECTION**
Oven does not go into cook cycle when COOK button is depressed.	1. Defective contacts on timer switch.	Test contacts.
	2. Primary-interlock switch is defective or out of adjustment.	Test switch.
	3. Defective thermal cut-out.	Test cutout.
	4. Defective COOK switch.	Replace.
	5. Defective cook relay.	Test relay.
	6. Open wiring between the above components.	Check wiring.
Oven goes into cook cycle, but timer does not time out.	1. Defective timer motor.	
	2. Open wiring in circuit to timer motor.	Check wiring.
Oven goes into cook cycle, but stirrer motor inoperative.	1. Defective stirrer motor.	Test motor.
	2. Open wiring in circuit to stirrer motor.	Check wiring.
Oven goes into cook cycle, but little or no heat is produced in oven load.	1. Shorted HV circuit between voltage-doubler circuit and magnetron.	Measure plate current.
	2. Defective power transformer.	Test transformer.
	3. Defective rectifier.	Test.
	4. Defective HV capacitor.	Test.
	5. Defective magnetron.	Measure power output.
Oven seems to be operating, but no heat nor COOK light comes on.	Secondary-interlock switch is defective or out of adjustment.	Test secondary-interlock monitor unit.
Oven goes into cook cycle but shuts down before end of cycle.	Thermal cutout opened.	Test cutout.
Power-source fuse blows when the COOK switch is depressed.	1. Defective power transformer.	Test.
	2. Secondary circuit is power transformer is shorted.	Check wiring.
DEFROST indicator comes on, but defrost cycle does not function.	1. Defective defrost motor.	Test.
	2. Defective DEFROST switch.	Test.
Both the indicator and defrost cycle do not function.	Defective selector switch.	Test defrost motor and switch. Check selector switch.

directions, or if an infinite resistance is read in both directions, the rectifier is probably defective and should be replaced.

HV Transformer. Disconnect the primary input terminals to measure the resistance with an ohmmeter. On the R × 1 scale the resistance of the primary coil should be less than 1Ω, and the resistance of secondary coil should be approximately 50Ω.

HV Capacitor. If the capacitor is open, no HV will be available to the magnetron. Disconnect the input leads and check for a short or open between the terminals by using an ohmmeter.

Checking on a high-ohms scale—if the HV capacitor is normal—the meter will indicate continuity for a short time, then indicate an open circuit once the capacitor is charged. If this is not the case, check the capacitor for a short between the terminals. If it is shorted, replace the capacitor.

Caution: Discharge the HV capacitor before checking.

Other Tests

Heater Transformer Test. The heater transformer can be checked with a voltmeter.

Caution: HV is present during the cook cycle, so extreme caution should be observed. Discharge the HV capacitor before touching any oven component or wiring. When checking low voltage, keep meter leads and hands away from the HV terminals.

With 120V AC applied to the primary winding of heater transformer, the voltage across the secondary winding should be approximately 3.2V AC. If the input voltage is normal but no secondary voltage is present, replace the heater transformer.

Primary-Interlock Switch Test. The primary-interlock switch can be checked with an ohmmeter. Isolate the switch; then connect the meter leads to the COM and N.O. terminals of the switch. The meter should indicate an open circuit with the door opened, and a closed circuit with the door closed. In case improper operation is indicated, make the necessary switch adjustment or replace the defective switch.

Secondary-Interlock Switch Test. Isolate the switch. Connect the ohmmeter leads to the COM and N.O. terminals of the switch. The meter should indicate an open circuit with the

Table 11-2. Troubleshooting Guide—Sharp Model 8200

OFF CONDITION		
PROBLEM	**POSSIBLE CAUSE**	**TEST OR CORRECTION**
Home fuse blows when power cord is plugged into wall receptacle.	Shorted wire in power cord or wire harness.	Replace cord or check wiring.
Oven and menu lights do not light with door opened. (Timer is at 0.)	1. No power in home.	Check wall outlet.
	2. Open wire in power cord	Replace same or repair wiring.
	3. Defective light bulb.	Replace.
	4. Defective secondary-interlock switch (lower latch switch).	Test switch.
	5. Open or loose wire connection to the above components.	Check wiring.
	6. Blown fuse.	See **Blown Fuse** procedure.
IDLE CONDITION		
PROBLEM	**POSSIBLE CAUSE**	**TEST OR CORRECTION**
With the timer set and door closed, oven and menu lights do not light.	1. Defective timer switch contacts.	Test timer switch.
	2. Open or loose wire connection.	Check wiring.
COOKING CONDITION		
PROBLEM	**POSSIBLE CAUSE**	**TEST OR CORRECTION**
Fan motor does not operate with COOK switch depressed.	1. Defective fan motor.	Replace.
	2. Defective primary-interlock switch.	Test switch.
	3. Defective COOK switch.	Replace.
	4. Defective oven thermal cutout.	Test cutout.
	5. Open or loose wire connection to the above components.	Check wiring.
Turntable motor does not operate, but fan motor operates.	1. Defective turntable motor.	Replace.
	2. Defective magnetron thermal cutout.	Test cutouts.
	3. Open or loose wire connection to the above components.	Check wiring.
Browner indicator light does not light and no heating occurs, but turntable motor operates.	1. Defective microwave timer switch contacts.	Test timer switch.
	2. Defective browner timer switch contacts.	Test timer switch.
	3. Open or loose wire connection to the above components.	Check wiring.

Table 11-2. cont.

PROBLEM	POSSIBLE CAUSES	TEST OR CORRECTION
Browning element does not operate, but the browner indicator lights.	1. Defective browning element. 2. Open or loose wire connection.	Check element. Check wiring.
Browning cycle operates normally, but microwave cooking cycle does not operate properly. (Variable-cooking mode selector is at FULL POWER position.)	1. Defective microwave timer switch contacts. 2. Defective vari-switch. 3. Defective heater transformer. 4. Defective HV components. 5. Open or loose wire connection to the above components.	Test switch. Test. Test. Check power output and HV components. Check wiring.
Microwave cooking cycle operates normally when the mode selector is set at FULL POWER position, but not in any other mode.	1. Defective vari-motor. 2. Open or loose wire connection. 3. Defective variable cooking mechanism.	Test motor. Check wiring. Check.
Oven goes into microwave (browning) cycle, but microwave (browner) timer does not time out.	1. Defective microwave (browning) timer. 2. Open or loose wire connection.	Test timer switch and motor. Check wiring.
Oven goes into cook cycle, but extremely uneven heating is produced in oven load.	Turntable motor does not operate.	Replace the motor or check wiring.
Oven goes into cook cycle, but the microwave or browner indicator does not light.	1. Defective light. 2. Open or loose wire connection.	Replace. Check wiring.
Oven goes into cook cycle but shuts down before end of the cycle.	1. Defective magnetron (oven) thermal cutout. 2. Open or loose wire connection.	Test cutout. Check wiring.

door opened and closed circuit with the door closed. When the meter leads are connected to the COM and N.C. terminals of the switch, it should indicate a closed circuit with the door opened and an open circuit with the door closed. In case improper operation is indicated, make the necessary switch adjustment or replace the defective switch.

Monitor Switch Test. Isolate the monitor switch. Connect the ohmmeter leads to the COM and N.C. terminals of the switch. The meter should indicate a closed circuit with the door opened, and an open circuit with the door closed. In case improper operation is indicated, replace the defective switch.

Blown Fuse. If the fuse in the primary-monitor switch circuit is blown when the door is opened, check the primary-interlock switch and monitor switch before replacing the blown fuse. If the fuse is blown because of improper switch operation, replace the defective switch and fuse at the same time. Replace just the fuse if the switches operate normally.

Caution: Replace the fuse with only a 15A slow-blow fuse (QFS-B0001 YBEO).

Microwave & Browner Timer Switches. Advance the timer and then return to zero. Check continuity between the terminals. If an open circuit is indicated at an advanced position or a closed circuit is indicated at zero, replace the timer.

Note: The microwave timer terminals between 1 and 2 will be a closed circuit if all is normal.

Microwave & Browner Timer Motors. Disconnect the wire leads from the timer motor to check for 120V. If voltage is present but the timer motor does not turn, replace the timer. If there is no power, check for loose or broken wiring.

Thermal Cutout Test. A continuity check across the thermal cutout should indicate a closed circuit unless the temperature of the thermal cutout reaches approximately 220°F (104°C). The thermal cutout resets automatically at approximately 175°F (79°C). An open thermal cutout indicates overheating of the magnetron assembly (or oven cacity). Check for restricted air flow to the magnetron (or oven cavity) through the opening in the chassis, especially the cooling duct and air intake.

Vari-Switch Test. The vari-switch is activitated by the lever and cam. When the variable-cooling mode selector is set at the FULL POWER position, the cam is placed far away from the cam lever. In this condition the vari-switch is turned on by a spring action of the cam lever.

The vari-switch can be checked with an ohmmeter. When the meter leads are connected to the COM and N.O. terminals of the switch, it should indicate a closed circuit with the cam lever released and an open circuit with the cam lever depressed. If improper operation is indicated, make the necessary adjustment on the mode selector or replace the defective switch.

Vari-Motor Test. The vari-motor rotates at 2 RPM. Check if 120V is present between the motor terminals. If voltage is present but the motor does not turn, replace the motor. If no voltage is present, check for loose or broken wire connections.

Browning Element. Disconnect the heater leads and measure the resistance with an ohmmeter. On the R × 1 scale, the resistance between the heater leads should be approximately 10Ω.

Disconnect the heater leads to measure the insulation resistance with a 500V, 100M insulation-resistance meter. The insulation resistance between heater lead and cavity should be more than 0.5M.

DOOR & INTERLOCK ADJUSTMENT

When moving or loosening the door assembly in such a case as door replacement, the following adjustment criteria should be observed:

1. The door-latch heads must smoothly catch the latch hooks through the latch holes.
2. Deviation of the door alignment from the horizontal line of the upper sash is to be less than 1 mm (0.04 inches).
3. The door is to be positioned with its face depressed toward the cavity face plate.
4. Install the outer case and check the microwave leakage around the door with an approved survey meter.

Model 8200 Latch Assembly Adjustment

The primary-interlock switch, secondary-interlock switch, and monitor switch are fixed in position; therefore, individual switch adjustment is not needed. In case the switches do not operate properly due to a misadjustment, the entire latch assembly should be adjusted in the following manner.

1. Loosen the two screws holding the latch assembly to the mounting bracked (Fig. 11-9).
2. With the door closed, adjust the position of the latch assembly by moving it back and forward, or up and down. The in-and-out play of the door allowed by the

Fig. 11-9. Latch adjustment—Model R-8200.

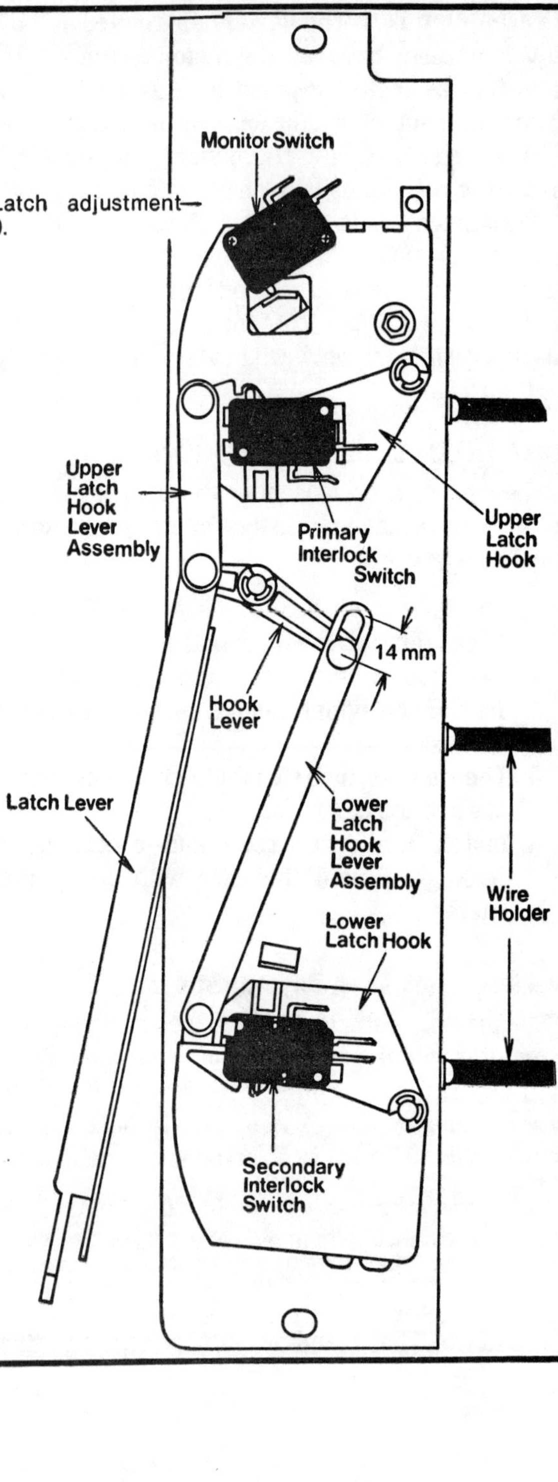

upper and lower latch hooks should be less than 0.5 mm (0.02 inches). The vertical position of the latch assembly should be such that there is 14 mm (0.56 inches) between the upper portion of the lower-latch hook-arm slot and the center of hook-lever pin with the door closed (Fig. 11-9).

3. Secure the screws firmly.

After the adjustment, make sure of the following points:

1. The in-and-out play of the door remains less than 0.5 mm (0.02 inches) at latched position. First check the upper-latch hook position, pushing and pulling the upper portion of the door toward the oven face. Then check the lower-latch hook, pushing and pulling the lower portion of the door toward the oven face. Both results (play of the door) should be less than 0.5mm (0.02 inches).
2. The distance between the upper portion of the lower latch hook arm slot and the center of the hook lever pin should be 14 mm (0.56 inches) at latched position.
3. The primary- and secondary-interlock switches should interrupt the circuit before the door can be opened.
4. The monitor switch should come on when the door is slightly opened.
5. Install the outer case and check the microwave leakage around the door with an approved survey meter.

Model 6300 Front-Latch Switch Test & Adjustment

When the thumb latch on the door handle is depressed, an audible click should be heard. When the thumb latch is released with door closed, it should activate the switches. If the thumb latch does not activate the switch when the door is closed, loosen the mounting-bracket screws and move the switch arm upward. Make sure the center barrier of latch switch comes to the slit of latch head (Fig. 11-10). This adjustment will allow the switch to be activated when the door is closed. The switch must release and interrupt the cook cycle before the door can be opened.

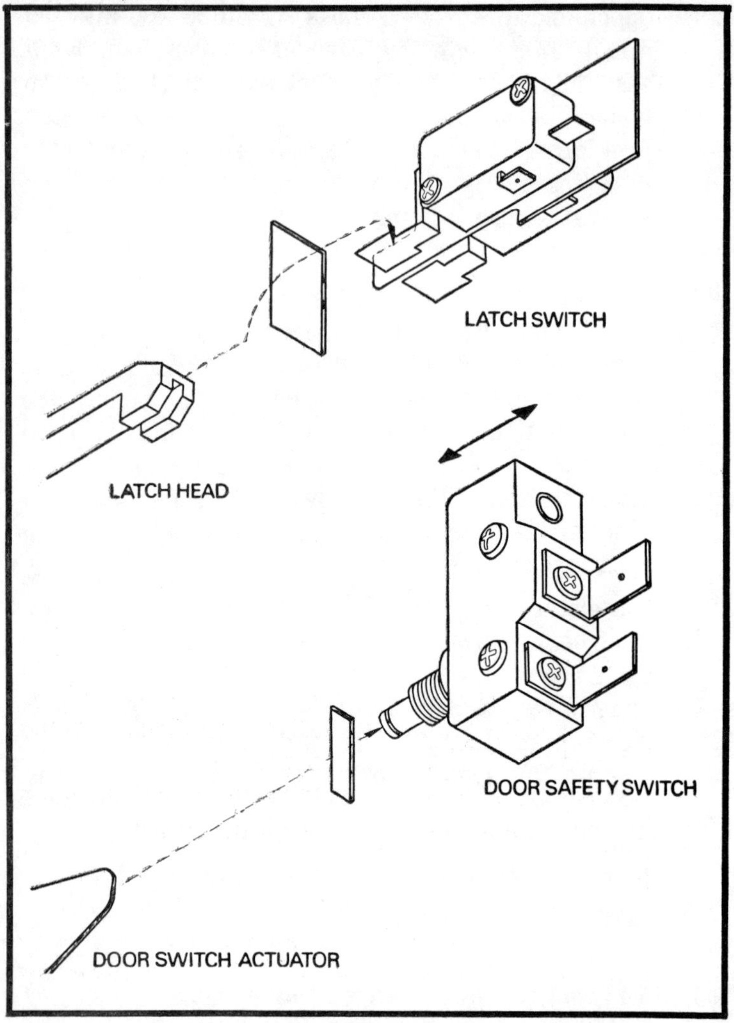

Fig. 11-10. Latch-switch assembly—Model R-6300.

The switch can be checked with an ohmmeter by connecting the meter leads to the COM terminal and the N.O. terminal of the latch switch. With the door open, the meter should read an open circuit. When the door is closed, the switch is closed, so the meter should indicate continuity.

Model 6300 Door-Safety Switch Adjustment

After the first audible click given by the front-latch switch, another audible click should be heard when the door is opened slightly. This is the door-safety switch being deactivated as the door cam arm moves away from the switch. The door safety switch can be adjusted by loosening the two mounting screws at the right-hand base of the oven. In case adjustment is required, make sure that the switch is cut off by the cam arm within an eighth-inch of the door opening at right edge from closed position. If the switch is defective or out of adjustment, the oven will appear to operate normally except for no cook-light or stirrer motor operation and no heat in the cooking load.

The door-safety switch can be checked with an ohmmeter. The meter leads connected to the COM and N.O. terminals should indicate an open circuit with the door open, and a closed circuit with the door closed.

Model 6560 Primary-Interlock Switch Adjustment

If the thumb latch does not activate the switch or the play of the door is excessive when the door is closed, the latch adjustment should be performed in the following manner.

1. Loosen the two screws holding the latch receptacle to the bracket. Adjust the position of the receptacle so the play of door is less than 0.5 mm (0.02 inches) at the latched position, then tighten the screws.
2. Loosen the two screws holding the switch to latch receptacle and move the switch arm upward.

Model 6560 Monitor Unit Adjustment

The whole secondary-interlock monitor can be adjusted by loosening two screws holding the unit to the base at the left of the oven and proceeding as follows.

1. Put the spacer of 5 mm (0.2 inches) thickness between oven face and the door right-center edge.
2. Shut the door.
3. Fix the monitor unit by securing it loosely (preparatory to adjustment) at the position where the secondary-interlock switch turns just on.
4. Change the spacer thickness to 8 mm (0.32 inches) between the cavity face and the door's right-bottom edge.

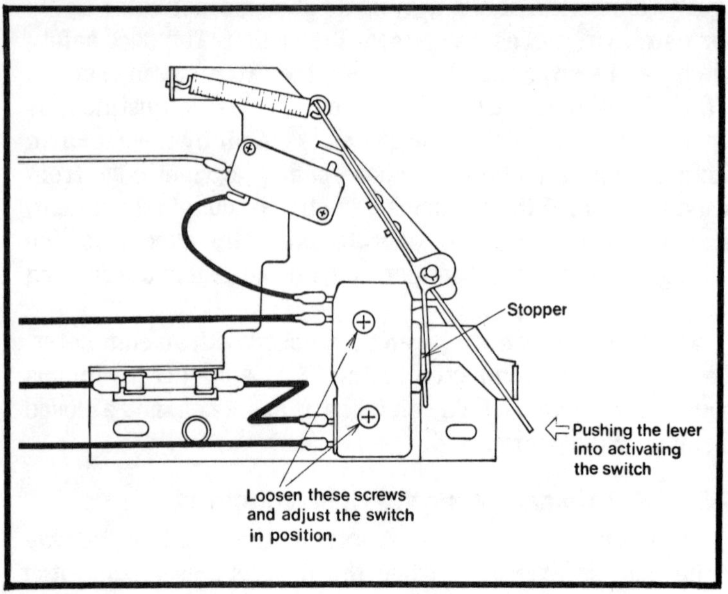

Fig. 11-11. Adjustment of secondary interlock—Model R-6560.

Model 6560 Secondary-Interlock Switch Adjustment

Since the secondary-interlock switch is installed on the mounting with a factory adjustment, there is no need to further adjust it. However, if the secondary-interlock switch cannot be activated even with the lever pushed to the end of the stopper, it should be adjusted in the following manner. This adjustment will allow the switch to be activated when the door is closed. The switch must release and interrupt the cook cycle before the door can be opened.

The switch can be checked with an ohmmeter by connecting the meter leads to the COM terminal and the N.O. terminal of the primary-interlock switch. With the door open, the meter should read an open circuit. When the door is closed, the switch is closed and the meter should indicate continuity. Under this condition, make sure that the secondary-interlock switch is off. It should turn on when you push the lever. In case the switch is out of position repeat steps 2, 3, and 4 of the previous procedure until it functions properly. Make sure that the monitor switch turns on within a door movement of 30 mm (1.1875 inches) at free edge. Tighten the screws; then glue each with screw-lock material.

Loosen the two screws holding the switch to the mounting. Adjust the position of the switch so as to properly activate the switch while holding the lever against the stopper; then tighten the screws (Fig. 11-11).

Model 5000 Front-Latch Switch Adjustment

When the thumb latch on the door handle is depressed, an audible click should be heard. When the thumb latch is released with the door closed, it should activate the switches. If the thumb latch does not activate the switches when the door is closed, loosen the mounting bracket screws and move the switch arm upward. Make sure the center barrier of latch switch comes to the slit of the latch head (Fig 11-12). This adjustment will allow the switch to be activated when the door is closed. The switches must release and interrupt the cook cycle before the door can be opened.

The switch can be checked with an ohmmeter by connecting the meter leads to the COM terminal and the N.O. terminal of the latch switch. With the door open, the meter should read an open circuit. When the door is closed, the switch is closed, so the meter should indicate continuity.

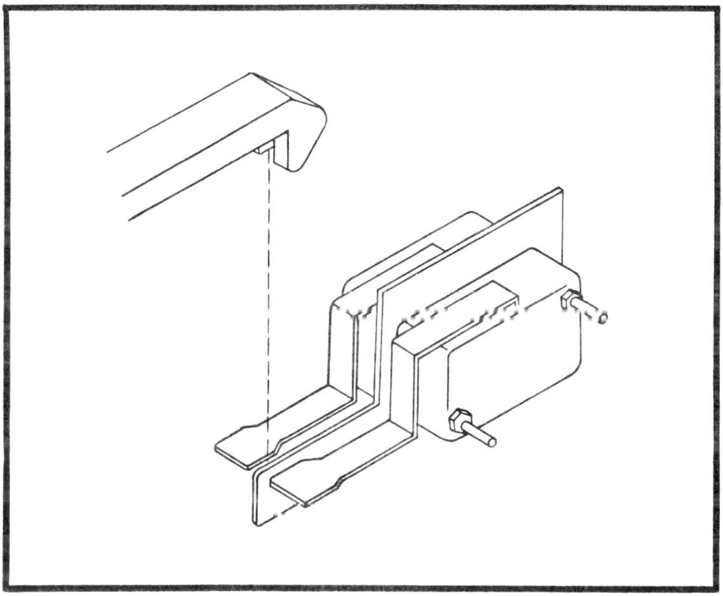

Fig. 11-12. Front-latch-switch adjustment.

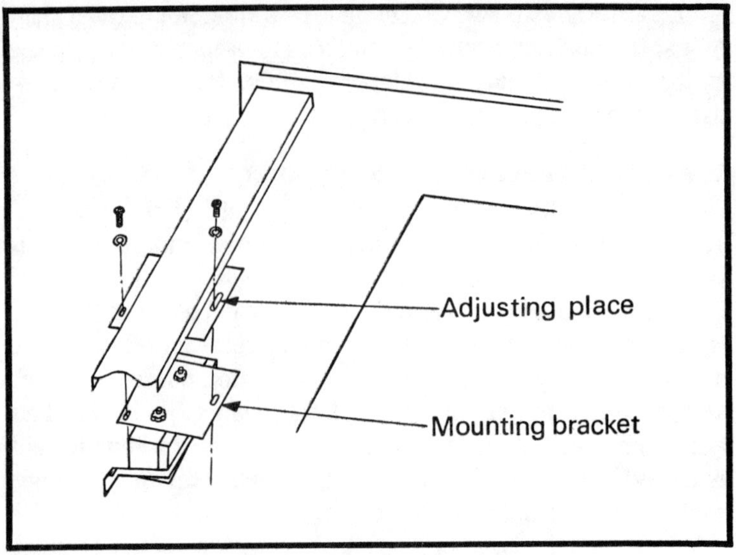

Fig. 11-13. Door-safety-switch adjustment.

Model 5000 Door-Safety Switch Adjustment

After the first audible click given by the front-latch switch, another audible click should be heard when the door is opened slightly. This is the door-safety switch being deactivated as the door hinge arm moves away from the switch. The door-safety switch can be adjusted by loosening the two mounting screws at the left chassis support as per Fig. 11-13. In case an

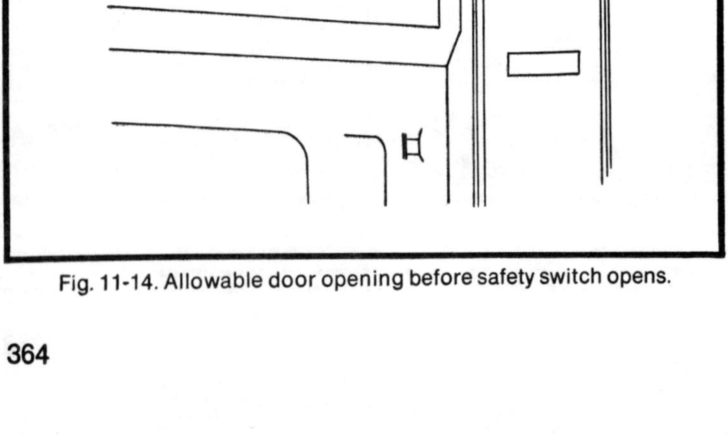

Fig. 11-14. Allowable door opening before safety switch opens.

adjustment is required, make sure that the switch is cut off by the cam actuator within 30 mm of door opening at right edge from closed position, as shown in Fig. 11-14, and tighten the screws.

The door-safety switch can be checked with an ohmmeter. The meter leads connected to the COM and N.O. terminals should indicate an open circuit with the door open and a closed circuit with the door closed.

Chapter 12
Servicing Tappan Microwave Ovens

This chapter will cover models made by Tappan Company[1] from 1970 to the present. The Mansfield, Ohio, appliance manufacturer pioneered the home microwave oven in 1955. Today it sells a wide range of microwave ovens for the home and institutional markets under its own name and, on the West Coast, under the O'Keefe and Merritt label. Here we shall be concerned with the popular models made by this company.

HOW MODELS 56-1235, -5235, -2343, -6343, -3453, -7453, -3454, & -7454 WORK

Figures 12-1 through 12-4 illustrate the operating conditions described here. To put the oven in the cook condition, the door and latch must be securely closed. This will activate two interlock switches. One interlock is activated by the door closing, and the other is activated by the door latch. The desired cooking time must be selected with the dial timer, which will close the switch contacts leading to the surge relay.

For the first few milliseconds of operation, current flows through the 5Ω resistor to the primary side of the transformer. The surge relay will energize, and the 5Ω resistor is then bypassed.

---

1. All illustrations in this chapter are courtesy of Tappan Company.

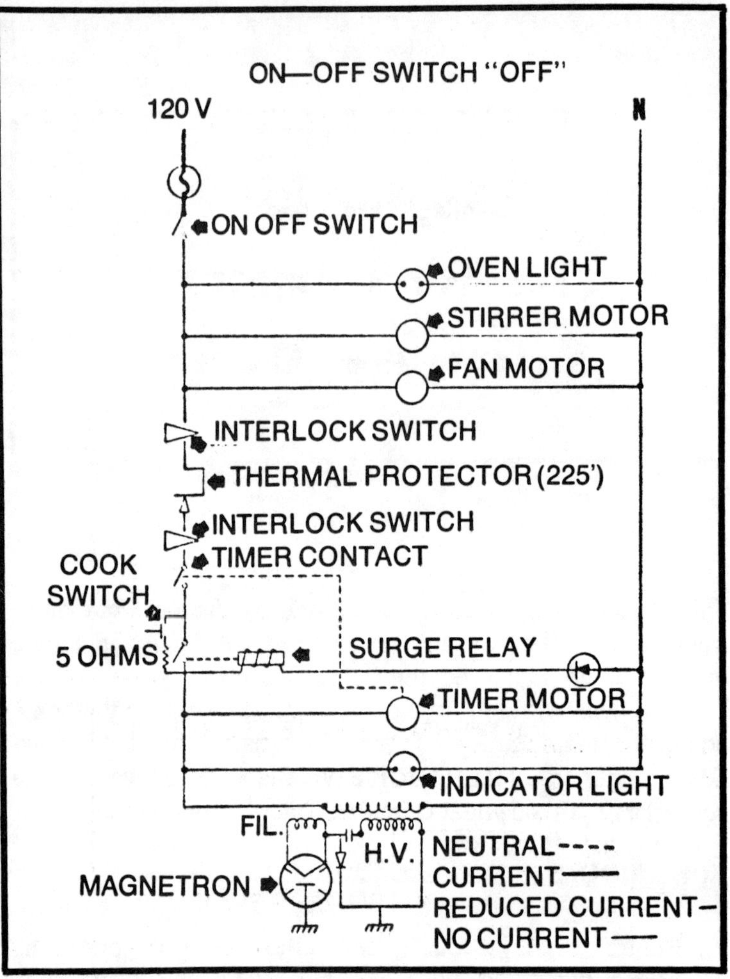

Fig. 12-1. Tappan circuit with ON-OFF switch at OFF.

The relay contacts close, providing power to the timer motor, which energizes the timer. This power will also close the relay contacts, providing a holding circuit for energizing the surge relay coil after the START switch is released.

The 120V AC input to the primary winding of the power transformer is converted to approximately 1800V AC output from the HV secondary winding. At the same time, the filament winding of the power transformer provides approximately 4V AC to heat the magnetron filaments.

The 1800V AC output from the power transformer's HV secondary is sent to a half-wave voltage-doubling circuit comprised of a single diode and one capacitor, where approximately 3500V DC (peak) is provided to the magnetron tube.

The −3500V DC applied to the cathode of the magnetron tube causes it to oscillate at the 2.4 GHz cooking frequency.

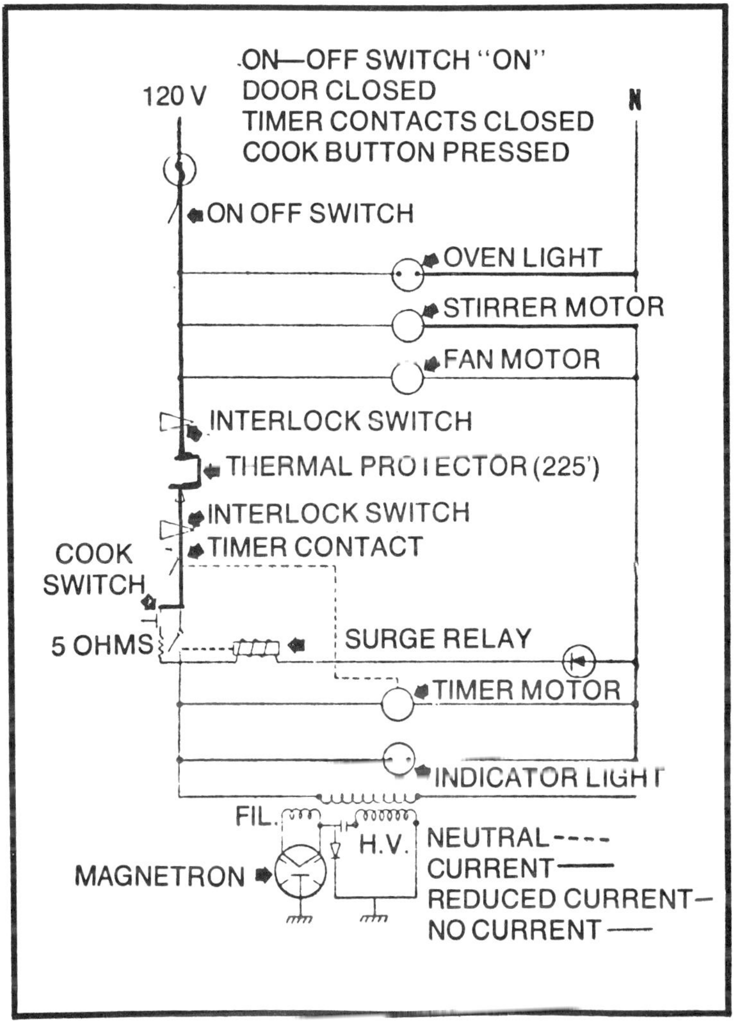

Fig. 12-2. Tappan circuit with ON-OFF switch at ON, door closed, timer contacts closed, COOK button pressed.

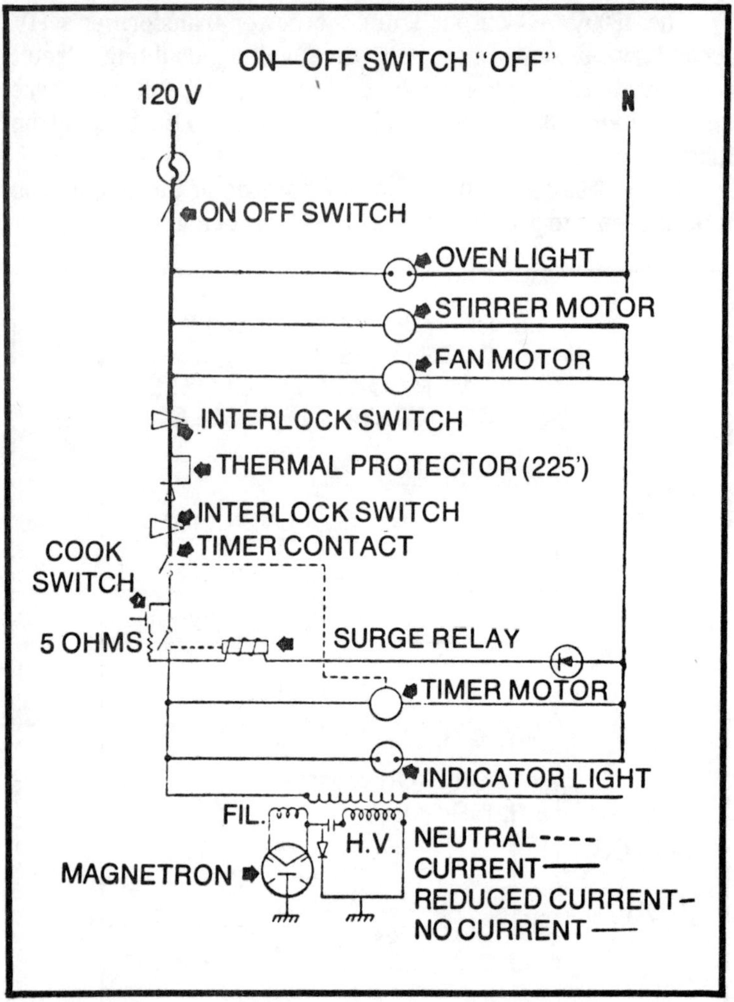

Fig. 12-3. Tappan circuit with ON-OFF switch at ON and door closed.

The RF energy produced by the magnetron tube is channeled through a waveguide, past the stirrer blade, and into the cavity, where the load is placed to be heated. When the timer has completed the selected cooking time, the timer switch opens, and the surge relay coil is de-energized. With the surge relay contacts open, the circuits are open to the cook light and power transformer. The cook light goes out, and the RF energy is no longer produced. A bell signals the end of the cook cycle.

TROUBLESHOOTING AND TESTING THE 56-1235, -5235, -2343, -6343, -3453, -7453, -3454, -7454

When you are troubleshooting a microwave oven, it is helpful to follow a logical program of checking. Such a program is given in Table 12-1. Many of the possible causes of trouble given in this table will require that a specific test of a component be made. These tests are given next.

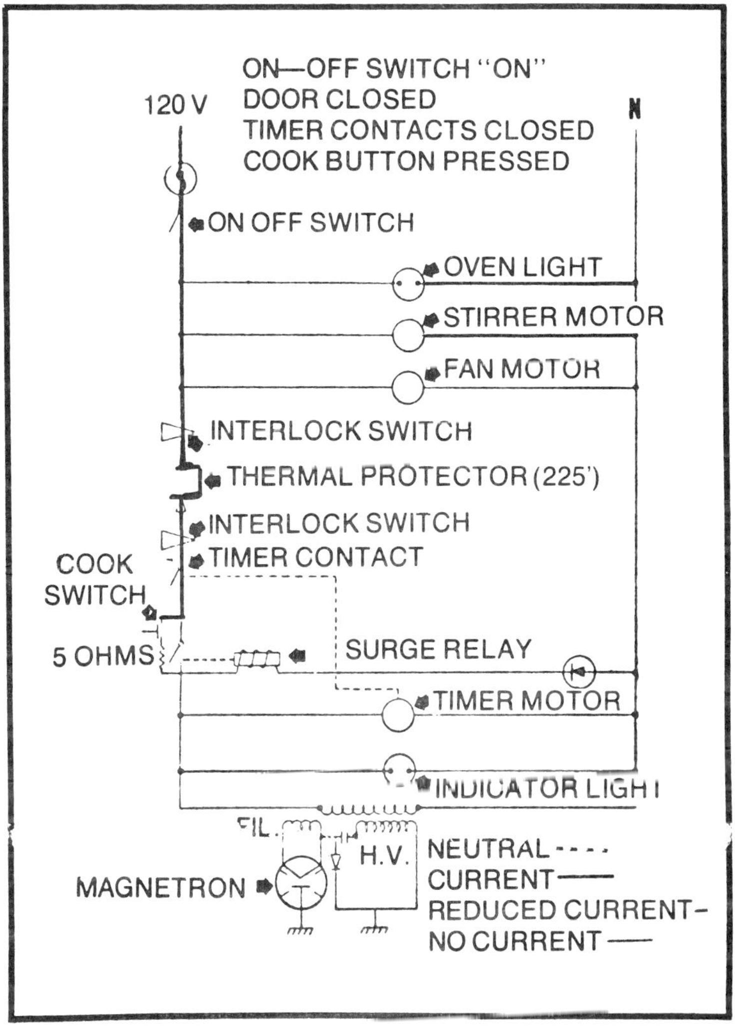

Fig. 12-4. Tappan circuit with ON-OFF switch at ON, door closed, timer contacts closed, COOK button released.

Table 12-1. Troubleshooting Guide—Tappan Models 56-1235, -5235, -2343, -6343, -3453, -7453, -3454, -7454

OFF CONDITION		
PROBLEM	**POSSIBLE CAUSE**	**TEST PROCEDURE OR CORRECTION**
Line fuse blows when power cord is plugged into wall receptacle.	1. Shorted or damged ON-OFF switch.	Replace switch.
	2. Shorted wire in power cord or wire harness.	Replace cord or check wiring.
ON CONDITION		
PROBLEM	**POSSIBLE CAUSE**	**TEST PROCEDURE OR CORRECTION**
Oven completely inoperative with switch turned on.	1. No power form service entrance.	Check wall outlet.
	2. Open wire in power cord or wiring harness.	Replace same or repair wiring.
	3. Defective ON-OFF switch.	Replace.
Oven cavity light does not illuminate.	1. Defective bulb.	
	2. Open wiring in circuit to light.	Check wiring.
Stirrer motor inoperative	1. Defective stirrer motor.	Replace stirrer motor.
	2. Open wiring in circuit to stirrer motor.	Check wiring.
Fan motor inoperative	1. Defective fan motor	Replace
	2. Open wiring in circuit to stirrer motor.	Check wiring.
COOK CONDITION		
PROBLEM	**POSSIBLE CAUSE**	**TEST PROCEDURE OR CORRECTION**
Surge relay does not pull in.	1. Defective contacts on timer switch.	Replace timer.
	2. Interlock switches defective or not properly adjusted.	See section on primary- or secondary-interlock switch.
	3. Defective surge resistor.	See **Surge Resistor Test**.
	4. Defective COOK switch.	See ON-OFF, COOK switch sections.
	5. Defective surge relay.	See **Surge Relay Test**.
	6. Open wiring between the above components.	Check wiring.
	7. Thermal cutout opened.	Replace thermal protector.
	8. Defective start switch (model 56-3454, 56-7454).	See OFF-START and DEFROST switch sections.
Surge relay pulls in, but no indicator light.	1. Defective indicator light.	Replace.
	2. Open wiring between timer contact and timer motor.	Check wiring.
Heat produced in oven load, but indicator light does not illuminate.	1. Defective indicator light.	Replace.
	2. Open wiring to indicator light.	Check wiring.

Table 12-1. cont'd.

PROBLEM	POSSIBLE CAUSE	TEST PROCEDURE OR CORRECTION
Indicator light indicates cycle, but little or no heat is produced in oven load.	1. Shorted HV circuit between voltage-doubler circuit and magnetron.	Check HV components. Check wiring.
	2. Defective HV transformer.	See **Power Transformer Test**.
	3. Defective diode.	See **Diode Check**.
	4. Defective capacitor.	See **Capacitor Check**.
	5. Defective magnetron.	See **Magnetron Filament Test**.
Oven goes into cook cycle but shuts down before end of cycle.	Open thermal cutout.	Replace thermal protector.
DEFROST CONDITION MODELS 56-3454 & 56-7454		
PROBLEM	POSSIBLE CAUSE	TEST PROCEDURE OR CORRECTION
Surge relay does not pull in.	⟶	See same problem, cook condition.
Oven works normally in cook, will not operate in defrost.	1. If DEFROST light not lit, panel DEFROST switch defective.	Replace panel DEFROST switch.
	2. If DEFROST light is lit, defrost timer defective.	Replace defrost timer.
Oven works normally in cook, with no change in defrost operation.	1. If DEFROST light not lit, panel DEFROST switch defective.	Replace control panel DEFROST switch.
	2. If DEFROST light is lit, defrost timer defective.	Replace defrost timer.

HV Components Tests

Because cooking times can be adjusted to compensate for power output, the magnetron tube should be replaced only if the water temperature rise test (described in Chapter 3) indicates a power output well beyond the normal limits. This test is only accurate if the line voltage is 120V while the oven is operating.

Caution: HV is present during the cook cycle, so extreme caution should be observed. When tests are completed, always unplug the oven from its power source and discharge the capacitor.

1. Unplug the oven from its power source.
2. After removing the outer wrapper, discharge the capacitor by shorting the two capacitor terminals together with an insulated screwdriver, then short each terminal to ground (Fig. 12-5).

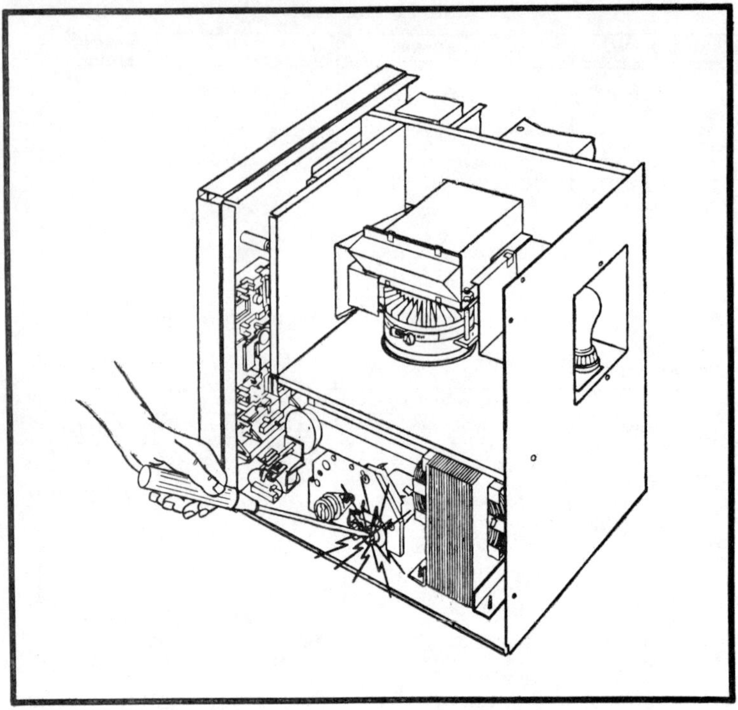

Fig. 12-5. Discharging the HV capacitor.

3. Visually check the HV wiring for proper connections. Refer to wiring diagram for proper wiring sequence.
4. After placing a cooking load in the oven cavity, apply power to the oven; put it into a cook cycle.
   Note: Do not touch any oven components or wiring during oven operation.
5. To check for filament operation, observe (through the small hole in the waveguide) the magnetron filament to see if it is glowing. The filament should begin to glow a second or two after the oven is put in a cook cycle. If the filament glow can be seen through the waveguide hole during the cook cycle, go to step 6. If no filament glow is seen in the waveguide during the cook cycle, turn the oven off and refer to *Power Transformer Test* and *Magnetron Filament Test*.
6. If the filament glow is seen during the cook cycle, turn the oven off, unplug the power cord, and discharge the

capacitor with an insulated screwdriver.
Disconnect the HV leads from the diode terminals.

7. With the ohmmeter set on the highest scale, measure the resistance between the diode terminals.
   Important: The meter must have at least a 9V battery to perform the check.
   Reverse the meter leads and again measure the resistance between each diode terminal. This test determines the front-to-back resistance of the diode. Depending on the polarity of the meter leads, an infinite resistance should be indicated in one direction and a deflection of the meter needle towards the zero mark should be indicated in the other direction. If normal indications are given as described, connect the diode and go to step 8. If a low resistance or infinite resistance is indicated in both directions, replace the diode; the diode is either shorted or opened.
8. Connect the HV leads to the magnetron assembly. Disconnect the leads to the capacitor. With an ohmmeter set on the highest scale, measure the resistance between the two capacitor terminals. The meter should indicate continuity between the terminals for a *short* time and should indicate an open circuit once the capacitor is charged. Reversing the meter leads should give the same indication. If normal indications are given as described, connect the diode and go to step 9. If the ohmmeter indicates continuity between the capacitor terminals at all times, or if no meter deflection occurs at all, the capacitor should be replaced.
   Note: The ohmmeter should be on a high scale when making a capacitor check.
9. With an ohmmeter set on the R × 1 scale, measure the resistance between HV terminal of the power transformer and chassis ground. The meter should indicate approximately 50Ω. If approximately normal readings are given, go to step 10. If extremely low resistance is given or if extremely high (or infinite) resistance is indicated with the above test, replace the power transformer.

10. If all operations appear normal in steps 1 through 9, replace the magnetron assembly; then check the oven for proper operation.

Magnetron Filament Test

In normal cook operation, a visible glow from the heated tube filament can be seen through the small holes in the waveguide during a cook cycle.

If no filament glow is seen during a cook cycle, turn the oven off, unplug the power cord, and discharge the capacitor with an insulated screwdriver. Disconnect the HV leads from the magnetron's filament terminals. Measure the resistance across the two filament terminals with the ohmmeter set on the R × 1 scale. If the normal resistance of less than 1Ω between magnetron terminals is indicated on the meter, make a power transformer test.

If high resistance or infinite resistance is indicated between the magnetron terminals, replace the magnetron assembly.

Diode Check

Disconnect the two leads to the diode. With the ohmmeter set on the highest scale, measure the resistance across the two diode terminals. Reverse the meter leads and again observe the resistance reading.

A normal diode will indicate infinite resistance in one direction and a deflection of the meter needle towards the zero mark is indicated in the opposite direction. If the diode check is normal, proceed to *Magnetron Check*. If continuity is indicated in both directions or if an infinite resistance is read in both directions, the diode is probably defective and should be replaced.

Note: Meters with less than a 6V or 9V battery are usually not adequate for checking the front-to-back resistance of the diode. The meter should be checked with a known-good diode before judging a diode to be defective because of an infinite resistance reading in both directions.

Power Transformer Test

An operational test can be made to determine the amount of filament voltage produced by the power transformer.

Caution: Approximately 1800V AC is present at the HV terminal of the power transformer during a cook cycle. Observe caution at all times.

1. With the oven's power cord unplugged and outer case removed, discharge the capacitor with an insulated screwdriver.
2. Remove the HV lead that connects the capacitor to the HV terminals.

   Note: During testing, keep meter, hands, etc. well away from the HV terminals of the power transformer.
4. Apply power to the oven. Put the oven in a cook cycle to get a filament voltage reading. A normal indication should be approximately 3.8V AC. Unplug the oven before disconnecting the meter leads.
5. If a normal 3.8V AC reading was indicated on the meter, go to step 6. If no voltage was indicated on the meter, set the meter to the proper scale. Check the 120V AC primary input to the power transformer during a cook cycle. If the input voltage is normal but no filament voltage is present, replace the power transformer.
6. If filament voltage is normal but the magnetron filaments do not glow, unplug the power cord, discharge the capacitor, install the HV lead, and refer to the *Magnetron Filament Test*.

HV Winding Continuity Check

One side of the power transformer's HV secondary is connected to the oven's chassis ground. The other end of the HV secondary has one terminal marked *HV*.

Since approximately 1800V AC is present at the HV secondary terminal during a cook cycle, il is not recommended that HV measurements be attempted. Normally, a continuity check of the HV secondary will be sufficient to determine the condition of the secondary winding. Proceed as follows:

1. With the oven unplugged, discharge the capacitor and remove the HV lead from the power transformer's HV secondary terminal.

2. With an ohmmeter set on the R × 1 scale, measure the resistance between the HV terminal of the power transformer and chassis ground. The reading should indicate around 50Ω.
3. If extremely low resistance is indicated or if extremely high (or infinite) resistance is indicated with the continuity test, replace the power transformer.

Capacitor Test

If the capacitor is open, no HV is available to the magnetron. A shorted capacitor normally causes high line current, which should trip the wall circuit breaker or blow the line fuse.

An ohmmeter can be used to check for shorted or a open capacitor. Proceed as follows:

1. Unplug the oven, discharge the capacitor, and remove the leads from the capacitor terminals.
2. With an ohmmeter set on the highest scale, measure the resistance between the two capacitor terminals. The meter should indicate continuity between the terminals for a short time and should indicate an open circuit once the capacitor is charged. Reversing the meter leads should give the same indication.
3. If the ohmmeter indicates continuity between the capacitor terminals at all times, or if no meter deflection occurs at all, the capacitor should be replaced.

Magnetron Assembly Test

To test for an open filament, isolate the magnetron filament leads. The meter should indicate less than 1Ω.

To test for a shorted magnetron, connect ohmmeter leads between the filament leads and chassis ground. This test should indicate an infinite resistance. If there is little or no resistance, the magnetron is grounded and must be replaced.

Primary-Interlock Switch

When the door latch is closed, an audible click can be heard; the primary-interlock switch is activated. When the latch is open, the switch is then deactivated. To test the switch for proper operation:

1. Disconnect power to the unit, remove the wrapper, and discharge the capacitor.
2. Remove the two leads to the primary interlock.
3. With the ohmmeter set on the R × 1 scale, take a continuity reading between the two terminals. With the latch closed, the reading should be zero; with the door latch open, a reading of infinite resistance should be indicated.
4. If the switch does not operate as described, replace it. *Note:* If interlock switch is replaced, it must be adjusted. See *Door and Interlock Adjustments* for proper adjustment procedures.

Secondary-Interlock Switch

After the first audible click given by the primary-interlock switch, another audible click will be heard when the door is open slightly. This is the secondary-interlock activating. To test it, proceed as follows:

1. Disconnect power to the unit. Remove the wrapper and discharge the capacitor.

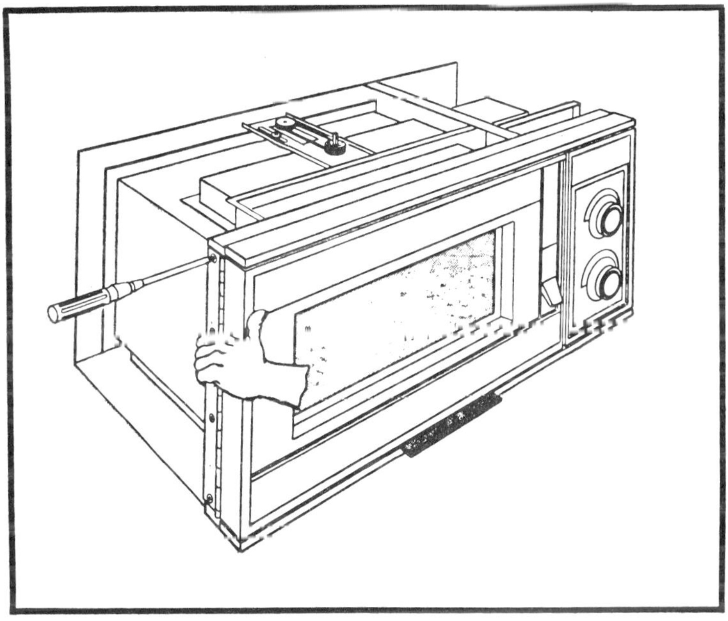

Fig. 12-6. Illustration for hinge adjustment.

2. Remove the two leads to the secondary-interlock switch.
3. With the ohmmeter set on the R × 1 scale, take a continuity reading between the two terminals. With the door closed, the reading should be zero; with the door open, an open circuit should be indicated.
4. If the switch does not operate as described, replace the switch.

   Note: If interlock switch is replaced, it must be adjusted. See *Door & Interlock Adjustments* for the proper adjustment procedure.

   Important: If, at any time, either the oven or house fuse blows due to an oven fault, both switches must be replaced. This is because the short-circuit currents will have damaged these important switches.

Surge Resistor Test

1. Disconnect power to the oven, remove the wrapper, and discharge the capacitor.
2. The surge resistor can be checked by removing the terminals leading to the surge relay and cook switch.
3. With an ohmmeter set on the R × 1 scale, measure the resistance between the two leads. The meter should indicate 5Ω; the resistor will then be operating properly.
4. If the indication on the meter is less than 5Ω or is infinity, replace the resistor. The resistor must be ordered from Tappan's Sentinel Service parts division. The resistor is especially designed not to ignite if high current is present.

Surge Relay Test

1. Disconnect power to the oven, remove the wrapper and discharge the capacitor.
2. The surge resistor can be checked by removing the two leads, one on the top terminal and one on the bottom terminal of the surge relay.
3. With an ohmmeter set on the R × 1 scale, measure the resistance between the top and bottom terminals of

the surge relay. The meter should indicate about 1300Ω. The surge relay should operate properly.

Note: If an open is indicated in one direction, reverse the leads and 1300Ω should be indicated. This is due to a diode that is placed in series with the winding of the relay.

4. If the indication on the meter is less than 1300Ω or much more than 1300Ω, the relay should be replaced.

Thermal-Protector Test

Unless overheating exists, a continuity check across the thermal-protector contacts should indicate a closed circuit. The thermal protector opens at approximately 225°F and resets automatically at approximately 185°F.

Models 56-3453 & 56-7453 ON-OFF & COOK Switch

1. Disconnect the power, remove the wrapper, and discharge the capacitor by placing an insulated screwdriver across both terminals, then from each terminal to ground.
2. Remove the ON-OFF and COOK switches, then remove the wires, noting what terminals the wires are connected to.
3. Set an ohmmeter on the R × 1 scale and zero the meter.
4. Place one of the test leads on terminal 12 and the other on terminal 8.
5. Depress the ON push button and the meter should deflect to zero.
6. With the leads on the same terminals as in step 4, depress the OFF push button. The meter should now indicate infinite resistance.
7. Now remove both leads. Place one on terminal 3 and the other on terminal 4.
8. Depress the COOK push button and watch the indication. When the button is pushed, the meter will deflect to zero. Release the push button; the meter will indicate infinite resistance. This test will prove the switch is operating properly.

9. If the switch is defective, it must be replaced. If it is not defective, place the wires on the terminals and secure the switch back in the control panel.

MODELS 56-3454 & 56-7454 OFF-START & DEFROST SWITCH

1. Disconnect the power, remove the wrapper, and discharge the capacitor by placing an insulated screwdriver across both terminals, then from each terminal to ground.
2. Set an ohmmeter on the R × 1 scale and zero the meter.
3. Place one of the test leads on terminal 5 and the other on terminal 6.
4. Depress the OFF push button; the meter should indicate an open circuit. Depress the START push button; the meter should indicate a closed circuit.
5. Disconnect the leads from terminals 5 and 6. Connect them to terminals 7 and 8.
6. Depress the START push button; the meter should indicate a closed circuit. Release the START push button; the meter should indicate an open circuit.
7. Disconnect the leads from terminals 7 and 8. Connect them to terminals 1 and 4.
8. Depress the DEFROST push button; the meter should indicate an open circuit. With the DEFROST push button in the out (COOK) position, the meter should indicate a closed circuit.
9. Disconnect the lead attached to terminal 1; connect it to terminal 3 (remove all leads from terminal 3 prior to connecting meter).
10. Depress the DEFROST push button; the meter should indicate a closed circuit. With the DEFROST push button in the out (COOK) position, the meter should indicate an open circuit. This test will prove the switch is operating properly.
11. If the switch is defective, it must be replaced. If it is not defective, place the wires back on the terminals.

DOOR & INTERLOCK ADJUSTMENT

The door adjustment is very important to prevent any excessive leakage of microwave energy. The RF levels should be checked after all door, interlock-switch, and associated repairs are made. The following door adjustments must be followed to the letter for proper door adjustment:

Hinge-Side Adjustment

1. Disconnect the power, remove the wrapper and glass shelf, and discharge the capacitor by shorting across the terminals; then from each terminal to ground with an insulated screwdriver.
2. Close the door and depress the latch.
3. Place an eighth-inch shim between the bottom of the door and the bottom trim. For proper appearance, the door should align with the control panel trim.
4. Loosen the four screws on the hinge side until the door can easily be moved back and forth.
5. Squeeze the door against the front frame and, at the same time, tighten the four screws—the center ones first, then the outer ones.

Latch & Roller Assembly

To make a proper latch and roller adjustment, the roller assembly must be adjusted so that with the door latch depressed the roller assembly will seat firmly in the lower crook of the latch. If it does not:

1. Remove the control panel.
2. The two screws on the front frame must be loosened enough to slide the roller assembly up and down.
3. Close the door and depress the latch. Adjust the roller assembly until it is seating in the lower crook of the latch.
4. Tighten the two screws that secure the roller assembly. After the door has been properly adjusted, refer to the interlock adjustments before operating the oven.

Primary-Interlock Adjustment

Warning: Do Not attempt to adjust interlock switches with electronic oven energized. If this warning is not observed, high levels of microwave leakage may be present.

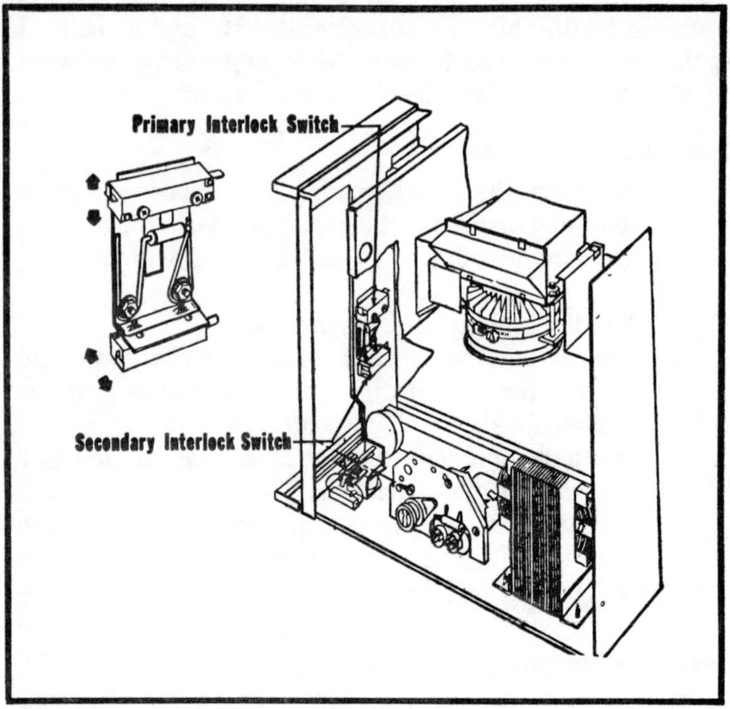

Fig. 12-7. Primary- and secondary-interlock adjustments.

1. Disconnect the power to the unit, remove the wrapper, and discharge the capacitor by placing an insulated screwdriver across the two terminals, then from both terminals to chassis ground.
2. Loosen the two screws that secure the upper (primary) interlock until the assembly can be moved freely up and down (Fig. 12-7).
3. Close the door and secure the latch.
4. Adjust the interlock down until the latch closes the switch. This can be heard by an audible click. Do not overadjust. Check it by operating the latch.
5. Secure the switch after proper positioning is accomplished.

   Note: The interlock switch *must not be* adjusted low enough to allow activating manually.
6. To test if the interlock is adjusted properly, insert an eighth-inch straight rod about 6 inches long into the latch opening on the front frame. Manually try to

activate the primary interlock switch (Fig. 12-8). If the switch cannot be activated, the interlock switch is adjusted properly. If the switch can be activated, the switch is not adjusted properly, and readjustments *must* be made. Refer to step 4.

Secondary-Interlock Adjustment

Warning: Do not attempt to adjust interlock switches with electronic oven energized. If warning is not observed, high levels of microwave leakage may be present.

1. Disconnect the power to the unit, remove the wrapper, and discharge the capacitor by placing an insulated screwdriver across the two terminals, then from both terminals to chassis ground.
2. Back out the two screws that secure the interlock until the assembly can be moved freely back and forth.
3. Close the door and secure the latch.
4. Move the switch towards the door until an audible sound of the switch closing can be heard. Do not overadjust. Check it by opening and closing the door.
5. Tighten and secure the switch with two screws.

MODELS 56-3453, -7453, -3454 & -7454 CONTROL PANEL DISASSEMBLY

1. Unplug the power cord, remove the outer wrapper, and discharge the capacitor.

Fig. 12-8 Testing interlock-switch adjustment.

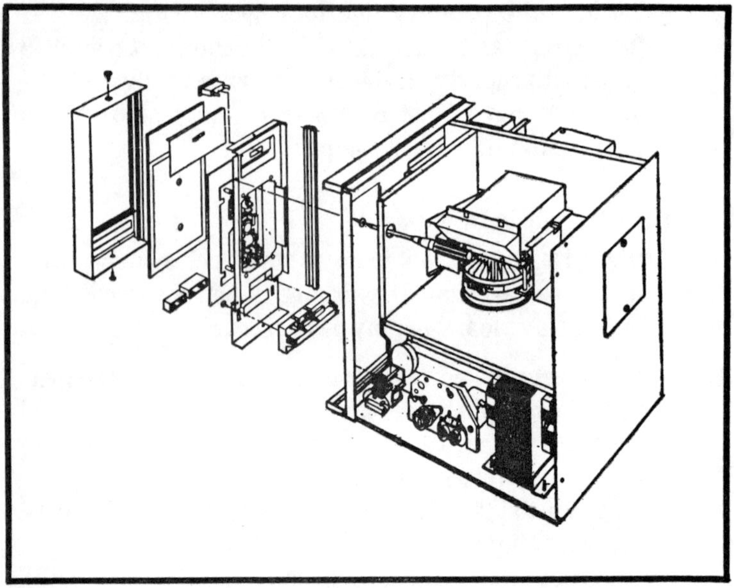

Fig. 12-9. Control panel of Tappan Models 56-3453, -7453, -3454, and -7454.

2. Remove the two screws shown in Fig. 12-9. (A screwdriver is shown near one of these.) One is attainable through the access hole in the air baffle; the other is also a common ground point.

   Note: The ground screw is not to be permanently removed.
3. Open the door and swing the control panel out about 2 inches and lift it out.
4. Remove the two dial knobs by pulling them off.
5. The trim can be removed by removing one screw in the top and bottom.
6. Lift the frame forward and lift off.
7. The glass can now be removed by lifting from the frame.
8. The cook light can be removed by removing the two wires, depressing the locking tabs, and pushing forward.
9. The ON-OFF and COOK switch can be removed by pulling off the push buttons and removing the two screws securing the switch.

10. The microwave sealing strips located on the control panel trim can be removed by pushing back; then lifting it off.
11. To remove the double rotary timer, use a flat-bladed screwdriver and carefully pry the timer from the mounting bracket. (See Fig. 12-10.) Replace wire for wire on the replacement timer.
12. To reassemble; reverse the procedure.

MODELS 56-1235, -5235, -2343 & -6343
CONTROL PANEL DISASSEMBLY

1. Unplug the power cord, remove the outer wrapper, and discharge the capacitor.
2. Remove the three mounting screws. One is attainable through the access hole in the air baffle, one is a common ground point, and one is between the two. (See Fig. 12-9 for proper location).

   Note: The ground screw is not to be permanently removed.

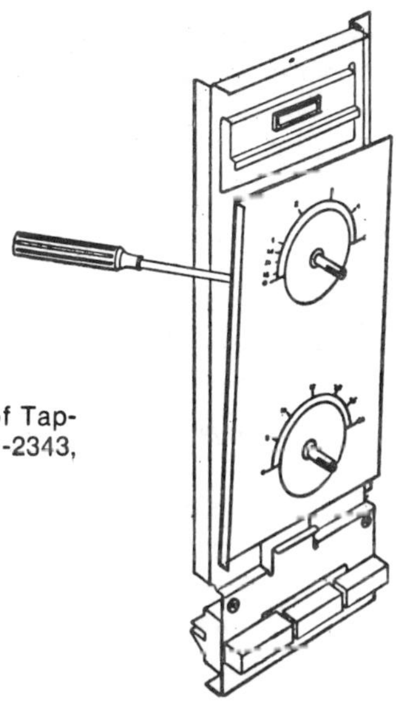

Fig. 12-10. Control panel of Tappan Models 56-1235, -5235, -2343, and -6343.

3. Open the door and swing the control panel out about 2 inches to lift it out.
4. Remove the dial knob by pulling it off.
5. The face panel can be removed from the assembly by removing four screws located on the corners of the assembly.
6. The cook light can be removed by detaching the two wires, depressing the locking tabs, and pushing backward.
7. The COOK switch (model 56-2343) can be removed by pulling off the push button and removing the two screws securing the switch. Replace them on the new switch, wire for wire.
8. To remove the timer, use a flat-bladed screwdriver and carefully pry the timer from the mounting bracket. (See Fig. 12-10.) Replace each head, wire for wire on the replacement timer.
9. To reassemble, reverse the above procedure.

HOW MODEL 56-1139 WORKS

The following paragraphs describe the circuit operation of this relatively early (1970) model Tappan oven for the various modes of operation. Refer to the schematic (Fig. 12-11) as you study these paragraphs.

Off Condition

1. ON-OFF switch is in the OFF position.
2. No components in the oven will operate.

Idle Condition

1. ON-OFF switch is in the ON position.
2. The following components are in operation:
    a. Dial light illuminates on the timer module.
    b. Cavity lights illuminate the cooking cavity.
    c. Blower motor forces cooling air through the air ducts surrounding the magnetron tube.
    d. Stirrer motor rotates the stirrer blade assembly at the top of the cavity.
    e. Time-delay relay heating element begins heating; contacts close after 5 seconds.
    f. Filament transformer begins heating the filaments of the magnetron tube.

Cooking Condition

1. The door is closed to activate both door interlock switches, i.e., front protective switch and rear protective switch.
2. The desired cooking time must be selected with the timer dial to close the switch contacts leading to the cook relay.
3. The START switch must be activated.
4. The following operations occur:
    a. The coil of the cook relay is energized.
    - Relay contacts T2 and L2 close to provide a current path to the timer motor, which begins to return to the OFF position of the timer. This also provides the holding circuit for energizing the coil of the cook relay when the START switch is released.
    - Relay contacts T1 and L1 close to provide a current path to the plate transformer and the COOK lights.

    b. The 110V AC input to the primary winding of the plate transformer is converted to approximately 180V AC on the secondary winding.
    c. The 1800V AC output from the plate transformer is sent to a doubling circuit comprised of a dual capacitor and two diodes. This 1800V AC is converted to approximately 3400V DC by the doubling circuit and sent to the magnetron tube.
    d. The − 3400V DC applied to the cathode of the magnetron tube causes the magnetron to oscillate at the 2.45 GHz cooking frequency.
    e. The RF energy produced by the magnetron tube is channeled through a waveguide, into the cavity feedbox, past the stirrer blade, and finally into the cavity, where the food is placed to be heated.
5. When the timer has timed out the desired cooking time:
    a. The timer switch opens and the coil of the cook relay is de-energized.

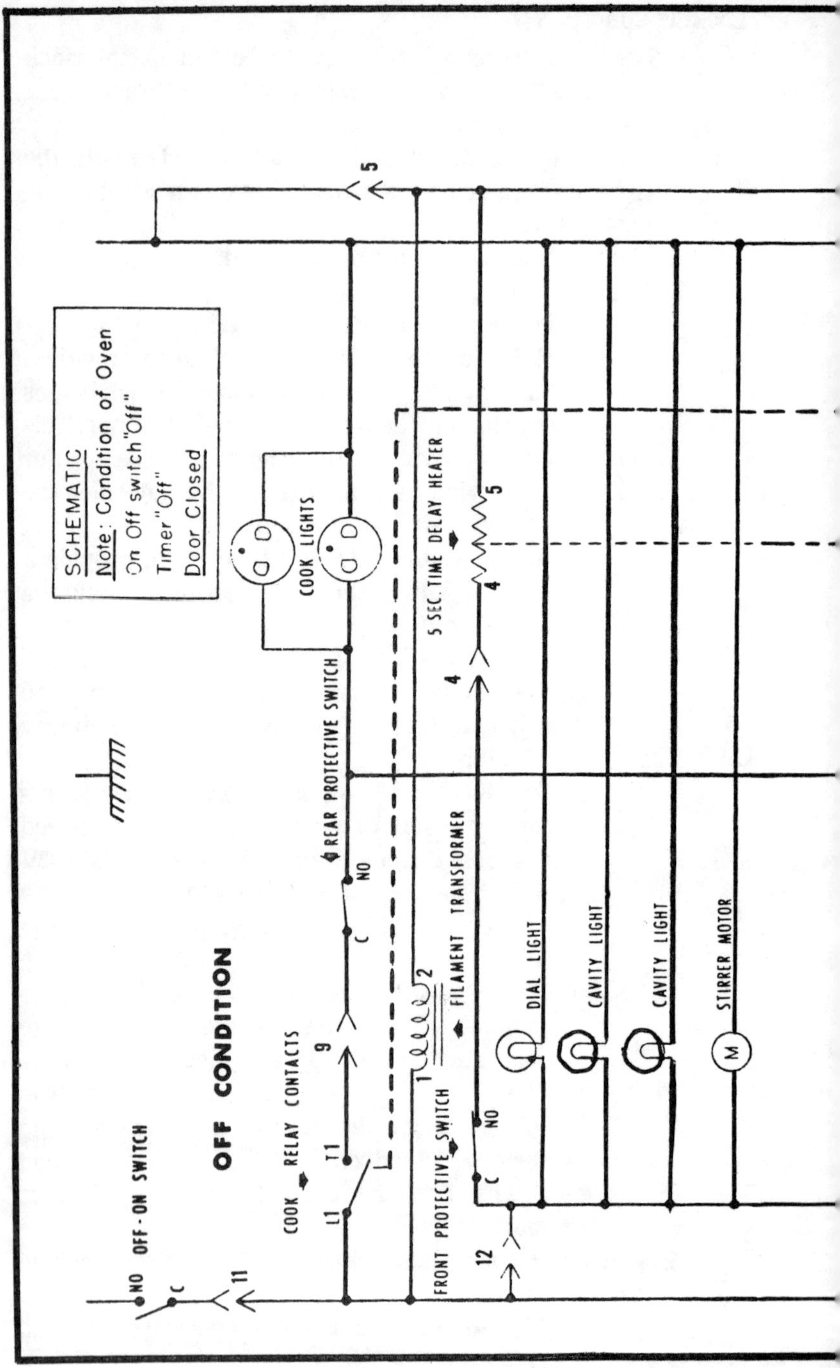
OFF CONDITION
SCHEMATIC
Note: Condition of Oven
On Off switch "Off"
Timer "Off"
Door Closed
OFF-ON SWITCH
NO
C
11
COOK RELAY CONTACTS
L1
T1
9
REAR PROTECTIVE SWITCH
C
NO
COOK LIGHTS
FILAMENT TRANSFORMER
1
2
FRONT PROTECTIVE SWITCH
12
4
5 SEC. TIME DELAY HEATER
5
DIAL LIGHT
CAVITY LIGHT
CAVITY LIGHT
STIRRER MOTOR
M

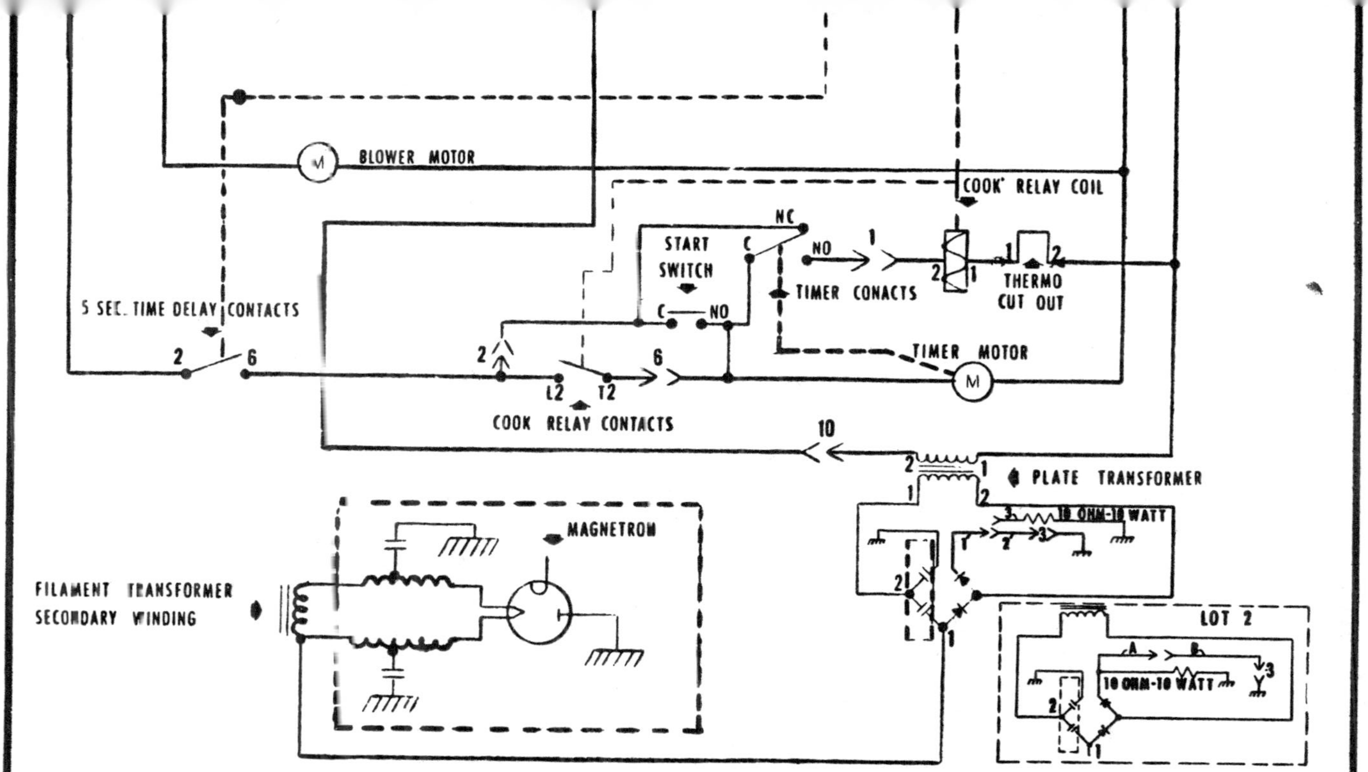

Fig. 12-11. Circuit of Tappan 56-1139 in OFF condition.

b. The contacts of the cook relay open.
- The COOK light goes out.
- Since the magnetron no longer has HV, no more energy is produced.

c. The timer bell rings for a short time to indicate the end of the cook cycle.
d. Basically, the oven has reverted to the idle condition.

MODEL 56-1139 TROUBLESHOOTING & TESTING

When you are troubleshooting a microwave oven, it is helpful to follow a logical sequence of testing. The troubleshooting sequence for Model 56-1139 is given in Table 12-2. The specific tests referred to in the troubleshooting table

Table 12-2. Troubleshooting Guide—Tappan Model 56-1139

OFF CONDITION		
PROBLEM	**POSSIBLE CAUSE**	**TEST PROCEDURE OR CORRECTION**
Line fuse blows when power cord is plugged into wall receptacle.	1. Shorted or damaged ON-OFF switch.	Replace.
	2. Shorted wire in power cord or wire harness.	Replace.
IDLE CONDITION		
PROBLEM	**POSSIBLE CAUSE**	**TEST PROCEDURE OR CORRECTION**
Oven completely inoperative when turned on.	1. No power from service entrance.	Check wall circuit breakers or fuses.
	2. Open wire in power cord.	Replace.
	3. Defective ON-OFF switch.	Replace.
Dial light does not illuminate on timer panel.	1. Defective bulb.	Replace.
	2. Open wiring between bulb and ON-OFF switch.	Check wiring.
Cavity light does not illuminate.	1. Defective bulb.	Replace.
	2. Open wiring between ON-OFF switch and bulb.	Check wiring.
Blower motor inoperative.	1. Defective blower motor.	Replace.
	2. Open wiring between blower motor and ON-OFF switch.	Check wiring.
Stirrer motor inoperative.	1. Defective stirrer motor.	Replace.
	2. Open wiring between stirrer motor and ON-OFF switch.	Check wiring.
Magnetron filament does not glow.	1. Open filament transformer.	Replace.
	2. Defective magnetron.	Replace.
	3. Open wiring from primary of filament transformer to ON-OFF switch or from secondary of filament transformer to tube connections.	Check wiring and connections.

Table 12-2. cont.

COOKING CONDITION

Caution: HV is present during the cook cycle. Observe caution.

PROBLEM	POSSIBLE CAUSE	TEST PROCEDURE OR CORRECTION
Oven will not go into cook cycle when START switch is activated.	1. Front protective switch defective or out of adjustment.	Replace or adjust.
	2. Defective 5-second time-delay relay.	Replace.
	3. Defective START switch.	Replace.
	4. Defective contacts on timer switch.	Replace timer motor assembly.
	5. Defective cook relay.	Replace.
	6. Defective thermal protector.	Replace.
	7. Open wiring between the above components.	Check wiring.
Line fuse blows when oven is placed into cook condition.	1. Overloaded circuit.	Check current-carrying capability of electrical outlet. Oven should be hooked to individual 15A circuit.
	2. Shorted component: a. START switch. b. Timer assembly. c. Cook relay. d. Rear protective switch. e. Plate transformer.	Repair or replace.
	3. Shorted wiring to above components.	Repair shorted wire.
Timer times out, but oven has no COOK light and does not heat.	1. Rear protective switch defective or out of adjustment.	Replace or adjust switch.
	2. Defective cook relay contacts L1 and T1.	Replace contacts.
	3. Open wiring.	Repair wiring.
Oven goes into cook cycle, but timer does not time out.	1. Defective timer motor.	Replace.
	2. Open wiring between cook-relay contact L2 and timer motor.	Check wiring.
Oven heats, but COOK light does not illuminate.	1. Defective COOK light.	Replace.
	2. Open wiring to COOK light.	Check wiring.
Oven goes into cook cycle, but unit heats very slowly or not at all.	1. Defective plate transformer.	Replace.
	2. Defective diode.	Replace.
	3. Defective capacitor.	Replace.
	4. Defective magnetron.	Replace.
	5. Defective filament transformer.	Replace.
	6. Open wiring between above listed components.	Check wiring.
	7. Shorted HV circuit between doubler circuit and magnetron.	Check for short circuit with ohmmeter.

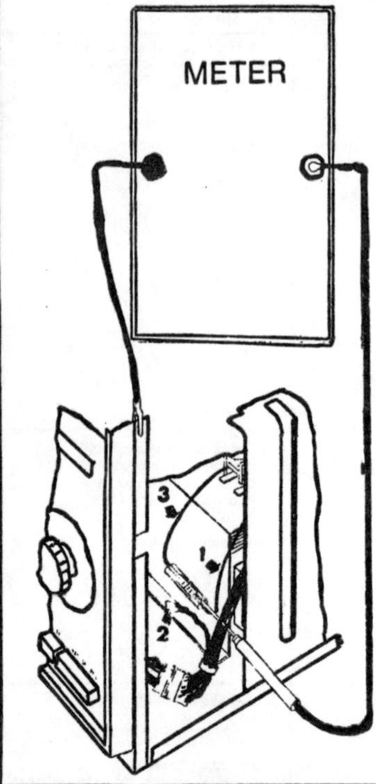

Fig. 12-12. Plate-current test—Tappan 56-1139, lot 1.

are given in the following paragraphs. Two different methods of testing the plate current are outlined. Refer to Figs. 12-12 and 12-13 to determine which wiring version is used in the unit under test.

Plate Current Test 1

This test should be made when the oven appears to operate normally except for little or no heating within the cavity.

Caution: HV is present during the cook cycle, so extreme caution should be observed. Do not hold leads or meter during the following test procedure. Use alligator clips to secure leads to test points.

1. Unplug the oven from its power source.
2. After removing the outer wrapper, remove the HV shield and discharge each capacitor section to the case with an insulated screwdriver or shorting bar.

3. Disconnect the orange wires marked *1* and *2* (Fig. 12-12).
4. Connect the orange wires marked *1* and *3*.
5. Using an ohmmeter, check the resistance from junction of wires *1* and *3* to chassis ground to verify the existence of the 10Ω metering resistor in the circuit.
6. Connect the positive lead of meter to junction of wires *1* and *3*, negative lead to chassis ground.
7. Set the meter on the proper scale to read at least 5V (5V DC equals 500 mA). The plate current should be between 400 and 450 mA (4.0 and 4.5V).
8. Connect the power cord, turn the oven on, and read the plate current.
9. After the test, unplug the oven, discharge the HV capacitors, and connect wires 1 and 2.
10. Replace the HV shield.

Plate Current Test 2

1. Unplug the oven from its power source.
2. After removing the outer wrapper, remove the HV shield and discharge each capacitor section to the case with an insulated screwdriver or shorting bar.
3. Disconnect orange wires shown as *A* and *B* in Fig. 12-13.
4. Using an ohmmeter, check the resistance from wire *A* to chassis ground to verify the existence of the 10Ω metering resistor in the circuit.
5. Connect the positive lead of meter to junction of wires A and negative lead to chassis ground. *Do not touch meter or meter leads during test.*
6. Set the meter on the proper scale to read at least 5V on the DC scale (5V DC equals 500 mA). The plate current should be between 400 and 450 mA (4.0 to 4.5V).
7. Connect the power cord, turn the oven on, and read plate current.
8. After the test, unplug the oven, and discharge the HV capacitors. Then connect wires *A* and *B*.
9. Replace the HV shield.

A Normal Reading of 4V DC Was Not Observed. The problem is most likely in one of two areas:

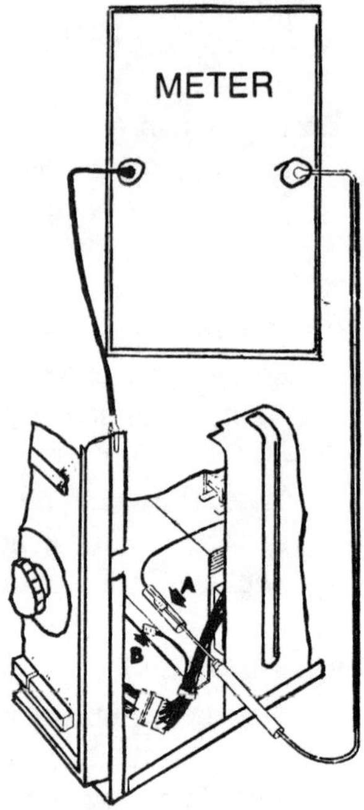

Fig. 12-13. Plate-current test—Tappan 56-1139, lot 2.

1. A short circuit to ground between the doubler circuit and the magnetron tube.
2. A defective magnetron tube.

Approximately One-Half the Normal Reading Was Observed. The diode may be defective. Check the front-to-back resistance of the diodes using an ohmmeter as follows:

1. Remove the power pack from the oven and isolate the diodes from the circuit.
2. Using the highest ohms scale of the meter, read the resistance across the two diode terminals.
3. Reverse the meter leads to the diode terminals and observe the resistance reading. A normal diode should read infinity or open in one direction and approximately 100K or more when the meter leads are reversed. If a short is indicated in both directions, or if

an open is read in both directions, the diodes are probably defective and should be replaced. Occasionally, a diode can short to ground through the base, and it should be replaced.

Note: Meters with less than a 9V battery are usually not adequate for checking the front-to-back resistance of the diodes.

No Voltage Is Indicated On the Test Meter. The following components should be checked.

- **Plate Transformer**—Unlock the wire leads to check for continuity between the two primary terminals with an ohmmeter. On the R × 1 scale, the resistance should be well below 10Ω. The same indication should be given when checking resistance between the secondary terminals.
- **Diode**—Refer to the previous procedure for testing diodes with an ohmmeter.
- **Capacitors**—If both capacitors were open, no HV would be available. An ohmmeter can be used to check for shorts between capacitors or open circuits. Using the high ohms scale, the meter should indicate continuity from ground to capacitors for a short time, then should indicate an open circuit once the capacitor is charged. The same indication should be given when measuring between terminals 1 and 2 of the dual capacitor.
- **Magnetron Assembly**—An open circuit between the doubler circuit and the magnetron cathode would given a zero reading in the plate current test. Check the wiring between the doubler circuit and the magnetrons filament leads. Also check the continuity between the two filament leads extending from the filter section. If an open circuit is indicated between the two filament leads, replace the magnetron. An open filament transformer or an open filament lead can give the same indication. Check for continuity between the primary leads and between the secondary leads of the filament transformer. The resistance of each winding should be well below 10Ω.

5-Second Time-Delay Test

Terminals 4 and 5 are the heater terminals of the time-delay relay. Resistance between 4 and 5 should be approximately 45Ω with the input leads removed. If an open circuit is indicated, replace the relay.

Front Protective Switch Test

When the thumb latch is depressed on the door handle, a click should be heard when the front protective switch is deactivated. When the thumb latch is released, the latch should activate the switch.

Some adjustment can be made on the switch by loosening its mounting bracket, then tightening the bracket after the adjustment. In some instances, the switch arm may have to be bent slightly.

The switch arm should be adjusted to allow the thumb latch to readily activate the front protective switch whenever the door is closed. However, the switch must release and interrupt the cook cycle before the door can be opened at all. The switch can be checked with an ohmmeter by connecting the meter leads to the common terminal and the N.O. terminal of the front protective switch. With the door open, the meter should read an open circuit. When the door is closed, the switch is closed so the meter should indicate continuity.

Rear Protective Switch Test

After the first click given by the front protective switch, another click should be heard when the door is opened slightly. This results from the rear protective switch being deactivated as the cam arm moves away from the switch.

The rear protective switch can be adjusted by loosening the two mounting screws at the left-hand base of the oven; then tightening the screws after the adjustment.

If this switch is defective or out of adjustment, the circuit is opened to the primary of the plate transformer and cooking light. The oven would appear to be operating normally except for a no COOK light indication, and the oven would not heat.

The meter leads connected to the COM and N.C. terminals should indicate a closed circuit with the door open, and an open circuit with the door closed.

Note: Isolate the components being checked from the circuitry when making continuity checks.

Other Tests

Timer Motor. A continuity check with an ohmmeter across the timer motor leads should indicate approximately 20Ω.

Cook Relay Coil. A continuity check for the coil of the cook relay should indicate approximately 10Ω.

Blower Motor. A continuity check across the blower motor leads should indicate well below 10Ω.

Stirrer Motor. A continuity check across the stirrer motor leads should indicate approximately 35Ω.

Thermal Protector. A continuity check across the thermal protector should indicate a closed circuit unless the temperature of the thermal protector is above 230°F. The thermal protector resets automatically at approximately 190°F.

HOW MODELS 56-1266, -5226, -10226 & -5026 WORK

The next section discusses the theory of operation of these models, which is directed specifically to the 56-1226 and 56-5226. Then the succeeding section covers the exceptions for models 56-1026 and 56-5026. Refer to the schematic in Fig. 12-14 as you read the explanations.

Pushing the ON switch applies power to oven lamp, stirrer, and fan. Until the door is closed, this is all that happens.

To put the oven in the cook condition, the door must be securely closed. This will activate three interlock switches. All three interlocks are activated by the door closing. The desired cooking time must be selected with the timer dial, which will close the switch contacts leading to the surge relay. Pushing the COOK switch then applies power to surge relay and surge resistor.

For the first few milliseconds of operation, current flows through the 5Ω resistor to the primary side of the transformer. The surge relay energizes and the 5Ω resistor is then bypassed. The relay contacts close, providing power to the timer motor, which energizes the timer. The relay contacts provide a holding circuit for the surge relay coil through the surge resistor after the COOK switch is released.

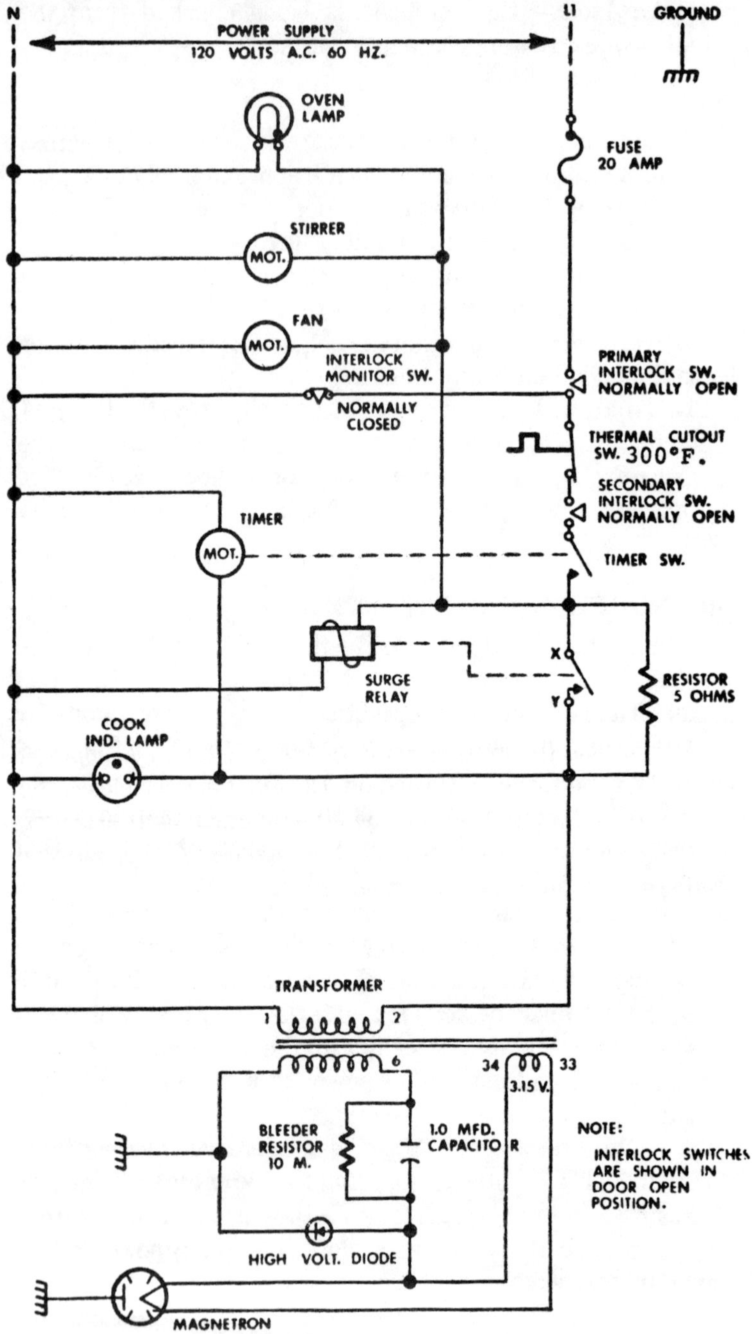

Fig. 12-14. Schematic diagram of Tappan Model 56-1026.

The 120V AC input to the primary winding of the power transformer is converted to approximately 200V AC across the secondary winding. At the same time, the filament secondary winding of the power transformer provides 3.15V AC to heat the magnetron filaments.

The 200V AC output from the power transformer's HV secondary is sent to a half-wave doubling circuit comprised of a single diode and one capacitor, where approximately 4000V DC (peak) is provided to the magnetron tube assembly.

The − 4000V DC apppied to the cathode of the magnetron tube causes the magnetron to oscillate at the 2.45 GHz cooking frequency.

The RF energy produced by the magnetron is channeled past the stirrer and into the cavity, where the load is placed to be heated. When the timer has completed the selected cooking time, the timer switch opens and the surge relay coil is de-energized. With the surge relay contacts open, the circuits are open to the COOK light and power transformer. The COOK light goes out and the RF energy is no longer produced. A bell signals the end of the cook cycle.

Model 56-1026 & 56-5026 Exceptions

Nothing on the oven is actually energized until the door is closed and the timer is turned on. Closing the door closes the interlocks and applies power up to the timer contacts. Turning on the timer applies power to the magnetron fan, stirrer motor, oven light, surge relay, and surge resistor.

MODELS 56-1226, -5226, -1026 & -5026 TROUBLESHOOTING & TESTING

When you are troubleshooting a microwave oven, it is helpful to follow a logical program of checking. Such programs are given for the above models in Tables 12-3 and 12-4. Many of the possible causes of trouble given in the tables will require that a specific component test be made. These tests are given next.

HV Components Test

This test is performed as given above for the other Tappan models, with the following exceptions.

Table 12-3. Troubleshooting Guide—Tappan Models 56-1226 & 56-5226

OFF CONDITION		
PROBLEM	**POSSIBLE CAUSE**	**TEST PROCEDURE OR CORRECTION**
Line fuse blows when power cord is plugged into receptacle.	1. Shorted or damaged ON-OFF switch. 2. Shorted wire in power cord or wire harness.	Replace switch. Replace cord or check wiring.
STANDBY CONDITION		
PROBLEM	**POSSIBLE CAUSE**	**TEST PROCEDURE OR CORRECTION**
Fuse blows.	Interlock problem	Check interlocks.
Oven completely inoperative with switch turned on	1. No power from service entrance. 2. Blown fuse. 3. Open wire in power cord or wiring harness. 4. Defective ON-OFF cook switch.	Check wall outlet. Check fuse. Replace same or repair wiring. Replace.
Oven cavity light does not illuminate (stirrer and blower operate.)	1. Open wiring in circuit to light. 2. Bad light bulb.	Check wiring. Change bulb.
Stirrer motor inoperative (cavity light and blower operate).	1. Defective stirrer motor. 2. Open wiring in circuit to stirrer motor.	Check stirrer motor. Check wiring.
Fan motor inoperative (cavity light and stirrer operate).	1. Defective fan motor. 2. Open wiring in circuit to fan.	Replace. Check wiring.
Fuse blows.	1. Shorted HV transformer. 2. Shorted HV capacitor. **Note**: A shorted diode or magnetron will not immediately blow fuse, but may damage the transformer by overheating.	Check transformer. Check capacitor.
Surge relay does not pull in.	1. Defective contacts on timer switch. 2. Interlock switches defective or not properly adjusted. 3. Defective surge resistor. 4. Defective COOK switch. 5. Defective surge relay. 6. Open wiring between the above components. 7. Thermal cutout opened.	Replace timer. See **Interlock Switch Test.** See **Surge Resistor Test.** See **ON-OFF & COOK Switches Test**. See **Surge Relay Test**. Check wiring. Replace thermal protector.
Surge relay pulls in, but no indicator light.	Defective indicator light.	Replace.

Table 12-3. cont.

COOK CONDITION		
PROBLEM	POSSIBLE CAUSE	TEST PROCEDURE OR CORRECTION
Relay pulls in and COOK light goes on when COOK button is pushed, but relay releases and light goes out when button is released.	Surge resistor open.	Replace.
Customer complains of slow or uneven cooking.	1. Stirrer not operating.	Check.
	2. Low magnetron output power.	Make water-temperature-rise check.
	3. Defrost cycler in defrost.	Check.
	4. Low line voltage.	Check.
Oven goes into cook cycle, but timer does not time out.	1. Defective timer motor.	Replace timer.
	2. Open wiring between timer contact and timer motor.	Check wiring.
Heat produced in oven load, but indicator light does not illuminate.	1. Defective indicator light.	Replace.
	2. Open wiring to indi-switch.	Check wiring.
Intermittent cook.	Defective ON/OFF cook switch.	Check and replace.
Inicator light indicates cycle, but little or no heat is produced in oven load.	1. Defrost control not set for cooking.	Set control for cooking.
	2. Shorted HV circuit between voltage-doubler circuit and magnetron.	Check HV components. Check wiring.
	3. Defective defrost control.	Replace.
	4. Defective HV transformer.	See **Power Transformer Test**.
	5. Defective diode.	See **Diode Check**.
	6. Defective capacitor.	See **Capacitor Check**.
	7. Defective magnetron.	See **Magnetron Filament Test**.
Oven goes into cook cycle but shuts down before end of cycle. Will start after a few minutes.	1. Magnetron thermal cut-out opened.	Replace thermal protector.
	2. Fan not operating.	Check fan.
	3. Bottom or back vents covered.	Check.

On models 56-1226 and 56-5226, make sure that the defrost feature is not turned on.

In checking for magnetron-filament operation, you should be able to see a glow from the filament by looking through the door toward the upper-right side of the cavity.

Table 12-4. Troubleshooting Guide—Tappan Models 56-1026 & 56-5026

OFF CONDITION		
PROBLEM	**POSSIBLE CAUSE**	**TEST PROCEDURE OR CORRECTION**
Line fuse blows when power cord is plugged into wall receptacle.	1. Shorted or damaged interlock switches. 2. Shorted wire in power cord or wire harness.	Replace switch. Replace cord or check wiring.
COOK CONDITION		
PROBLEM	**POSSIBLE CAUSE**	**TEST PROCEDURE OR CORRECTION**
Oven completely inoperative with timer turned on.	1. No power from service entrance. 2. Blown fuse. 3. Open wire in power cord or wiring harness. 4 Defective timer switch.	Check wall outlet. Check fuse. Replace same or repair wiring. Replace.
Oven cavity light does not illuminate (stirrer motor and fan operate).	1. Defective bulb. 2. Open wiring in circuit to light.	Replace. Check wiring.
Stirrer motor inoperative (cavity light and fan operate).	1. Defective stirrer motor. 2. Open wiring in circuit to stirrer motor.	Check stirrer motor. Check wiring.
Fan motor inoperative (stirrer and cavity light operate).	1. Defective fan motor 2. Open wiring in circuit to fan.	Replace. Check wiring.
Fuse blows.	1. Interlock problem 2. Shorted HV transformer. 3. Shorted HV capacitor. **Note**: A shorted diode or magnetron will not immediately blow fuse.	Check interlocks. Check transformer. Check capacitor.
Surge relay does not pull in.	1. Defective contacts on timer switch. 2. Interlock switches defective or not properly adjusted. 3. Defective surge resistor. 4. Defective surge relay. 5. Open wiring between the above components. 6. Thermal cutout opened.	Replace timer. See **Interlock Switch Test**. See **Surge Resistor Test**. See **Surge Relay Test.** Check wiring. Replace thermal protector.
Surge relay pulls in, but no indicator light.	1. Defective indicator light. 2. Defective timer	Replace. Replace.

Table 12-4. cont.

PROBLEM	POSSIBLE CAUSE	TEST PROCEDURE OR CORRECTION
Oven goes into cook cycle, but timer does not time out.	1. Defective timer motor.	Replace timer.
	2. Open wiring between timer contact and timer motor.	Check wiring.
Heat produced in oven load, but indicator light does not illuminate.	1. Defective indicator light.	Replace.
	2. Open wiring to indicator light.	Check wiring.
Indicator light indicates cycle, but little or no heat is produced in oven load.	1. Shorted HV circuit between voltage-doubler circuit and magnetron.	Check high-voltage components. Check wiring.
	2. Defective HV transformer.	See **Power Transformer Test**.
	3. Defective diode.	See **Diode Check.**
	4. Defective capacilor.	See **Capacitor Check.**
	5. Defective magnetron.	See **Magnetron Filament Test**.
Oven goes into cook cycle but shuts down before end of cycle. Will restart after a few minutes.	1. Magnetron thermal cut-out opened.	Replace thermal protector.
	2. Fan not operating properly.	Check fan.
	3. Bottom or back vents covered.	Check.
Customer complains of slow or uneven cooking.	1. Stirrer not operating.	Check.
	2. Low magnetron output.	Make water temperature rise check.
	3. Line voltage very low.	Check.

The resistance between the HV terminal of the power transformer and the oven's chassis ground should be approximately 77 Ω instead of 50Ω.

Magnetron Filament Test

This is the same as for the other models. The filament glow, however, should be observed by looking through the door toward the upper-right side of the cavity.

Diode Check

The diode check is the same as already given for the other Tappan models. If the diode checks out okay, proceed to the magnetron test.

Power Transformer Test

The testing procedure is the same as for the other models. The voltage readings, however, are different. For the models presently under consideration, the HV reading should be 2000V AC and the filament voltage should be approximately 3.15 V AC.

HV Winding Continuity Check

This is the same as for the other models. The voltage at the resistance from the HV terminal to ground is about 77Ω.

Capacitor Test

To make this test, proceed as explained earlier for the other Tappan models. Be sure the ohmmeter is set on a high-resistance scale when making the test.

Magnetron Assembly Test

To test for an open filament, isolate the filament leads and measure the resistance between them. It should be less than 1Ω.

To test for a shorted magnetron, connect an ohmmeter between the magnetron filament and oven ground. The ohmmeter should indicate infinite resistance. If it indicates little or no resistance, the magnetron is shorted and must be replaced.

Surge Resistor Test

Test this resistor as explained previously. Be certain to order a replacement from the manufacturer if replacement is indicated. This is a special resistor that will not ignite if subjected to high current.

Interlock Switch Test

These tests are as given under *Primary-Interlock Switch* and *Secondary-Interlock Switch*.

Solid-State Surge Relay Test

1. Disconnect the power to the oven, remove the wrapper, and discharge the capacitor.
2. Check the surge resistor (attached to relay).
3. Remove wires to the relay switch. With an ohmmeter set on the $R \times 1$ scale, measure the resistance between terminals of the surge relay switch. The meter should indicate an open. Manually depress the

relay. The contact inside switch should close to give you a shorted reading.

4. You will not be able to check the coil of the relay because the full-wave bridge between it and the terminal will give you an open reading on an ohmmeter. The best way to check it is to remove the leads to coil terminals, attach a line cord across terminals, and plug it in. If relay closes, the bridge and coil are good. With the relay energized, an ohmmeter check across the relay switch terminals should show them closed.

Safety Switch Test

Remove the fuse and replace it with the oven lamp. With the door open, set the unit up for a normal cooking operation.

1. If the bulb illuminates brightly, one of the interlock switches is stuck in the closed position. Replace all the switches.
2. If the bulb illuminates dimly, the switches are open (normal).

   Caution: When replacing these switches, use only Tappan replacement parts. Replacing them with any other part will void the warranty and may cause injury to the user. If, at any time, either the oven or house fuse blows due to an oven fault, all interlock switches must be replaced. This is because the short-circuit currents will damage these important switches.

Thermal Protector Test

Unless overheating exists, a continuity check across the thermal protector contacts should indicate a closed circuit. The thermal protector opens at approximately 300°F and resets automatically at 220°F.

ON-OFF & COOK Switches Test

1. Disconnect the power, remove the wrapper, and discharge the capacitor by placing an insulated screwdriver across both terminals and then from each terminal to ground.

2. Remove the wires from back of the switch, noting what terminals the wires are connected to.
3. Set a ohmmeter on the R × 1 scale and zero the meter.
4. Place one of the test leads on terminal 7 and the other on terminal 8.
5. Depress the ON push button; the meter should deflect to zero on the meter.
6. With the leads on the same terminals as in step 4, depress the OFF push button; the meter should now indicate an infinite resistance.
7. Now remove both leads. Place one on terminal 3 and the other on terminal 4.
8. Depress the COOK push button and watch the meter. When the button is pushed, the meter will deflect to zero. Release the push button; the meter will indicate infinite resistance. This test will prove the switch is operating properly.
9. If the switch is defective, it must be replaced. If it is not defective, place the wires on the terminals. Secure the switch back in the control panel.

56-1026 & 56-5026 Timer Test Models

1. Remove a wire from one side of the timer switch.
2. Connect an ohmmeter, set on the R × 1 scale, across the terminals of the timer switch.
3. Making sure oven is not plugged in; turn the timer on. The meter should indicate zero. If it does not, replace timer.

Models 56-1226 & 56-5226 Power-On Cycling Test

1. By observing the magnetron's filament glow through the door with the power control set at DEFROST, you can see the magnetron cycle on and off approximately each 15 seconds. You can also hear a change in transformer sound (hum) as the magnetron loads it down.

MODELS 56-1025, -1226, -5226, -5026 DOOR & INTERLOCK ADJUSTMENTS

The door adjustment on all microwave ovens is very important to prevent any excessive leakage. The RF leakage should be checked after all door, interlock switches, and associated repairs are made. See Chapter 3 for the proper testing procedures.

The following door adjustments must be followed to the letter for proper door adjustments.

Hinge Side Adjustments

1. Disconnect the power.
2. Remove the screws to slide the wrapper back.
3. Close the door.
4. The door should align with the control panel.
5. Loosen the four screws on the hinge side until the door can be easily moved back and forth.
6. Squeeze the door against the front frame and, at the same time, tighten the four screws—the center screws first.

Latch Assembly Adjustment

To make proper latch adjustment, observe that the assembly must be set so that with the door closed, the latch arms seat firmly on the interlock assembly. If they do not:

1. Remove the outer door panel and control trim.
2. Loosen the two screws holding latch to frame (Fig. 12-15) enough so that the assembly moves easily.
3. Close the door. The latch assembly must be adjusted so that the latch arms depress the upper and lower interlocks until they are closed.
4. Tighten the two screws that secure the latch assembly.

Proper adjustment is obtained when: 1) Closing the door energizes the upper and lower interlocks. 2) When the door handle push button is depressed, the door is easily unlatched.

Interlock Switch Adjustment

The interlock switches are not adjustable on these models. If you suspect one of the interlocks is not making, use the interlock test to check.

CONTROL PANEL COMPONENTS REMOVAL

1. Unplug the power cord, remove the outer wrapper, and discharge the capacitor.

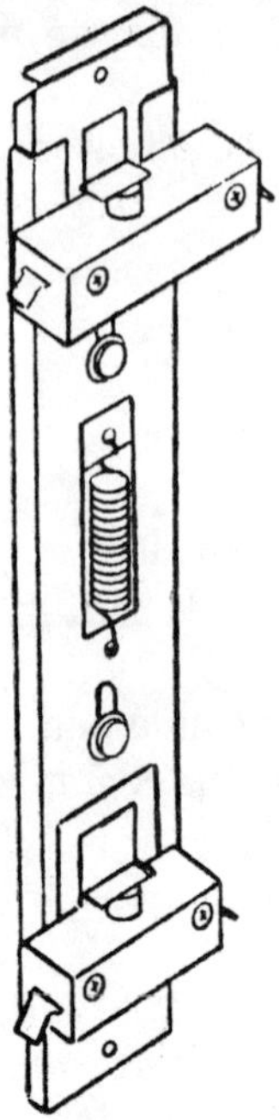

Fig. 12-15. Latch-assembly adjustment.

2. Remove the two screws that hold control-panel trim in place. (See Fig. 12-16 for location.)

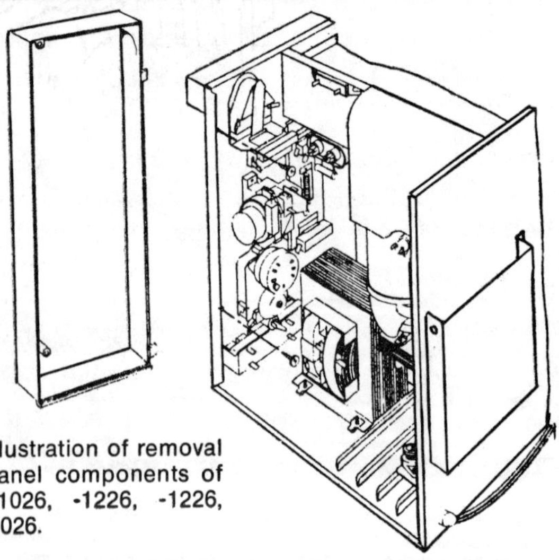

Fig. 12-16. Illustration of removal of control-panel components of Models 56-1026, -1226, -1226, -5226, and -5026.

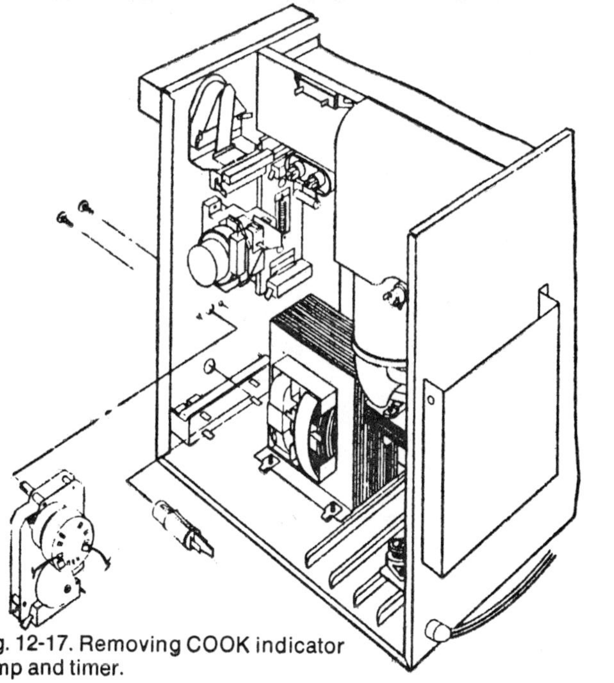

Fig. 12-17. Removing COOK indicator lamp and timer.

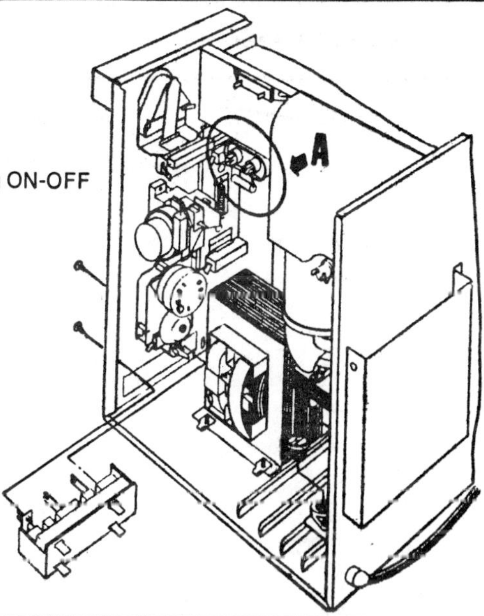

Fig. 12-18. Removing ON-OFF and COOK switch.

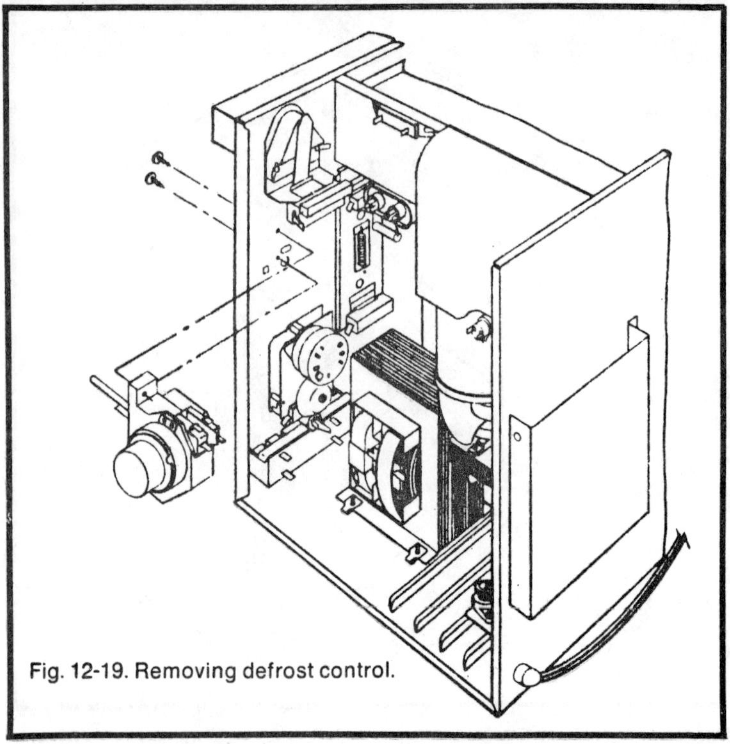

Fig. 12-19. Removing defrost control.

3. Pull off the knobs.
4. Open the door. Swing the trim out about 2 inches and lift off.
5. The COOK light can be removed by removing the wires and pushing it forward.
6. The ON-OFF and COOK switch can be removed by removing the two screws securing the switch, then replacing it with the new switch, wire for wire. (See Figs. 12-17 and 12-18.)
7. To remove the rotary timer, remove the two screws that hold it in place. Replace it wire for wire with a replacement timer.
8. To remove the defrost control on Models 56-1226 and 56-5226, remove the two screws holding it in place. Then remove the wires. Replace it wire for wire. (See Fig. 12-19.5

   Note: Screws that hold this control are special; do not lose them.

Chapter 13
Servicing White-Westinghouse Microwave Ovens

The countertop microwave ovens of White-Westinghouse[1] include a 500W economy model and single- and dual- timer 650W models.

Economy model KM320S has a cooking power output of 500W and a 12-minute timer. Next step up in the line covered is model KM420, which has a 650W output with a 14-minute timer. Rounding out the coverage is model KM520, which has a 650W output and 5- and 30-minute timers. About 70% of the items cooked in a microwave oven will cook in less than 5 minutes. The 5-minute timer of model KM520 with its markings at 15-second intervals, provides accurate timing of these foods. The 30-minute timer allows for cooking large items such as chickens and roasts.

HOW THEY WORK

This section describes the primary-circuit operation of each of the above models in turn, then the secondary circuit operation.

KM320S Primary Circuit

All microwave ovens built beginning August 7, 1974 must by federal law incorporate an interlock monitor circuit. The

1. All of the illustrations in this chapter are courtesy of White-Westinghouse Appliance Company.

monitored circuit involving the door-interlock switch is designated the *secondary* interlock. The start-stop latch switch actuated by the door latch and START-STOP buttons are designated as the *primary* interlock.

The door-interlock switch is mounted behind the front bulkhead in the upper-left corner and is actuated by the pivot arm attached to the door. This switch is normally open when the door is open, and closed when the door is closed.

The monitor-sensing switch is a snap-in type and mounts in a slot in the upper-left corner of the front bulkhead near the door's pivot arm. This switch is normally closed when the door is open and open when the door is closed.

The monitoring switch is connected between the normally open contact of the door-interlock switch and the time-delay mechanism (a thermal fuse inserted in a resistor). The time-delay mechanism is mounted on the rear oven bulkhead in front of the transformer. If the door-interlock switch fails to open when the door is opened, current will flow through the sensing switch and energize the resistor, causing the fuse to open. Ten seconds is the maximum time to open the fuse.

Since time must run out or the STOP button must be pushed before the door can be opened, power is interrupted to the other parts of the oven, preventing oven operation during the actuation time of the monitor circuit. You must correct the door-interlock switch or sensing-switch problem when replacing the fuse. (See *Component Testing*.)

Power is supplied to the oven through a heavy-duty, 115V, 3-lead, grounding-type cord. The ground lead is screwed to the oven cavity. (Refer to the schematic, Fig. 13-1.)

The *hot* lead goes to the thermal cutout and on to the SPDT interlock switch, which is operated by the door's pivot arm. Thus, when the door is open, there is no power supplied to the oven. The door-interlock switch is a normally open switch.

When the door is closed, the pivot arm closes the door-interlock switch. Two leads come from the interlock switch. One goes to the interlock-sensing switch, which is open when the door is closed. The second goes to one of the timer terminals, then from there to the solenoid on the START-STOP switch. If the START switch is pushed with the timer contacts open, you will energize the solenoid. This will cause solenoid chatter, since the unit is trying to shut off as you are trying to start it.

With the timer contacts closed, a circuit will be set up through the oven light to the other side of the line; the light will come on. Then push the START button; this will close the

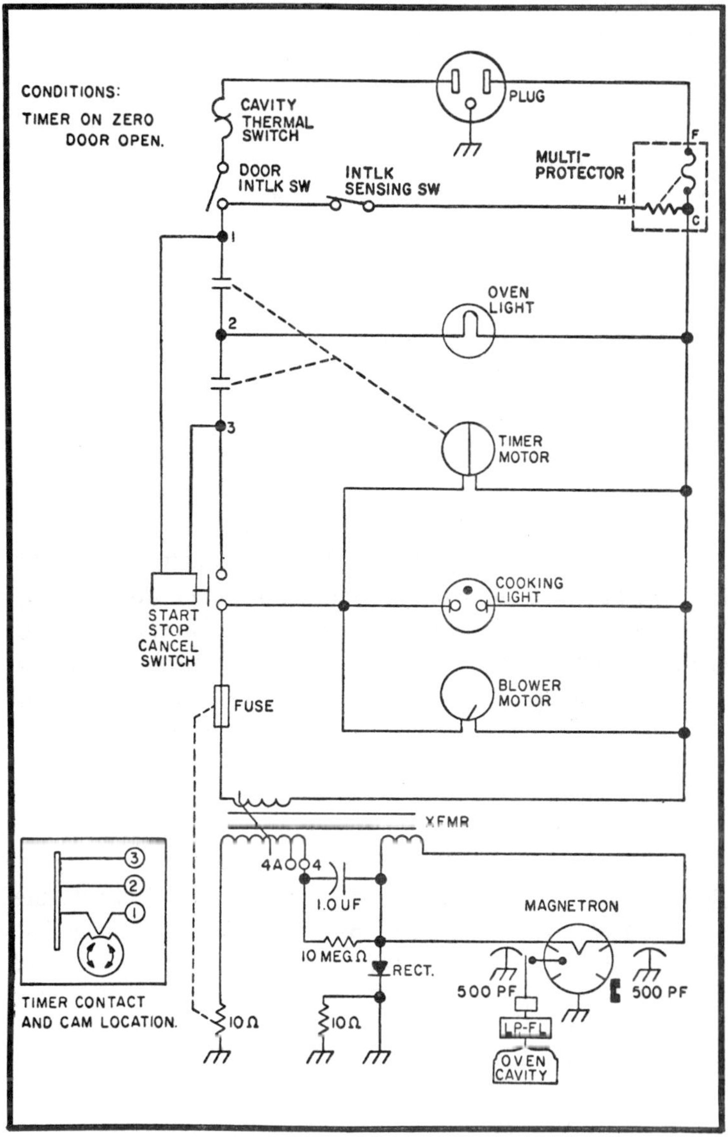

Fig. 13-1. Circuit of White-Westinghouse KM320S.

circuit through the timer motor, blower, cooking light, thermal fuse, primary of the transformer, and through the multiprotector to the ground side of the line.

KM420-1 Primary Circuit

To set up the primary circuit of the oven, it is first necessary to plug in the oven. You can then track your circuit from one side of the line to the other side of the 115V circuit. (See Fig. 13-2.)

With the door open, the oven light will be on because the circuit is from one side of line through the left interlock switch, to the oven light, and out the other side of the line.

When you close the door, all three interlocks are closed. This turns the oven light off unless the oven light switch is in the closed position. If the oven light switch is closed, there is a circuit from one side of the line through the oven light switch, through the oven light, and out the other side of the line.

With the door closed, two steps are required to energize the primary circuit. Pushing down on the latch lock switch closes the START switch. Now you have to turn the timer on to complete the circuit to the primary winding of the transformer. The circuit then goes from one side of line; through the left and latch interlock switches, START switch, right interlock switch, timer switch, cavity thermal cutout, relay, tube thermal cutout, and primary of the transformer; across the other side of the START switch; and out the other side of line. At the same time there is a circuit through the timer motor dial light, fan motor, and relay.

KM420-2 Primary Circuit

The primary circuit for this model has been changed slightly but is quite similar to the circuit for the *−1* version. The tube thermal cutout has been moved to a point between the line cord and the wiring to the oven-light switch and left interlock switch. There is a thermal fuse (secondary overload protector) located in the circuit where the original tube thermal cutout is pictured in Fig. 13-2.

KM420-3 Primary Circuit

These ovens, built after August 7, 1974, incorporate an interlock-monitor circuit. The monitored circuit, in this case

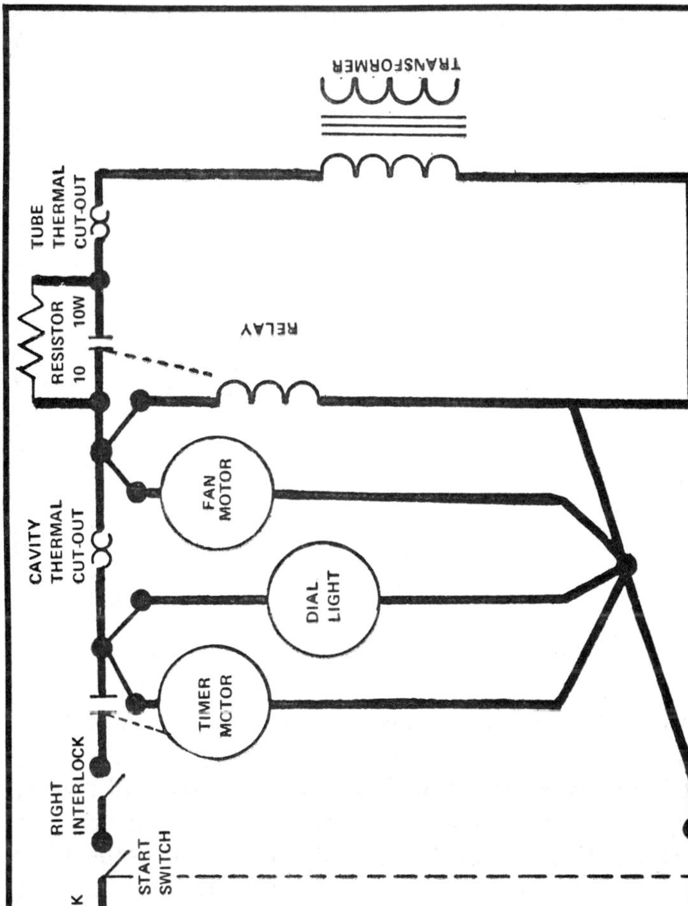

Fig. 13-2. Control circuit of White-Westinghouse KM420.

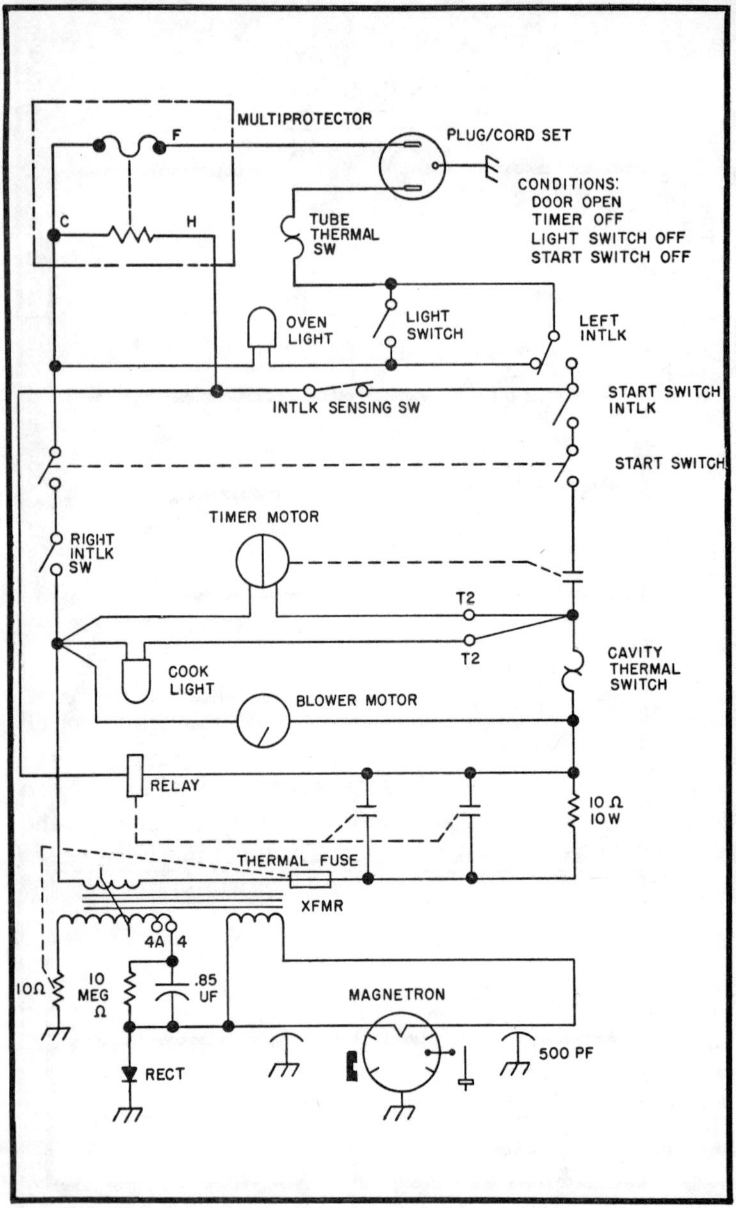

Fig. 13-3. Complete circuit of White-Westinghouse KM420.

the left-door interlock switch, is the secondary interlock. The primary interlock is the slide-latch bar switch (START switch), which is activated by the slide-latch bar.

To set up the primary circuit in this model it is first necessary to plug in the oven. You can then track your circuit from one side of the line to the other side of the 115V circuit. (See Fig. 13-3). There is the fuse in the multiprotector on one side of the line. One side of the relay connects with the interlock-sensing switch and goes through the multiprotector resistor before it reaches the fuse side of the line.

With the door open, the oven light will be on because the circuit is from one side of the line through the tube thermal cutoff, through the left interlock switch to the light, through the fuse, and out the other side of the line. If the light does not come on with the door open, push the oven light switch. If the light comes on, the left interlock switch is inoperative and must be replaced. If the light still does not come on, the fuse, thermal cutoff, or bulb may be open.

Assuming there is a light when the door is open, close the door and the light should go off. If not, push on the light switch. If this does not turn off the light, the left interlock switch needs adjusting or replacing. As the door is closed the door pivot arm actuates the left interlock switch arm, opening the interlock-sensing switch, and closing the N.O. contacts of the interlock switch. The right interlock and the latch interlock are closed at the same time.

With the door closed, two steps are required to energize the primary circuit. Pushing down on the latch bar locks the door as it closes the START switch. The timer must now be set for a timed operation to complete the circuit to the primary winding of the transformer.

KM520 Primary Circuit

To understand the circuit, begin at the line cord and go from one side of the line cord to the other side of the line cord as follows. (See Fig. 13-4.)

Trace from the line cord to the terminal board. Go from the terminal board to the oven light switch, through the switch to the oven light bulb, through the bulb to the other side of the line at terminal TM-1 on the 5-minute timer (T-5).

Trace from the terminal board to the door latch switch, through the switch to the oven light bulb, through the bulb to the other side of the line at terminal TM-1 on the 5-minute

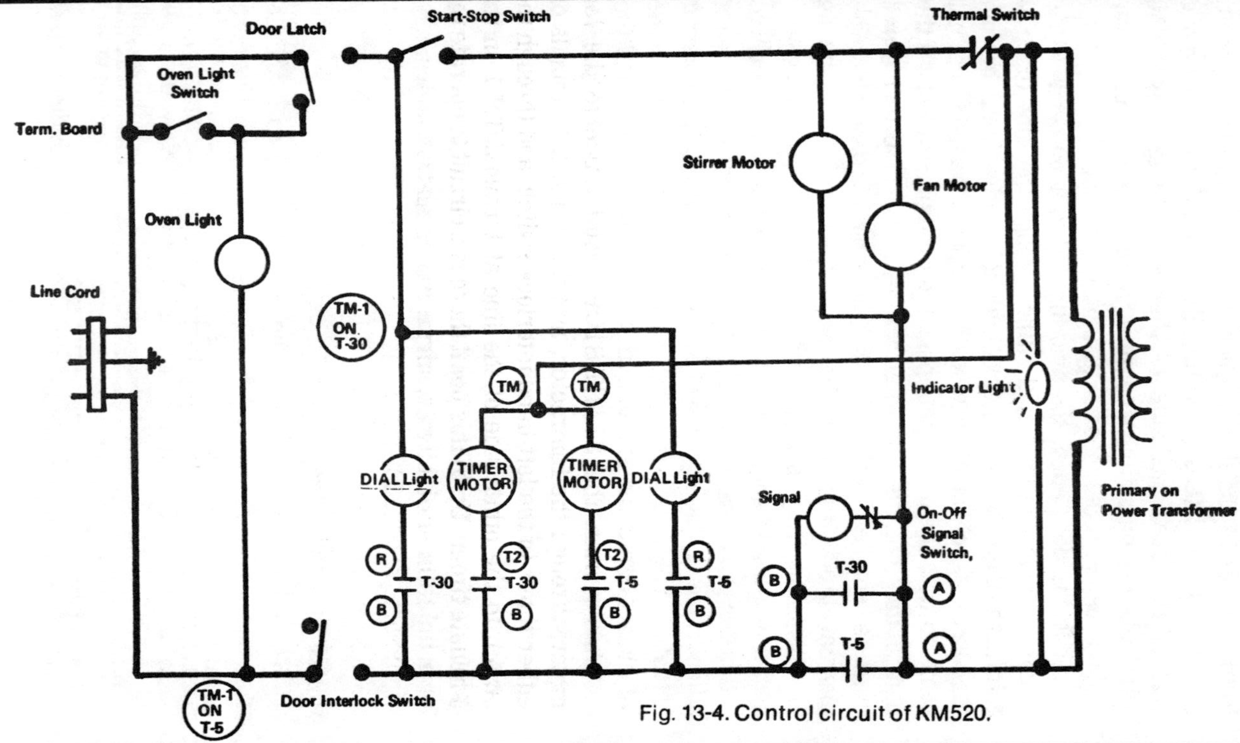

Fig. 13-4. Control circuit of KM520.

timer. So far, with the oven door open, the only electrical circuit is through the oven light bulb. The oven light switch is bypassed by the oven latch switch.

With the oven door closed and the door-latch switch closed, go the the START-STOP switch. Now continue into the control circuit.

From the door-latch switch to terminal TM-1 on the 30-minute timer (T-30), trace to the dial lights on each timer, through the dial lights and timer contacts *R* and *B* to the left-hand interlock switch, through the switch to the other side of the line at terminal TM-1 on the 5-minute timer (T-5).

With the oven door closed, the door-latch switch closed to the START switch, and the left-hand interlock switch closed, there is a circuit to control the operation of the oven light through the oven light switch. By setting either timer, the dial light will operate.

From the door latch switch trace to the START-STOP switch, through the START-STOP switch to the stirrer motor, through the motor and the *A* and *B* contact on either timer to the left-hand interlock switch, and through the switch to the other side of the line at terminal TM-1 on the 5-minute timer.

Again from the START-STOP switch trace to the fan motor, through the fan motor, and the *A* and *B* terminals on either timer to the left-hand interlock switch, and through the switch to the other side of the line at terminal TM-1 on the 5-minute timer. Up to this point there is a circuit to operate the oven light, timer dial lights, stirrer motor, and fan motor.

From the START-STOP switch trace through the thermal switch to terminal TM on either timer, through the timer motor and terminals T2 and *B* on the timer, to the left-hand interlock switch, and to the other side of the line at terminal TM-1 on the 5-minute timer.

Also from the thermal switch, trace to primary on the power transformer, through the primary, through the *A* and *B* terminals on either timer, to the left-hand interlock switch, and through the switch to the other side of the line at terminal TM-1 on the 5-minute timer.

Across the primary of the transformer is an indicator light to indicate the control circuit is completed. There is a buzzer and an on-off switch across the *A* and *B* contacts on either timer to indicate when the cooking operation has finished. The on-off switch can disconnect this signal if the customer so desires.

KM320S Secondary Circuit

The HV circuit in this unit is a voltage doubler. A simple explanation of its operation is as follows (refer to Fig. 13-1).

As the unit is started, the transformer supplies approximately 1700V AC to the circuit, which charges the oil-filled capacitor. As the AC reverses or flows in the other direction, the diode prevents the voltage in the capacitor from being discharged back through the diode. This voltage is then added to the approximately 1700V produced in the transformer winding on the negative side of the AC wave, raising the circuit voltage to approximately 3600V (peak). The diode used is both a one-way current-limiting device and a rectifier changing the voltage from AC to DC.

The resulting voltage of the circuit is not quite double that of the transformer output because there is loss in the circuit and the capacitor does not completely discharge.

The 10Ω resistor with the fuse inside it (the fuse is in the primary circuit) is for the purpose of protecting the transformer should a short occur in the secondary circuit. When a HV short to ground occurs the current in the secondary circuit of the transformer will increase. This increase in current heats up the resistor, which opens the fuse in the primary circuit.

Caution: The circuit's 3600V potential cannot be measured without special equipment. *Do not attempt to read secondary voltages*. To attempt to read these voltages serves no useful purpose in servicing the equipment and subjects you to an unnecessary hazard.

KM420 & KM520 Secondary Circuit

The secondary circuit is activated by the control or primary circuit. Off the transformer secondary, there are two voltages. (Fig. 13-5). The 3.1V is used to heat the filament of the magnetron tube. The second winding is 4000V and it is used for proper oscillation of the magnetron tube. The full-wave bridge rectifier gives a DC voltage. The rectifier's negative potential is placed on the cathode. By placing the negative

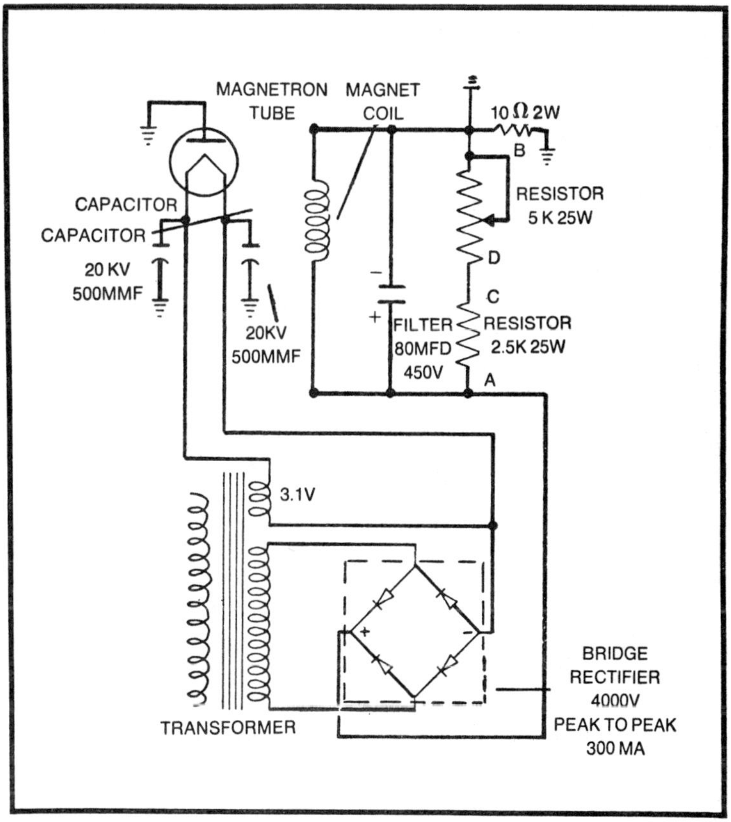

Fig. 13-5. High-voltage circuit of KM420 and KM520.

potential on the cathode, the anode (plate) is at 4000V positive with respect to the cathode.

The positive output of the rectifier goes through the control circuit of the magnetron tube. In the control circuit there is a magnetic coil which directs the electron flow through the magnetron tube in a circular motion to set up proper oscillation a 2.45 GHz. There is also a filter capacitor of 80 GF, (which helps filter the DC voltage on the magnetron tube) and a 7.5K (two resistors, one fixed at 2.5K and the other variable at 5K) adjustable resistor. Increasing the resistance of the variable resistor increases the magnetic field around the magnetron tube, which in turn reduces its plate current. From the top of the adjustable resistor, where the magnetic coil and the filter are tied, a 10Ω, 2W resistor is connected to ground. It is normally bypassed. This resistor is used for a ground circuit in measuring the current of the magnetron.

PRIMARY CIRCUIT TESTS

There are a few quick operating checks that can be made to pinpoint some of the difficulties that occur in the primary circuit. When operating the oven, *always be sure there is a load* in it. For the test listed below, the oven is plugged into a power source, a load is placed in the oven, the door is closed, the timer is turned to an operating position, and the LOCK-COOK latch is pushed down.

Oven Operates but Doesn't Cook. This indicates that the thermal fuse (located in the secondary circuit) is open or the transformer is open. Thermal fuse failure is generally caused by the failure of one or more parts in the secondary circuit. This circuit should be checked thoroughly before replacing the fuse and operating the oven.

Blower and Timer Operate, but Don't Light. Check wiring or replace oven light.

Blower and Dial Light Operate, but Timer Doesn't Advance. Check wiring or replace timer.

Timer and Dial Light Operate, but Blower Doesn't Operate. Check wiring, cavity thermal switch, and blower motor.

Oven Light Doesn't Operate. Check the wiring. Check the light switch or replace the oven light bulb.

All the other tests made to the primary circuit will have to be made with the power off. The oven cavity will have to be pulled out from the outer case and the air baffle lifted to get at some of the components. With one lead removed from the component, an ohmmeter can give you a continuity check across the terminals when the component is in an operating position.

SECONDARY CIRCUIT CHECKS

Before an attempt is made to check the resistance of the secondary circuit or to handle any service on this circuit, first unplug the oven and discharge the capacitors by running a screwdriver shank between the capacitors' leads and the filter assembly. Before making the resistance check, be sure the adjustment clamp is moved to the *top* of the adjustable resistor.

KM420 & KM520 Resistance Check

The resistance between points *A* and *B* in Fig. 13-5 should be 550Ω (magnet coil).

A to B is 450–550Ω. Check the position of the resistor clamp on the adjustable resistor.

A to B is 0–550Ω. Remove magnet coil lead from the ground connection, then check the resistance at points *A* and *B*. If the new reading is 7.5K, the coil is shorted to ground. If the new reading doesn't change, remove one lead of the 80 μF capacitor from the ground connection; then check the resistance at points *A* and *B*. If the reading is not 7.5K, the 80 μF capacitor is shorted to ground. If the reading doesn't change, check the resistors. Check the resistance between *A* and *C*; it should be 2.5K Check the resistance between *B* and *D* with adjustment clamp at the top of the adjustable resistor; it should be 5K. Replace any faulty resistor or resistors.

A to B is 7.5K. The magnetic coil is either open or disconnected from the circuit.

A to B is 600Ω. Remove the coil lead from the ground connection; then check the resistance at points *A* and *B*. If the new reading is above 7.5K, the resistor circuit is open. Check the resistance between *A* and *C*; it should be 2.5K. Check the resistance between *B* and *D* with the adjustment clamp moved to the top of the adjustable resistor; it should be 5K.

KM420 & KM520 Transformer Check

Figure 13-6 shows the resistance readings in the primary and secondary windings of the power transformer. In checking the primary winding across *A* and *B*, resistance should be below 2Ω. In checking across points *C* and *D* (3.1V AC filament winding), the resistance reading should be below 2Ω. In checking the secondary windings between points *E* and *F*, the resistance should be between 100 and 200Ω (1000V is developed on this winding). An additional check on the transformer is to take each lead and measure to ground for any high- or low-resistance leakage.

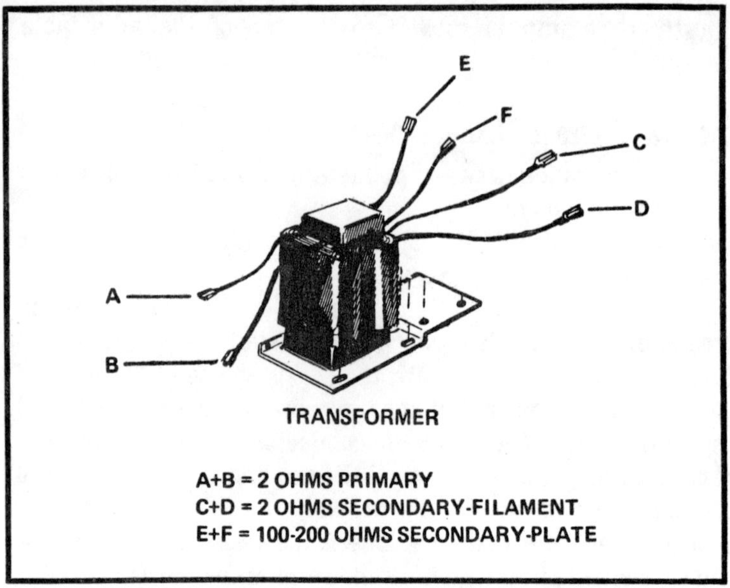

Fig. 13-6. Transformer connections—Models KM420 and KM520.

KM420 & KM520 RF Capacitors Check

In checking these capacitors, place the ohmmeter leads between the capacitor terminal where the filament transformer leads were disconnected and filter box assembly (Fig. 13-7). There should be no continuity reading. If a reading is obtained, it is an indication that the filter capacitor is shorted. When making this check use a high-resistance scale on meter.

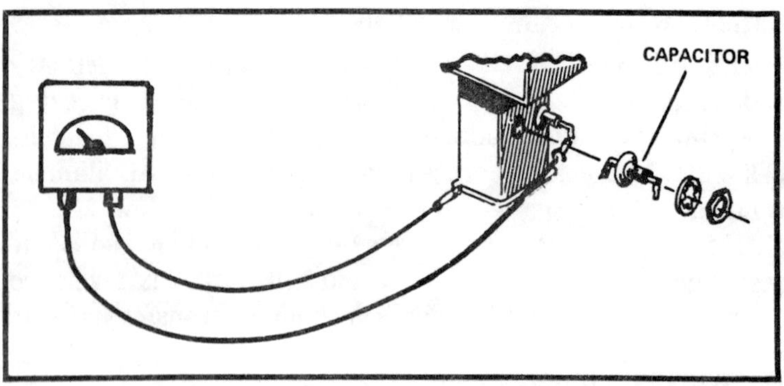

Fig. 13-7. Checking RF capacitors—Models KM420 and KM520.

KM420 & KM 520 Rectifier Check

In checking the bridge rectifier, set the ohmmeter to read on the highest scale. Check resistance between *A* and *B* of Fig. 13-8. Reverse leads and check again. There should be a considerable difference (minimum of one-half of full-scale deflection) between the two readings. If reversing the leads does not change the reading, replace the bridge rectifier.

Repeat this procedure between *A* and *C*, *B* and *D*, and *C* and *D*. The bridge may be checked in the unit, but one end of the AC wires must be removed. The AC wires are the secondary wires from the transformer which develops the 4000V AC.

KM420 & KM520 Magnetron Check

There is not much checking to be done on the magnetron tube. There are two resistance checks. One is to check the filament winding to see if it is open, and the other is to check for a ground at the cathode. There are some visible checks that can be made by looking directly at the magnetron tube, but this would require dissembling the magnetron assembly. To make a check of the filament winding on the magnetron, take a resistance reading across the filament leads. The resistance should be between 1 and 2Ω. In checking the resistance between the cathode of the magnetron and ground, there should be no continuity reading. Any resistance reading would indicate a high-resistance short to ground and a defective magnetron tube.

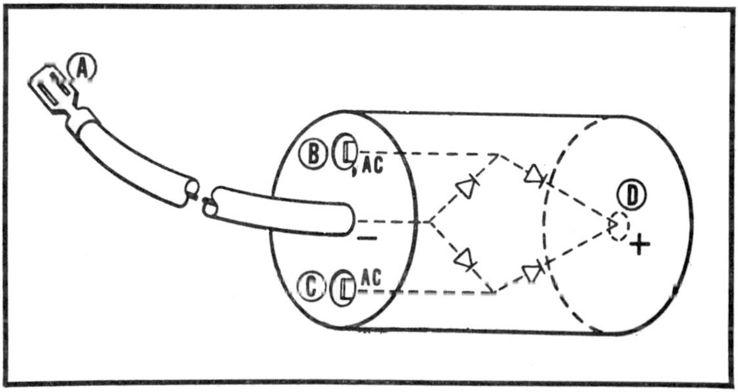

Fig. 13-8. Checking rectifiers—KM420 and KM520.

Magnetron Current Check for All Models

1. Unplug the oven.
2. Discharge the RF capacitors.
3. Disconnect ground wire from the grounding lug on the bulkhead.
4. Attach the ammeter leads between the disconnected end and the ground lug mounted on the bulkhead.
5. Set the ammeter on the highest scale to make the first check; then reduce the meter scale to read in the area of 300 mA, or 0.3A.
6 Check the meter leads for good connection and proper polarity.
7. Place a cooking load in the oven cavity (cup of water).
8. Plug in the oven. Start oven operation and record the milliampere reading. For proper oven operation (time-temperature relationship), the reading should be 300 mA for the KM520 and 275 mA for the KM320 and KM 420. An adjustment can be made to increase or decrease the reading.
9. Unplug the oven. To make the adjustment on the KM420 or KM520, loosen the screw holding the metal band on the 5K (or 7.5K) adjustable resistor. To *increase* milliamperes, slide the band down toward the bottom of resistor. To *decrease* milliamperes, slide the band up toward the top of resistor. In making this adjustment, the first position of the adjusting band should be half the distance from where the adjustable band was when the check was made—either towards the top or bottom of resistor, depending on what direction is needed for correction. After locating the new position, tighten screw on metal clamp and check the meter leads for good connection.

For the KM320, if the reading is lower than 275 mA, change the gray wire coming from the oil-filled capacitor to tap *4* on the transformer. If the milliampere reading is high, change the gray lead to tap *4A* on the transformer. If the reading continues to be high or low, the oil-filled capacitor may be changed to correct the problem.

KM320S Capacitor Test

If the capacitor is open, no HV is available to the magnetron. A shorted capacitor normally causes high current in the voltage-doubler circuit; this should cause the thermal fuse to open. An ohmmeter can be used to check for a shorted or open capacitor. Proceed as follows:

1. Unplug the oven, remove outer shell, discharge the RF capacitors and the oil-filled capacitor, remove the leads from the oil-filled capacitor, and disconnect one side of the resistor.
2. With an ohmmeter *set on the highest scale,* measure the resistance between the two capacitor terminals. The meter should indicate continuity for a short period of time and then should indicate an open circuit once the capacitor is charged. Reversing the meter leads should give the same results.
3. If the ohmmeter indicates continuity between the capacitor terminals at all times, or if no meter deflection occurs at all, the capacitor should be replaced.

KM320S Fuse Resistor

This assembly is located in the HV circuit (Fig. 13-9). It senses an increase in one leg of the voltage-doubler circuit and will open (the fuse portion) the circuit in the transformer primary. The *resistor* is connected between the transformer secondary and rectifier. The *fuse* is connected in the primary circuit. If there is a sudden increase in current in the secondary, the resistor will heat, opening the fuse. This will cause the oven to stop operating. Test the fuse or resistor with an ohmmeter as indicated in checking the primary circuit.

KM320S Rectifier

A special rectifier consisting of a single diode is used. The rectifier used will conduct high current in one direction with a low voltage drop but conduct a low current at a large voltage drop in the other direction. If the voltage becomes excessively high, this special diode will conduct high current for a short time and effectively protect the other components from

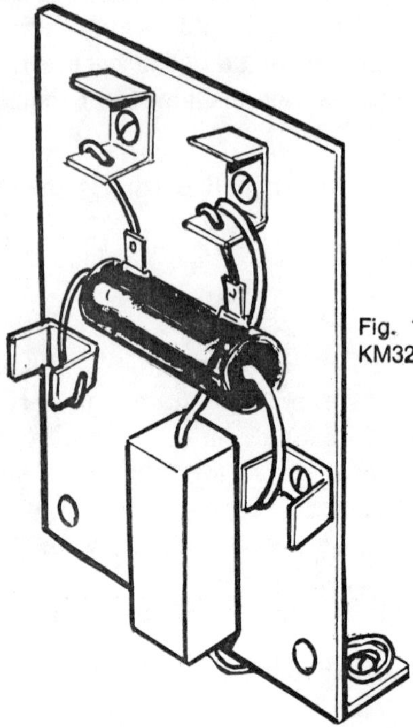

Fig. 13-9. Fuse resistor—Model KM320S.

shorting. To check the diode, set an ohmmeter to the R × 10K or higher.

Check the resistance between the terminals. Reverse leads and check it again. There should be a considerable difference (minimum of one-half of full scale) between the two readings, approximately a 10:1 ratio. If reverseing the leads does not change the reading, replace the diode. The diode may be checked in unit, but one lead must be removed.

KM320S RF Capacitor Check

Caution: The capacitors are charged to 3600V, and it is extremely important that they be grounded before handling. Capacitors will normally be grounded when the cover is removed through the cover clip. If not, place a screwdriver blade against the capacitor can, then to capacitor, to discharge it. To test it, proceed as follows.

1. Remove the capacitor cover.
2. Remove the capacitor can and gasket from the bottom of can.
3. Slip the side of the plastic protective cover between the ground clamp and the capacitor while checking the capacitors (Fig. 13-10).
4. Remove both leads from either capacitor.
5. Set an ohmmeter on the R × 10K scale and check from capacitor to ground. Any reading less than 100K indicates a faulty capacitor.
6. After checking or replacing the capacitors, dress the tube leads to ensure that they are not in such a position as to short to the capacitor box.

SWITCH ADJUSTMENTS

The following sections describe the adjustment of the interlocks and switches on the White-Westinghouse ovens.

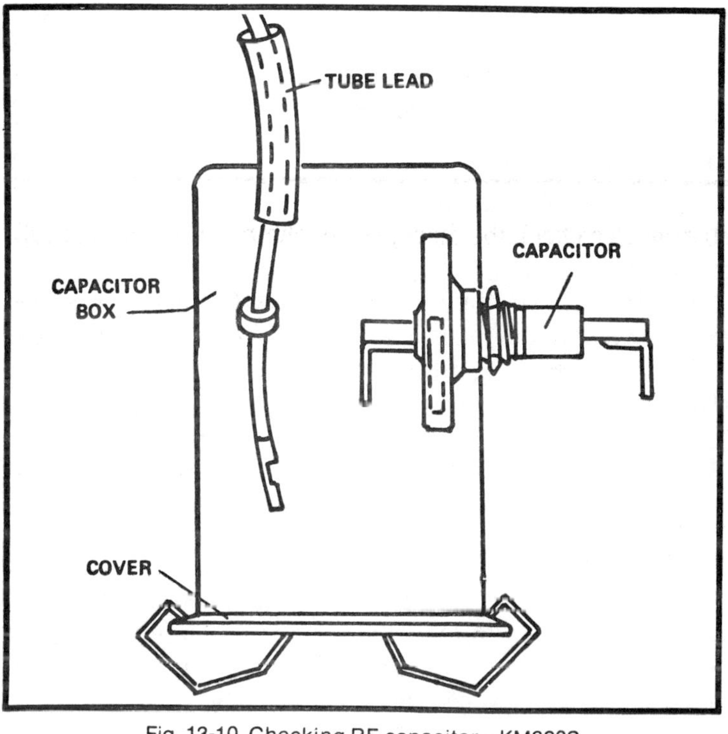

Fig. 13-10. Checking RF capacitor—KM320S.

Since different switches are used on different ovens, a different procedure must be used for each oven.

KM320S Door-Interlock Switch

With the unit unplugged, the door open, and the outer case removed, attach the ohmmeter leads to the common and normally open terminals. The ohmmeter should read infinity (open circuit).

Close the door, the ohmmeter should read zero (closed circuit). When, the door is opened, the door-interlock circuit must open before the door movement exceeds 0.3125 inches travel measured at the upper-right front of door from fully closed position. Adjustment is made by bending the switch arm.

KM320S Interlock-Sensing Switch

With the unit unplugged, the door open, and the outer case removed, attach the ohmmeter leads to the common and normally closed terminals. The ohmmeter should read zero (closed circuit).

Close the door. The ohmmeter should read inifinity (open circuit).

KM320S Switch Sequencing

With the unit unplugged, the outer case removed, the timer off, and the START switch off, attach the ohmmeter leads to the common terminal of door-interlock switch and normally closed terminal of interlock-sensing switch. The ohmmeter should read infinity (open circuit).

Open and close the door slowly. The ohmemter reading should be infinity (open circuit). This indicates proper sequencing of the door-interlock sensing-switch interlock. There is no sensing switch adjustment. Adjustment must be made on the door-interlock switch.

KM320S Start-Stop Switch

With the unit unplugged, the outer case removed, the door closed, and the START button depressed, visible contacts must be closed. The door must lock and return to the fully closed position after pull is exerted on door handle. If the door does not return to the closed position, the timer knob, control panel, and latch-switch mechanism must be removed. Adjustment is made by reforming the metal latch tab retaining the door strike.

With the STOP button depressed, visible contacts must open. The door must be unlocked.

KM42020-1, -2 Left Interlock

The left interlock is normally open when the door is open and will not allow power to flow through the unit. The switch may be checked with an ohmmeter before replacing the oven cavity in the outer cabinet, or it may be checked when assembled by observing the operation of the oven light. Opening the oven door will allow the oven light to become energized, which indicates that the switch is working.

Note: There have been cases of the grease shield interfering with the operation of the interlock-switch arm. This should be checked when making any adjustment to the switch arm.

KM420-3 Left Interlock & Sensing Switch

1. Pull both leads to the common terminals.
2. Place an ohmmeter across the two terminals.
3. There should be infinite reading at all times when door is opened or closed.
4. Adjust the switch bracket to make any adjustments to the switches.

KM420 Latch Interlock

Place a water load in the unit. To check the operation of the latch interlock, close the door, push the latch down, and turn the timer knob to energize the timer. Timer will advance when interlock is operating properly. If the interlock is inoperative, the unit will not function. An ohmmeter check of the switch with the door closed shows continuity and should show an open circuit with an eighth-inch of door opening.

KM420 Right Interlock

1. Place a water load in the unit.
2. Close the door and engage the latch switch.
3. Energize the timer.

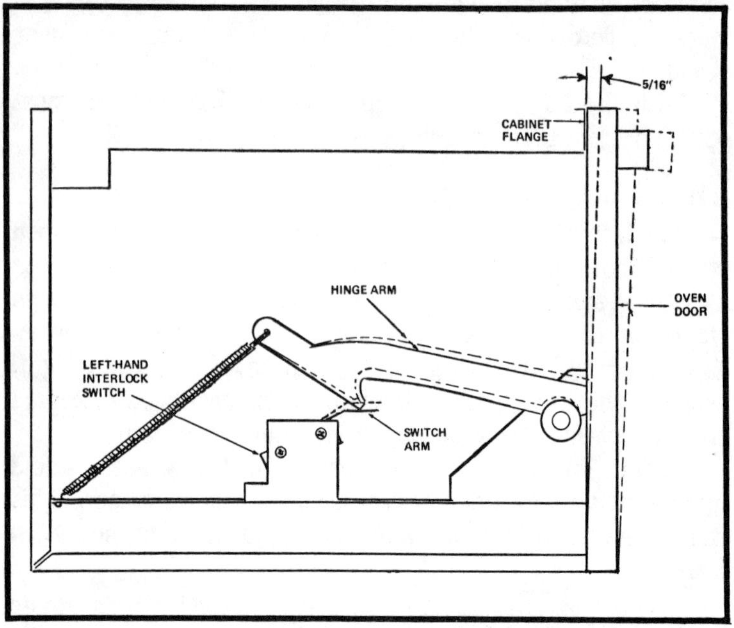

Fig. 13-11. Adjustment of interlock-switch arm in relation to hinge arm—KM520.

4. Grasp the door handle, pull out on the bottom of the handle to de-energize the switch.
5. If the switch is functioning properly the unit will stop operating. After complete service on a unit, check the operation of all interlock switches in accordance with the procedure outlined avove.

Figure 13-11 shows the adjustment of the interlock switch arm in relation to the hinge arm to open the circuit at a minimum of 0.3125 inch opening of the oven door. This will ensure that the control circuit is open anytime the oven door has traveled at least 0.3125 inches from the cabinet flange.

KM520 Pushbutton Assembly

Figure 13-12 shows the relationship between the reset button and spring assembly and the START push button. With the oven door open and the reset button in the out position, there is pressure on the back side of the START push button. With pressure on the back of this button and if an attempt is made to start the oven's operation, a canceling effect is caused

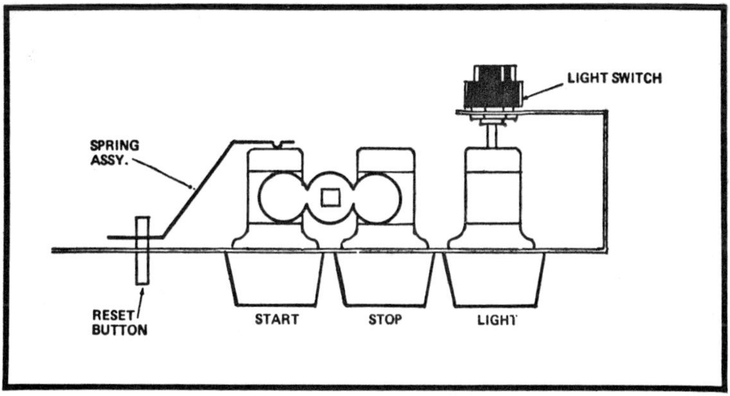

Fig. 13-12. Relationship of reset button and spring assembly to START push button with door open.

by the reset button and spring assembly, causing the start button to return to the out position. This prevents the operation of the oven.

Figure 13-13 shows the oven door closed, which applies pressure to the reset button, which in turn moves the spring assembly attached to the reset button away from the back side of the START pushbutton. This allows full travel of the START button, which in turn will close the START switch. With the spring assembly recessed away from the START pushbutton,

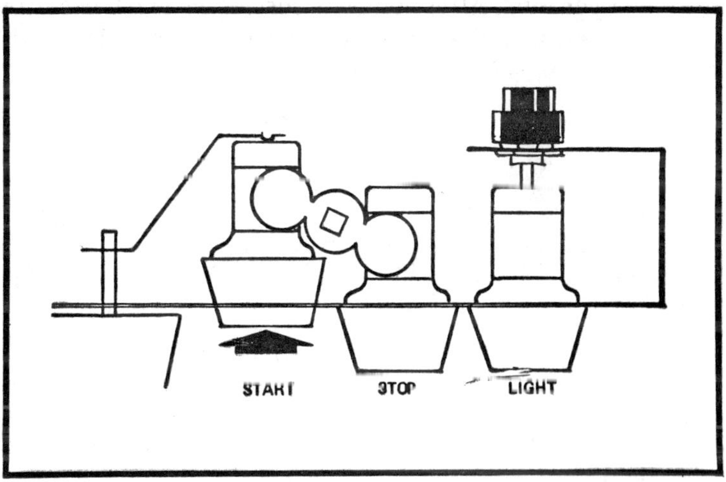

Fig. 13-13. Relationship of reset button and spring assembly to START pushbutton with door closed.

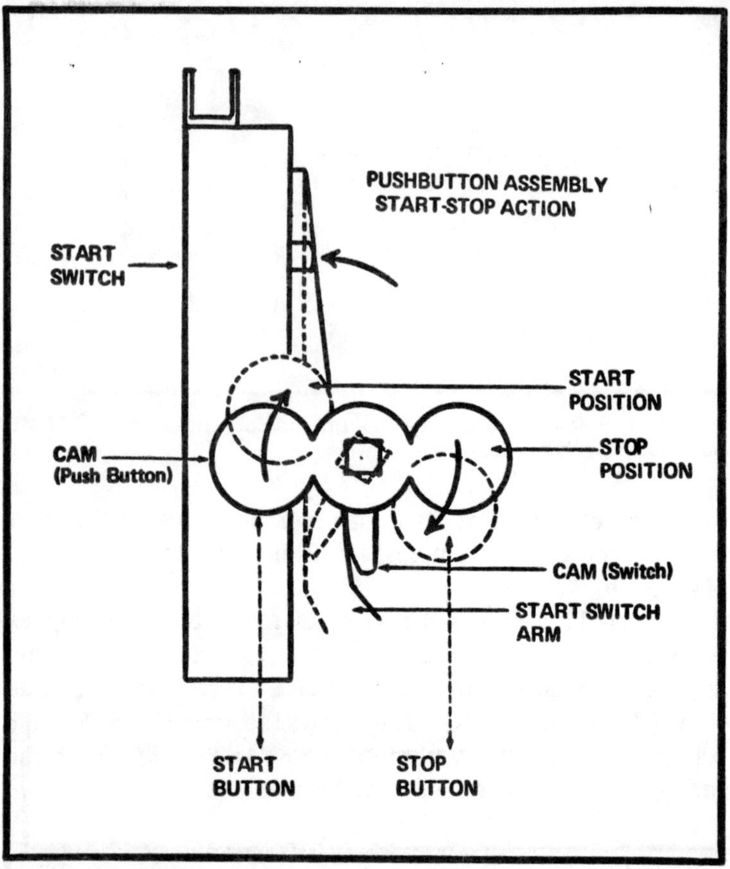

Fig. 13-14. Position of switch cam relative to pushbutton cam when switch cam is in on or off position.

the START pushbutton can now remain in the in (start) position.

Figure 13-14 shows where the switch cam is, while in the *on* or *off* position, in relation to the push button cam.

When the START button is depressed, it causes the pushbutton cam to rotate the switch cam, which in turn applies pressure to the START switch arm. This pressure closes the START switch.

An adjustment can be made between the cam switch and the switch arm to allow proper operation of the START switch. With the loosening of the switch mounting screws, the switch can be correctly positioned.

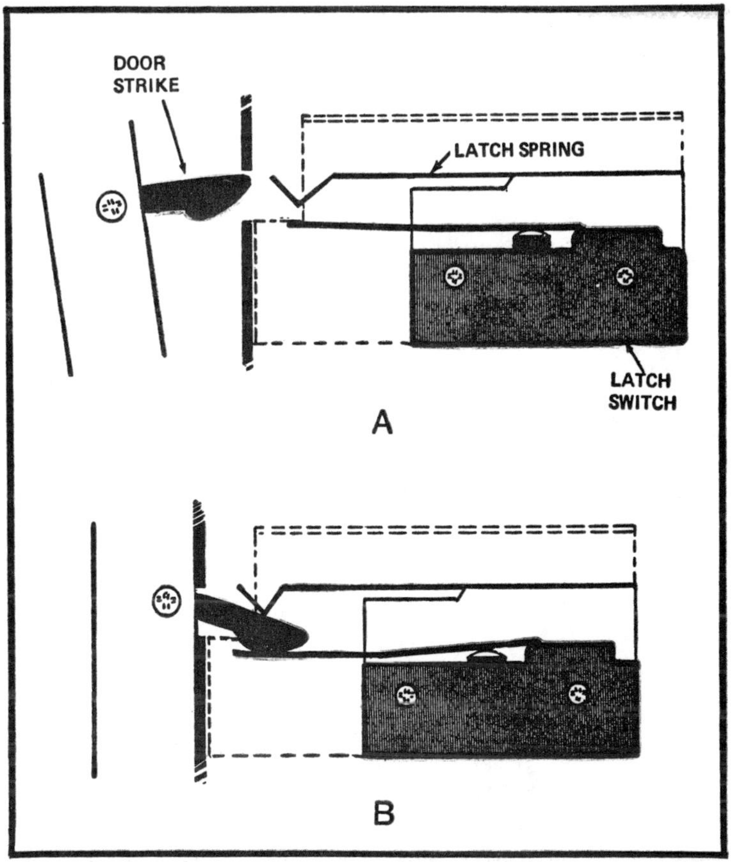

Fig. 13-15. Latch-switch assembly with oven door open (A) and closed (B).

KM520 Latch Switch

Figure 13-15 shows the latch switch in the open position and closed position. Part A shows the oven door and its latch with the oven door open and shows the position of the interlock or latch switch and the latch spring. Part B shows the oven door closed and the latch reaching through the oven liner opening and pressing on the latch switch. The door latch pressing against the door-latch switch closes the latch switch. This will ensure proper oven door closing before the control circuit is complete.

Index

Index